Contested Sustainability Discourses in the Agrifood System

The industrial agrifood system is in crisis regarding its negative ecological, economic, and social externalities: it is unsustainable on all dimensions. This book documents and engages competing visions and contested discourses of agrifood sustainability.

Using an incremental/reformist to transformation/radical continuum framework for alternative agrifood movements, this book identifies tensions between competing discourses that stress food sovereignty, social justice, and fair trade and those that emphasize food security, efficiency, and free trade. In particular, it highlights the role that governance processes play in sustainability transitions and the ways that power and politics affect sustainability visions and discourses.

The book includes chapters that review sustainability discourses at the macro and meso levels, as well as case studies from Africa, Australia, Canada, Europe, South America, and the USA.

Douglas H. Constance is Professor of Sociology at Sam Houston State University, Huntsville, Texas, USA.

Jason Konefal is Associate Professor of Sociology, at Sam Houston State University, Huntsville, Texas, USA.

Maki Hatanaka is Associate Professor of Sociology at Sam Houston State University, Huntsville, Texas, USA.

Books in the Earthscan Food and Agriculture Series

The Commons, Plant Breeding and Agricultural Research
Challenges for Food Security and Agrobiodiversity
Edited by Fabien Girard and Christine Frison

Agri-environmental Governance as an Assemblage
Multiplicity, Power, and Transformation
Edited by Jérémie Forney, Chris Rosin and Hugh Campbell

Agricultural Development and Sustainable Intensification
Technology and Policy Challenges in the Face of Climate Change
Edited by Sekhar Udaya Nagothu

Environmental Justice and Soy Agribusiness
Robert Hafner

Farmers' Cooperatives and Sustainable Food Systems in Europe
Raquel Ajates Gonzalez

The New Peasantries
Rural Development in Times of Globalization
(Second Edition)
Jan Douwe van der Ploeg

Sustainable Intensification of Agriculture
Greening the world's food economy
Jules N. Pretty and Zareen Pervez Bharucha

Contested Sustainability Discourses in the Agrifood System
Edited by Douglas H. Constance, Jason Konefal and Maki Hatanaka

For further details please visit the series page on the Routledge website: www.routledge.com/books/series/ECEFA/

Contested Sustainability Discourses in the Agrifood System

Edited by Douglas H. Constance,
Jason Konefal and Maki Hatanaka

Routledge
Taylor & Francis Group
LONDON AND NEW YORK

earthscan
from Routledge

First published 2018 by Routledge

2 Park Square, Milton Park, Abingdon, Oxfordshire OX14 4RN
52 Vanderbilt Avenue, New York, NY 10017

Routledge is an imprint of the Taylor & Francis Group, an informa business

First issued in paperback 2020

British Library Cataloguing-in-Publication Data
A catalogue record for this book is available from the British Library

Library of Congress Cataloging-in-Publication Data
Names: Constance, Douglas, editor. | Konefal, Jason, editor. | Hatanaka, Maki, editor.
Title: Contested sustainability discourses in the agrifood system / edited by Douglas H. Constance, Jason Konefal and Maki Hatanaka.
Description: Abingdon, Oxon ; New York, NY : Routledge, 2018. | Series: Earthscan food and agriculture series | Includes bibliographical references and index.
Identifiers: LCCN 2018008821 | ISBN 9781138063099 (hardback) | ISBN 9781315161297 (ebook)
Subjects: LCSH: Food industry and trade. | Sustainable agriculture. | Agricultural industries.
Classification: LCC HD9005 .C623 2018 | DDC 338.1–dc23
LC record available at https://lccn.loc.gov/2018008821

ISBN: 978-1-138-06309-9 (hbk)
ISBN: 978-0-367-58936-3 (pbk)

Typeset in Goudy
by Wearset Ltd, Boldon, Tyne and Wear

We dedicate this book to the distinguished environmental sociologist Frederick H. Buttel (1948–2005).

Contents

Contributors

Editors

Douglas H. Constance is Professor of Sociology at Sam Houston State University in Huntsville, Texas, USA. His research area is the social impacts of the globalization of the industrial agrifood system and alternative agrifood systems. He has numerous journal articles, book chapters, and books on these topics. His most recent co-edited book is *Alternative Agrifood Movements: Patterns of Convergence and Divergence* (2014) by Emerald Press. He is past Editor in Chief of the *Journal of Rural Social Sciences*. He is also past President of the Southern Rural Sociological Association (2003), the Agriculture, Food, and Human Values Society (2008), and past Vice-President of the Rural Sociological Society (2013). Dr Constance currently serves as the Quality of Life Representative and Chair of the Administrative Council of the United States Department of Agriculture Southern Region Sustainable Agriculture Research Education Program (USDA/SSARE).

Jason Konefal is an Associate Professor in the Department of Sociology at Sam Houston State University, USA. His research examines the relationship between political economic structures and practices and opportunities for social change. Specifically, he is interested in how neoliberalization is affecting governance processes and possibilities for equality, justice, sustainability, and democracy in food and agriculture. Dr Konefal's publications have appeared in the *Journal of Rural Studies*, *Agriculture and Human Values*, and *Organization & Environment*.

Maki Hatanaka is Associate Professor of Sociology at Sam Houston State University, USA. Her research examines economic globalization and efforts to regulate it in the areas of food, agriculture, and the environment. In particular, her interests focus on forms of non-state sustainability governance, particularly standards, metrics, certification, and accreditation. Her research has been published in numerous edited volumes and journals, including *Science*, *Journal of Rural Studies*, *Food Policy*, *World Development*, *Agriculture and Human Values*, and *Sociologia Ruralis*.

Contributing authors

Margaret Bancerz is a PhD candidate in Policy Studies at Ryerson University in Toronto, Canada. She holds a Masters in Environmental Studies from York University, and an Honours Bachelor of Arts in International Studies from Glendon College, York University. Her research focuses on wicked policy issues, specifically food policy, and how multi-stakeholder initiatives can be used to work through policy conundrums.

Mary A. Beckie holds an interdisciplinary doctorate in Sustainable Agriculture and Rural Development from the University of Saskatchewan, Canada, and is Associate Professor and Director of Community Engagement Studies in the Faculty of Extension, University of Alberta, Canada. Her academic work focuses on sustainable community development, alternative agri-food systems, local government, and citizen engagement. She has conducted research on organic agriculture and farmer marketing in Canada and Europe, agroecology and permaculture in Cuba, and food security initiatives in Sri Lanka and India. She is the Western Canada lead for FLEdGE (Food: Locally Embedded, Globally Engaged), a study on sustainable regional food systems.

Daniel Bornstein is a PhD student in the Department of Community and Environmental Sociology at University of Wisconsin-Madison, USA. He is currently studying how scientists and managers utilize ecological indicators to evaluate progress of the restoration of the Florida Everglades, a programme aimed at restoring the ecosystem to its historical conditions while ensuring adequate water supply for agriculture and urban development. His past work, based in The Gambia, has focused on the country's attempt to comply with food safety standards, as well as on the introduction of high-yield rice varieties to smallholder farmers.

Livia Boscardin is a sociologist at the University of Basel, Switzerland, specialized in human-animal studies. Her PhD thesis on 'Sustainable Exploitation. The Political Ecology of the Livestock Revolution' was funded by a scholarship of excellence of the Swiss National Science Foundation. Livia was the coordinative assistant of the doctoral programme 'Law and Animals: Ethics at Crossroads' of the University of Basel and a Visiting Scholar at the Animal Studies Initiative, NYU. Today, she is on the Committee of the European Association for Critical Animal Studies (EACAS).

Arantxa Capdevila is a political scientist holding a PhD on Audio-Visual Communication. She is Associate Professor at the Department of Communication Studies of the Rovira i Virgili University, Spain. Her main research line is based on the study of political communication processes by applying rhetorical-argumentative methods and the analysis of persuasive messages. Her last research project (as IP) was titled 'The Role of Metaphor in the Social Perception and definition of Conflicting Issues: Institutions, media and the citizens,' financed by the Spanish Ministry of Economy and Competitiveness.

Robert Magneson Chiles is an Assistant Professor of Rural Sociology in the Department of Agricultural Economics, Sociology, and Education and the Department of Food Science and a Research Associate in the Rock Ethics Institute at The Pennsylvania State University, USA. His current scholarship focuses on the measurement and governance of agricultural greenhouse gas emissions from the standpoint of climate justice, ethical obligations towards animals in the developing world, and the community development potential of plant-based protein offerings. He received his PhD from the University of Wisconsin-Madison and his BA from Stanford University.

Marina Di Masso is an environmental sociologist, and postdoctoral researcher at the Sustainable communities, Social innovation and Territory research group (SoPCI) of the Universitat de Vic – Universitat Central de Catalunya (UVic-UCC), Spain. Her major research interests are transformative processes of social change, with a focus on the agrifood system, particularly on critical perspectives of the global agrifood system, alternative food networks, and food sovereignty as a transformative paradigm. Recent research and personal interests include wider alternative socioeconomic practices, with a focus on feminist economics and social and solidarity-based economy. She is a member of the Agroecology and Alternative Food Systems Chair of the UVic-UCC.

Josep Espluga-Trenc is a sociologist, Associate Professor at the Department of Sociology at the Universitat Autònoma de Barcelona (UAB), Spain, and researcher at the Institute of Government and Public Policies (IGOP). He is also a member of the Research Group for Agriculture, Livestock and Food under Globalization (ARAG-UAB), at the same university. His major research field is environmental and health sociology, with special focus on the social perception of risk. He is Co-director of the Postgraduate Program 'Promoting Agroecological Farming at the Local Level,' and coordinates the Minor on Sustainable Development and Global Citizenship of the UAB.

Michael Gertler is Associate Professor in the Department of Sociology, and Fellow in Community and Co-operative Development, Centre for the Study of Co-operatives, at the University of Saskatchewan, Canada. He teaches rural and environmental sociology, sociology of agriculture and food, and the sociology of communities and community development. In field studies in Canada, the USA, Costa Rica, and Mongolia, he has focused on agricultural restructuring and organizational innovations that support sustainable development. His current research is on the organic subsector, grazing co-ops, and the restructuring of community pasture systems in Saskatchewan. He is a board member of Public Pastures–Public Interest.

JoAnn Jaffe is Professor of Sociology and Social Studies at the University of Regina, Saskatchewan, Canada. She is editor of the interdisciplinary journal, *Prairie Forum*, and a past President of the Rural Sociological Society. She has conducted field research in the Caribbean, North America, Central America,

East Africa, and Israel. Jaffe also frequently works in partnership with community groups. She was a review editor for the Global Volume of the International Assessment of Agricultural Knowledge, Science, and Technology for Development (IAASTD, 2006–2009) and an author and co-editor of *Farm Communities at the Crossroads* (2003) and *Contesting Fundamentalisms* (2004).

Les Levidow is a Senior Research Fellow at the Open University, UK, where he has been studying agrifood-environmental issues since the late 1980s. His research topics have included the following: sustainable development, bioeconomy, agroecology, waste-conversion technologies, agrifood-energy innovation, the water-energy-food nexus, institutional capacity-building and learning, European integration, regulatory expertise, scientific uncertainty, and the precautionary principle. He is co-author of two books: *Governing the Transatlantic Conflict over Agricultural Biotechnology: Contending Coalitions, Trade Liberalisation and Standard Setting* (Routledge, 2006); and *GM Food on Trial: Testing European Democracy* (Routledge, 2010). He is also Editor of the journal *Science as Culture*.

Allison Marie Loconto is a Research Fellow at the National Institute for Agricultural Research (INRA) and the Institute for Research, Innovation and Society (IFRIS); and she is a Visiting Expert at the Food and Agriculture Organization of the United Nations (FAO). She holds a PhD in Sociology from Michigan State University, an MA in International Affairs and Development from the American University and is currently the President of the Research Committee on Agriculture and Food (RC40) in the International Sociological Association (ISA). Her research interests include governance by standards, innovations in certification systems, responsible innovation, and the governance of transitions toward sustainable agriculture. Recent research has been published in the *ANNALS of the American Academy of Political and Social Science*, the *Revue Française de Sociologie* and *Agriculture and Human Values*.

Athena Moseley earned her Bachelor of Arts at Sam Houston State University, USA, with a Major in Sociology and a Minor in Political Science. She is currently a Marketing Manager for The Moseley Team at RE/MAX Hometown where she enjoys compiling analytics and experimenting with campaign strategies.

Marta G. Rivera-Ferre is a Ramón y Cajal Research Fellow and Assistant Professor at the University of Vic – Central University of Catalonia, Spain. She coordinates the research line 'Sustainable communities, social innovations and territories' within the group 'Societies, policies and inclusive communities'. She is a member of the editorial board of the *International Journal of Sociology of Agriculture and Food* and the *Journal of Sustainable Development*. She has participated as lead author in the UN-led assessments: the International Assessment of Agricultural Knowledge Science and Technology for

Development (IAASTD) and the rural chapter of the Intergovernmental Panel on Climate Change (IPCC).

Kiah Smith is a sociologist and Research Fellow in the School of Social Science, The University of Queensland, Australia. Her research spans food security and food justice, climate change, resilience, ethical trade, gender, livelihoods and green economy. She currently works on an ARC Discovery project exploring financialization, agribusiness, land and food security in Australia. Previously, Kiah worked as a research analyst with the United Nations Research Institute for Social Development; she is also a *Future Earth* Fellow, member of the ISA RC40 and RC09, TASA and the Agrifood Research Network, and co-convenor of the Brisbane Fair Food Alliance (a civic food collective). She is the author of *Ethical Trade, Gender and Sustainable Livelihoods*.

Jennifer Sumner teaches in the Adult Education and Community Development Program at OISE/University of Toronto, Canada. She is the author of *Sustainability and the Civil Commons: Rural Communities in the Age of Globalization* (University of Toronto Press, 2007), Editor of *Learning, Food and Sustainability: Sites for Resistance and Change* (Palgrave-Macmillan, 2016) and Co-editor of *Critical Perspectives in Food Studies* (Oxford University Press, 2017).

Jessica Thomas is a PhD student at LISIS (Interdisciplinary Laboratory on Sciences, Innovations and Societies), a research unit of the French National Institute for Agricultural Research (INRA). Her research interests are the analysis of the political, scientific and professional motivations driving the '*agroecological transition*.' Through a socio-historical approach she questions how agroecology has become the new paradigm of French agricultural development. This research analyses how agroecological public policies impact the interactions between decision-makers, scientists, teachers and professionals, as well as the processes of production and diffusion of knowledge and innovations for agricultural development.

Paul B. Thompson holds the W. K. Kellogg Chair in Agricultural, Food and Community Ethics at Michigan State University, USA, where he is also a Professor in the Departments of Philosophy, Agricultural, Food and Resource Economics and Community Sustainability. His 2015 book *From Field to Fork: Food Ethics for Everyone* (Oxford University Press) won the prestigious Book of the Year Award from the North American Society for Social Philosophy. Thompson's work in this volume was supported by his USDA/Michigan Agbioresearch Hatch Project 'Theory Development for Sustainability,' Project Number MICL02324.

Acronyms

AKS	Agricultural Knowledge System
CAD	Canadian dollar
CAPI	Canadian Agri-Food Policy Institute
CBoC	Conference Board of Canada
CFA	Canadian Federation of Agriculture
COTA	Canada Organic Trade Association
CWB	Canadian Wheat Board
EEIG	Economic and Environmental Interest Grouping
EOR	Ecologies of risk
FNSEA	National Federation of Farmers' Unions
FSC	Food Secure Canada
GM	Genetically modified
GMO	Genetically modified organism
LIIB	Low indirect impact biomass
MSI	Multistakeholder initiative
NC	Neo-conventional
NGO	Non-governmental organization
PFPP	People's Food Policy Platform
POGI	Prairie Organic Grain Initiative
RSB	Roundtable on Sustainable Biomaterials
USD	United States dollars
WWF	World Wildlife Fund
ZT	Zero-till

Part I

Introduction

1 Contested sustainability discourses in the agrifood system

An overview

Douglas H. Constance

> … contemporary unsustainable agro-food systems have an enormous amount of staying power and a capacity to maintain their dynamics despite socio-environmental costs and risks and despite the depletion of the 'natural capital' of agroecosystems. Thus, what we need to understand social scientifically is how it can be the case that unsustainable agriculture is capable of being sustained for lengthy periods, if not indefinitely.
>
> (Frederick H. Buttel, 2006: 214)

It is now generally accepted that that the industrial agrifood system is unsustainable, as stated by Buttel (2006). Agriculture in the Anthropocene Era is the leading contributor to climate change (Godfray and Garnett, 2014; Kuyper and Struik, 2014). The climate impacts of the Anthropocene make it imperative that we change the way food is produced, distributed, and consumed (Campbell *et al.*, 2017; IPES-Food, 2016; Rockstrom *et al.*, 2017). The problem is well understood, the solutions are much more difficult and highly contested (Almås and Campbell, 2012; Holt Gimenez and Shattuck 2011; Scoones, 2016).

This edited volume critically investigates the topic of contested sustainability discourses in the agrifood sector through a collection of papers presented at the 2016 International Sociological Association and International Rural Sociological Association meetings in Vienna and Toronto, respectively. The sessions were sponsored by Research Committee 40 on Agriculture and Food of the ISA. This book documents and engages these competing visions and contested discourses of agrifood sustainability through 14 substantive chapters, and Introduction and Conclusion.

Concerns about the sustainability of industrial agriculture were first raised systematically with the Brundtland Report in 1987 as part of a more general concern regarding sustainable development (Velten *et al.*, 2015). Since that report the term sustainability has come into "play" as competing factions and interests manoeuvre to capture the definition (Buttel, 2006; Constance, 2010; Scoones, 2016). The ecological crisis was the first to manifest, followed by social and economic crises regarding food production and consumption (Magdoff *et al.*, 2000; Constance *et al.*, 2014). The productivist model based on intensive, specialized monoculture combined with intensive, concentrated livestock production created a metabolic rift – the geographic separation of the nutrient/waste cycle. This model was coordinated by the nation-State through the USDA and commodity programmes, the Land Grant Universities through public research on mechanization, genetics and breeding, chemical inputs, and powerful commodity groups linked to agribusiness corporations (Hightower, 1978). The institutionalization of Cochran's (1993) treadmill of production rendered U.S. 'agriculture ecologically-unsustainable (and economically and socially unsustainable)' (Buttel 2006, 216). This industrial model was diffused globally as part of development projects, with similar outcomes: environmental degradation, social stratification, and corporate domination. And, although the industrial model produced vast quantities of food, global hunger remained a serious problem. The economic and food shocks of 2008–2009 accelerated sustainability concerns. The industrial agrifood system was unsustainable, and calls for the transition to a new paradigm grew louder (IPES-Food, 2016). To complicate the scenario, by the year 2050 the world population is predicted to be nine billion people who need to be fed without expanding the agricultural land base, and while reducing negative environmental impacts such as greenhouse gas emissions, water logging, and chemical contamination (UNFAO, 2009; Feed the Future, 2015; Gardner, 2009).

In response to this realization, two competing visions have emerged as the path forward to sustainably feed the world: the sustainable intensification approach and the agroecology approach (Levidow, 2015). These two transitions paths are the outcome of historically competing visions and contested discourses regarding the preferred model of the agrifood system (Thompson, 2010a; McMichael, 2014). Agroecology strives to maximize diversity through holistic systems that create long-term fertility. The discourse focuses on honouring indigenous cultures and appropriate technologies that support a decentralized agrifood system aligned with concepts of ecological resilience, food sovereignty, fair trade and social justice (Altieri, 2002; IAASTD, 2008; Fernandez *et al.*, 2013; Gliessman, 2013; IPES-Food, 2016). Although the sustainable intensification discourse began with agroecology roots (Pretty, 1997), its current application is dominated by the Green Revolution focus on high-technology solutions applied to specialized monocultures designed for growing more food on less land with more efficient use of resources. This discourse to feed the world is embedded in food security, comparative advantage, and free trade positions (Garnett *et al.*, 2013; Struik *et al.*, 2014; Petersen and Snapp, 2015; Rockstrom *et al.*, 2017).

The tension between the food security and the food sovereignty visions aligns with the transition paths. The food security path is based on neo-productivist, high-tech solutions using all available tools and technologies, including intellectual property, as the new paradigm to meet the challenge of feeding the world with sustainable intensification (Marsden, 2013; McMichael, 2014). The food sovereignty path is based on agroecology and a social justice framework. The food security path is patterned on consequentialist philosophy grounded in utilitarian assumptions about agrifood science and rurality. The agroecology path employs rights-based rhetoric to support its social justice agenda. The food security path is incremental with reforms to "green" to the existing system, while the food sovereignty path pushes for transformative change to the system (Thompson, 2010b; Holt Gimenez and Shattuck, 2011; Constance *et al.*, 2014). Where the current system promises to sustainably intensify, the agroecologists prefer to intensify sustainability. The agroecologists warn that sustainable intensification is an oxymoron at least (Eckard, 2015), and more probably a "wolf in sheep's clothing" (FOE 2012). The neo-productivists promise their green solution can feed the world, while the low-tech agroecology approach cannot. While food sovereignty advocates argue for a transition path informed by deep agroecology, conventional agriculture proponents have countered with food security discourses focusing on ecological modernization, sustainable intensification, and climate smart agriculture (Levidow, 2015).

Although there are other recent books that deal with aspects of the crisis of and alternatives to the industrial agrifood system (Duncan, 2015; Marsden and Morley, 2015), existing works neither systematically document the philosophical and practical discourses behind the two emergent transition paths, nor document the political economic contestations as each path advances its agenda as the solution to feed the world sustainably. More specifically, the agroecology and food sovereignty paths are now well documented (Wittman *et al.*, 2010; Claeys 2015; Trauger 2015; IPES-Food, 2016). The primary contribution of this book is that it adds the sustainable intensification position to the discourse and reveals the contested aspect of sustainability discourses, practices, and policy across geographic regions and levels of contestation. Once the contest is over, and a certain discourse prevails, the metrics, standards, practices, and policies become 'recipes for reality' (Busch, 2011), and thereby structure and institutionalize the governance of the agrifood system. Guthman's (2004) work on organics is a good example of this process. In this book we provide examples of the contest over Busch's 'recipes for reality.'

The macro framework: agrifood regimes

The meta-discourse on contested agrifoods transitions examined in this book is contexted within the agrifood regimes literature in the sociology of agriculture and food. The agrifood regimes literature is useful to provide a meta-frame for the contested sustainability discourses. Adapted from the French Regulationist School (Aglietta, 1979; Lipietz, 1987; Boyer 1990), regimes approaches organize history into stable global trade arrangements and their symbiotic, rule-based, governance mechanisms (Friedmann and McMichael, 1989). Contradictory relations within each regime generate crises, decay, and a period of transition to the next regime. Agrifood regimes analysis supports a structured method of analysis of the global political economy of agrifood production and consumption by examining the specific historical geo-politics that underpin the stable periods of capital accumulation and the transitional periods between regimes (Campbell and Dixon, 2009). Our point is that the characteristics of the next regime are 'in play' as powerful groups use their resources to advance their discourse regarding the preferred transition path to global sustainable agriculture.

The First 'Frontier' Regime (Settler-Colonial Regime: 1870s–1930s) was dominated by Great Britain as the core countries extracted natural resources from the peripheral countries to support industrial development at home. The Second 'Fordist' Regime (Agricultural Surplus Regime: 1950s–1970s) was dominated by the United States as it used its agricultural surplus as a food weapon in strategic postcolonial Cold War states. (Friedmann and McMichael 1989; see Campbell and Dixon 2009). Three similar perspectives on the contested transition to the Third 'Free Trade' Regime are useful for our purposes: the Corporate Food Regime (McMichael, 2005; 2009); the Neoliberal Food Regime (Pechlaner and Otero, 2010; Otero, 2011); and the Corporate Environmental Regime (Friedmann, 2005; 2009).

In the 1970s the 'Globalization Project' emerged as a corporate counter-response to the social-democratic regulatory and civil-rights successes of the Fordist era. Neoliberalism provides the ideological underpinnings of globalization as TNCs drive the process of economic restructuring, which transforms agrifood supply chains in favour of capital over subordinate groups. This flexible accumulation system is diffused around the globe as new territories in the global South are integrated into supply chains that service the global North. In this market-driven system intensification maximizes yield and generates surpluses and global exports, which then marginalizes the productive capacities of less-intensive methods elsewhere, accelerating the treadmill of production as all farms and regions are pushed to adopt intensification methods or be absorbed by those who do.

Agrifood TNCS, funded by transnational finance capital and supported by global governance organizations such as the WTO, IMF, and World Bank, diffuse the Globalization Project and its 'Food from Nowhere' model globally within nation-states, who through neoregulation, facilitate and mediate the process. Resistance movements such as La Via Campesina challenge the Corporate Food Regime and call for food sovereignty, agroecology, and the rights of indigenous peoples. Through neo-regulation, the state and supra-state actors both construct and mediate the contested formulation of public policies that constrain or enable the actions of TNCs and social movement groups and thereby restructure the political economy of the region and the world.

As the legitimation crisis of the indistural model and Corporate Food Regime accelerated, the dominant regime responded to charges of unsustainability with a new market model referred to as the 'Corporate-Environmental Food Regime.' This 'green regime' shifts agro-industrial production methods towards reducing some harmful environmental effects, fulfilling consumer demand for 'green' products, and conventionalizing alternative agricutural forms such as organics, animal welfare, and fair trade. This convergence of corporate greening and mainstream environmental social movements generates a system whereby global agrifood retailers service a subdivided system of class diets of quality-audited foods for the transnational rich and standardized products for poor consumers. While criticized for the market-based approach, growing corporate domination, and neoliberal individualism, this 'Food from Somewhere' variant provides valuable ecological services to the industrial agrifood system (Campbell, 2009).

Grounded in the regime's analytical framework, Holt Gimenez and Shattuck (2011) provide a typology of alternative agriculture social movement (AAM) organizations to help analyse the competing discourses on the practices and policies of the nascent regime in general, and the contested discourses in the book chapters. The typology ranges from 'Neoliberal' positions at one pole, to 'Reformist' to 'Progressive' to 'Radical' at the other pole. Neoliberal positions call for techno-scientific solutions to unsustainability with incremental changes to the dominant system. This view stresses free market fundamentalism and economic liberalism. The vision is to expand global markets by replacing inefficient producers with efficient agro-export enterprises (land grabs), using all available technologies (GMOs), organized by TNCs (food enterprises) and supported by governments and major donor organizations. Reformist positions stress food security through mainstreaming green certification schemes such as organics or fair trade and sustainable roundtables. This discourse focuses on market-led reforms in coordination with global donors such as the Gates Foundation and international development organizations such as the World Bank and Consultative Group on International Agricultural Research (CGIAR).

The Progressive position focuses on food justice through agroecological production embedded in local communities with workers' rights and solidary economies. In this view food is right, it should be healthy instead of disease-causing, and governments should enforce and protect that right against powerful market forces. Community Supported Agriculture (CSA), food policy councils and foodsheds are key organizational structures. Radical positions such as La Via Campesina call for transformative system change to agroecology and food sovereignty. In this view the WTO-orchestrated, corporate-dominated system based on GMOs and copyrights must be dismantled and replaced with a democratization and re-peasantization of the agrifood system because it – industrial agriculture – is killing the planet.

The book proceeds from two main premises. First, the existing agrifood system is unsustainable and in crisis. Second, the new regime is "in play" as institutional actors jockey to advance their discourse and vision as the better path forward to feed the world sustainably. The main objective of this book is to reveal the specific discourses and actors who are competing to define the future structure of the global agrifood system. In doing so, we answer Buttel's (2006) call for more social-science research on the peculiarities of sustaining the unsustainable.

The book structure

This edited book consists of an Introduction (Part I), fourteen substantive chapters (Parts II–IV), and a Conclusion (Part V). The substantive chapters are divided into three sections: Part II includes four framing chapters; three that focus on the problematics of sustainable intensification in relation to agroecology, and one that proposes the concept of the Civil Commons as the 'answer to the problem.' Part III includes six chapters that mix broader theoretical engagements with empirical studies. Part IV engages the discourse on contested governance of the agrifood with four chapters, moving from national to global initiatives.

The concluding chapter first situates the substantive chapters within the two transitions framework utilizing the Incremental/reformist to Transformation/radical continuum for alternative agrifood movements, then it synthesizes the contribution of the book regarding the contested futures of the agrifood system.

Part II: framing the contested discourse

In Chapter 2 entitled 'Sustainable intensification: agroecological appropriation or contestation?' Levidow traces the global promotion of sustainable intensification (SI) to raise yields with less environmental harm through a broad 'toolkit' including agroecological methods. In the European context, agri-intensification has diverse forms and policy agendas. SI has been advocated to help keep farmers on the land by making their cultivation methods more market-competitive, while conserving biodiversity elsewhere; this approach complements a land-sparing strategy. By contrast, a different intensification agenda promotes biodiverse agroecosystems, complementing a land-sharing strategy. The latter corresponds with an alliance of farmers and civil society organizations (CSOs) promoting agroecology to transform the dominant agro-food regime. In their efforts towards supportive policies, such alliances have gained larger budgets for agroecological methods in the EU's R&D programmes. But their efforts at 'greening the CAP' have resulted in rules which still subsidise higher-yield practices without necessarily benefiting agri-biodiversity. Levidow concludes that by recognizing these tensions, practitioners can better develop strategies for intervening in various agri-policy arenas, even where SI remains implicit.

In Chapter 3 entitled 'Sustainable intensification as a sociotechnical imaginary' Thompson applies the concept of the sociotechnical imaginary to evaluate sustainable intensification. The sociotechnical imaginary was developed in science studies to analyse how individuals, groups and cultures weave implicit technical commitments into their visions of the future. Sustainable intensification has implicated agricultural technology within an envisioned future in several different ways since the idea was introduced in the 1990s. He traces the history of these changes to the current situation, where the potential for genuine change in the direction of agricultural science and technological innovation has become much less visionary and transformative than it was in the early stages. While earlier imaginaries incorporated an explicit recognition of ethics as an element of envisioning the future of food systems, the current sociotechnical imaginary has disciplined this dimension of sustainable intensification, resulting in a very conservative vision that does not challenge the trajectory of agricultural technology that emerged under industrial capitalism.

In Chapter 4 entitled 'Agrifood discourses and feeding the world: unpacking sustainable intensification' Constance and Moseley ground the sustainable intensification versus agroecology discourse within a historical perspective.

Throughout its modern history agriculture in the United States has been characterized by two competing visions grounded in the 'agrarian ethic' versus the 'industrial ethic'. The current manifestation of these competing visions is the 'sustainable intensification' versus 'agroecology' discourse to feed the world in the face of climate change and population growth. The term sustainable intensification was first used in the 1990s in reference to modernizing agriculture in developing countries to generate local and regional foods in order to meet growing population pressures. At that time the discourse and practice on sustainable intensification was informed by agroecology; it was to be based on appropriate technologies and sensitive to cultural norms. In 2006 the UNFAO announced that sustainable intensification was the new paradigm that would feed the earth's population of nine billion in 2050. It was defined as growing more food on the same amount of land with fewer environmental externalities. The FAO called on governments and NGOs to implement the paradigm, marshalling all available technologies. Sustainable intensification is now criticized, labelled as an oxymoron with numerous ethical concerns. They trace the archaeology of the competing discourses, with attention to the topic of sustainable intensification and agrifood transition futures.

In Chapter 5 entitled 'Sustainability as the civil commons: laying the groundwork for sustainable agriculture' Sumner proposes the concept of the global commons as the answer to the problems of unsustainable industrial agriculture. As powerful interests compete to control the meaning of sustainability, it is crucial that it not be co-opted to support the continuation of corporate-controlled, chemical-based, monoculture agriculture. She has developed an understanding of the term 'sustainable' that provides the basis for a counter-hegemonic discourse: sustainability involves the civil commons – co-operative human constructions that protect and/or enable universal access to life goods. The more a society builds the civil commons, the more sustainable it becomes; the more it encloses the civil commons, the less sustainable it becomes. Following this logic, sustainable agriculture would entail agriculture that builds the civil commons. Agriculture that increases exports for entrepreneurs, enriches transnational corporations, functions as a sales site for GMOs or degrades the land, water and air would not qualify as sustainable agriculture. But agriculture that preserves the soil, involves community and national control of food production or contributes to nutritious school lunch programs would qualify.

Part III: contested discourses in theory and practice

In Chapter 6 entitled 'Zero Hunger discourse: neoliberal, progressive, reformist or radical?' Smith explores the framing of sustainability through the discourse of Zero Hunger at multiple levels: national – Brazil; regional – West Africa; and global – UNFAO/Standards Development Goals (SDGs). Historically, Zero Hunger emerged as a policy platform in Latin America: it is state-centric with high involvement of civic actors, redistribution via smallholder farmer cooperatives, social protection programmes and an emphasis on agroecology. Zero

Hunger West Africa appears to blur food sovereignty, food self-sufficiency, right to food, food justice, and food and nutrition security discourses. At the global level, submissions to the SDG process reveal contestation by civic actors around themes of production, markets, nutrition, rights, justice and participation. The chapter applies an analytical framework developed from the typologies of Holt Gimenez and Shattuck (2011) and Clapp and Dauvergne (2011), who argue for the existence of four main positions shaping food/environmental restructuring today: neoliberal, progressive, reformist/institutionalist, and radical.

In Chapter 7 entitled 'Greenwashing the animal-industrial complex: sustainable intensification and the livestock revolution' Boscardin focuses on sustainability discourses in the livestock industry. Animal agriculture emits more greenhouse gases than the global transport sector and is the main driver of biodiversity loss. The chapter examines how the neo-productivist 'Livestock Revolution' – a projected upsurge in the consumption of animal foods of around 70 per cent by 2050 – should be attained by sustainable intensification. In addition, the perspective of critical animal studies posits the violent commodification of animals as a political problem. The chapter discusses how the colonial animal-industrial complex annihilates billions of animal lives, crosses planetary boundaries, and undermines social justice. Rather than a 'natural development benefitting the poor,' the Livestock Revolution is a form of directed change in global nutrition serving capital interests. The chapter shows how this 'sustainable', 'modern', high-tech form of animal exploitation deepens the ecological, societal, and food crises.

In Chapter 8 entitled 'Are food quality schemes an alternative to the conventional food system? Reflections on the EU metaphors on agrifood quality regulation' Espluga-Trenc, Di Masso, Rivera-Ferre, and Capdevila engage the contested discourse on food quality in Europe. The environmental, economic and social impacts of the current hegemonic agrifood regime, based on mass food production and global market oriented distribution, have led governments and state agencies to increasingly promote alternative policies prioritizing food quality over quantity. Some of these policy measures try to highlight distinctive features based on geographical origin of foods (e.g. appellations of origin) or production systems (e.g. organic farming) in order to understand how food quality is conceptualized and to assess its transformative potential. In this chapter, the authors present an analysis of the metaphors used in one of the main European legal texts on this subject (EU Regulation 1151/2012). Results show how this agrifood quality policy is mainly oriented to protect those producers at risk of being excluded from the current hegemonic agrifood regime, but this protection is basically understood in economic terms, while other environmental and social domains are almost absent from the policy aim.

In Chapter 9 entitled 'Discourses on sustainability in the French farming sector: the redefinition of a consensual and knowledge-intensive 'agroecology' Thomas contextualizes the institutionalization of agroecology in France. She presents an analysis of diverse discourses in the context of the implementation of a national programme supporting agroecology since 2013. The chapter reveals that these public policies are an emblematic example of the constrained political strategies undertaken by reformist governments in countries where agriculture is highly intensified. The first part of the chapter presents an analysis of the French government's discourse, and some of its interventions, supporting sustainability of the agrifood system, and what it ultimately produces in terms of issue and solution-framing. The aim is to discuss the normativity and performativity underpinning these institutional practices. The second part draws attention to specific references to valued knowledge, and knowledge production processes, to support the agroecological transition. By unfolding debates on knowledge between various stakeholders of the French agricultural sector, this section reveals the actors' choices and hybridizations underpinning knowledge valuation activities.

In Chapter 10 entitled 'Duelling discourses of sustainability: neo-conventional and organic farming on the Canadian Prairies' Gertler, Jaffe, and Beckie document the tensions between two farming systems in Canada. Crop production on the Canadian Prairies is large-scale, export-oriented, and risky. In this context, they investigate the competition between two approaches to sustainability: neo-conventional and organic farming. The former features reduced tillage and herbicide resistant, geneticallymodified crops – especially canola. It is championed by an agro-food complex that includes agribusiness, governments, universities, and purpose-built promotional organizations. The organic agro-food complex is smaller but enjoys broad support from consumers and environmental organizations. They explore how each has grown but also why the organic sector has not kept pace with growing demand due to the underfunding of organic research, interaction of technologies and land markets, barriers to social reproduction, and how, despite mutual borrowing, this competition remains cultural and political as well as technical. They conclude that agroecological innovation in both camps is constrained by commercial logics and that there is need for further advances in environmental and social sustainability.

In Chapter 11 entitled 'Contested sustainability discourses as lived experience: conflicted feelings towards meat in consumers' narratives and life stories' Chiles adds the issue of consumer perspectives on meat consumption as a contested discourse. He argues that many of the sustainability discourses that circulate in academic and popular media pay insufficient attention to how socio-environmental controversies deeply perplex ordinary consumers. He uses narrative analysis to examine the contested sustainability discourses that surround meat consumption in the life stories of three focus group participants. Their narratives collectively illuminate the key discourses which have shaped normative understandings about meat throughout American history: scarcity, abundance, comfort, guilt, pride, tradition, and apprehension. The purpose of this chapter is to unpack these tensions and unravel the different threads which tug on the hearts, minds, and stomachs of modern consumers.

Part IV: contested agrifood governance

In Chapter 12 entitled 'Shifting visions of sustainability in United States agriculture: a case study of the role of multi-stakeholder governance' Konefal and Hatanaka investigate the role of multi-stakeholder governance in contested agrifood discourses. Given the lack of consensus as to what counts as sustainable agriculture, a key component of agricultural sustainability transitions is developing a vision of sustainability and associated tools that can be used to foster changes in agricultural practices. They examine the role of multi-stakeholder governance in this process. They present case studies of two leading multi-stakeholder initiatives for U.S. agriculture: Field to Market and the Stewardship Index for Specialty Crops. They find that multi-stakeholder governance is generating a shift in the meaning of sustainability in U.S. agriculture. Specifically, whereas sustainability has historically been associated with alternative agrifood movements in the U.S., the visions of agricultural sustainability being advanced by Field to Market and the Stewardship Index for Specialty Crops are more in alignment with the emerging notion of sustainable intensification. Most notably, Field to Market and SISC are advancing a "neutral" approach that does not judge particular farming systems or practices, but rather emphasizes quantitative measurements and eco-efficiencies. Hence, their research indicates that in the context of the U.S., multi-stakeholder governance is codifying sustainability in ways that are largely accommodative of the existing structure and practices of U.S. agriculture.

In Chapter 13 entitled 'Understanding the challenge of problem definition in multistakeholder initiatives: lessons from sustainability policy frames in Canadian non-state food strategies' Bancerz applies the contested sustainability frame to four non-state food strategies in Canada. In 2017 the Canadian federal government began a multi-stakeholder process to develop a food policy for Canada. This chapter serves as a cautionary example for policymakers and demonstrates the potential challenges experienced in multi-stakeholder processes. One of the biggest contributors to the collapse of such processes is the failure to develop a clear problem definition of the policy issue being discussed. To show this, critical frame analysis is applied to examine four Canadian non-state food strategies and how they problematize sustainability in the Canadian food system. Sustainability as a consensus frame is a concept that generally resonates with everyone but can also be framed differently by actors leading to differing policy solutions. In order to develop successful food policy for Canada, a clear and common understanding of sustainability in the food system and as part of the food policy problem is vital.

In Chapter 14 entitled 'Standardizing "unused" land: the politics of indicators in land classification' Bornstein investigates the problem of indirect land use change associated with the expansion of biofuel production on previously forested lands, with special attention to carbon emission issues. He uses a case study of a biofuel certification scheme aimed at resolving this issue. The multi-stakeholder Roundtable on Sustainable Biomaterials has promoted biofuel cultivation on 'unused' farmland to avoid deforestation. This study traces the development of the RSB's technical indicator for unused land, illuminating the contested process of land classification. Initially, scientific experts were attentive to how local conditions may complicate distant categorizations of land as available for plantations. Yet ultimately, the Roundtable abandons a definition of unused land that takes into account the future possibilities of local land use patterns. The outcome is a classification of unused land that is favourable to the industry players seeking certification.

In Chapter 15 entitled 'Justifying the standardization of sustainability impact' Loconto investigates the role of meta-governance organizations regarding contested agrifood sustainability. Sustainability standards are increasingly being used to encourage, measure and prove the adoption of sustainable practices. Yet it is unclear what measures have been developed to 'prove' the 'responsibility' of standards development organizations in 'impacting' sustainable practices. By drawing upon the French school of pragmatic sociology, Loconto argues that the ISEAL Impacts Code is a 'good' form of meta-governance that enables the use of sustainability standards to prove the responsibility of corporations to make sustainability impacts. However, the 'goodness' of this form appeals to multiple orders of worth, which illustrates that there are aspects of sustainability impact, particularly its forms, that remain contested.

Part V: conclusion

In Chapter 16 – the Conclusion – entitled 'Fault lines in agricultural sustainability: contestation, cooptation, reform, and transformation' Konefal discusses the key fault lines and tensions in agricultural sustainability discourses. Building on the analyses presented in the preceding chapters, seven divergences in contemporary discourses on agricultural sustainability are identified: technology, social innovation, experts, democratization, markets, nature, and the extent of change. The effects of governance and institutionalization on sustainability discourses and their enactment are examined. Current governance processes are shown to have a moderating effect on sustainability transitions.

References

Aglietta, M. (1979). *A Theory of Capitalist Regulation*. New Left Books, London.

Almås, R. and Campbell, H. (eds) (2012). *Rethinking Agricultural Policy Regimes: Food Security, Climate Change and the Future Resilience of Global Agriculture*. Emerald Group Publishing Limited, Bingley, UK.

Altieri, M.A. (2002). Agroecology: The Science of Natural Resource Management for Poor Farmers in Marginal Environments. *Agriculture, Ecosystems and Environment* 93, 1–24.

Boyer, R. (1990). *The Regulation School: A Critical Introduction*. Columbia University Press, New York.

Busch, L. (2011). *Standards: Recipes for Reality*. MIT Press, Cambridge, MA.

Buttel, F.H. (2006). Sustaining the Unsustainable: Agro-food Systems and the Environment in the Modern World. In: P. Cloke, P., Marsden, T., and Mooney, P.H. (eds), *Handbook of Rural Studies*. Sage Publications, London, 212–229.

Campbell, B.M., Beare, D.J., Bennett, E.M., Hall-Spencer, J.M., Ingram, J.S.I., Jaramillo, F., Ortiz, R., Ramankutty, N., Sayer, J.A. and Shindell, D. (2017). Agriculture Production as a Major Driver of the Earth SyNstem Exceeding Planetary Boundaries. *Ecology and Society* 22, 4, 8.

Campbell, H. (2009). Breaking New Ground in Food Regime Theory: Corporate Environmentalism, Ecological Feedbacks and the 'Food from Nowhere Regime. *Agriculture and Human Values* 26, 4, 309–319.

Campbell, H. and Dixon, J. (2009). Introduction to the Special Symposium: Reflecting on Twenty Years of the Food Regimes Approach in Agri-food Studies. *Agriculture and Human Values* 26, 4, 261–265.

Claeys, P. (ed.) (2015). *Human Rights and the Food Sovereignty Movement: Reclaiming Control*. Routledge, London.

Clapp, J. and Dauvergne, P. (2011). *Paths to a Green World: The Political Economy of the Global Environment*. The MIT Press, Cambridge, MA.

Cochran, W. (1993). *The Development of American Agriculture*. University of Minnesota Press, Minneapolis.

Constance, D.H. (2010). Sustainable Agriculture in the United States: A Critical Examination of a Contested Process. *Sustainability* 2, 1, 48–72.

Constance, D.H., Renard, M.C. and Rivera-Ferre, M. (eds) (2014). *Alternative Agrifood Movements: Patterns of Convergence and Divergence*. Emerald Group Publishing Ltd, Bingley, UK.

Duncan, J. (ed.) (2015). *Global Food Security Governance: Civil Society Engagement in the Reformed Committee on World Food Security*. Routledge, London.

Eckard, R. (2015). Sustainable Intensification: Oxymoron or Unavoidable Imperative. *The Age*, 10 June. Available at: www.theage.com/au/action/printArticle?id=6586069 (Accessed 10 May 2016).

Feed the Future (2105). *Feed the Future: The U.S. Government's Global Hunger and Food Security Initiative*. Washington, D.C. Available at: https://feedthefuture.gov/progress 2015/ (Accessed 22 May 2017).

Fernandez, M., Goodall, K., Olson, M. and Mendez, V.E. (2013). Agroecology and Alternative Agri-food Movements in the United States: Toward a Sustainable Agri-food System. *Agroecology and Sustainable Food Systems* 37, 1, 115–126.

FOE (Friends of the Earth) (2012). *Wolf in Sheep's Clothing: An Analysis of the Sustainable Intensification of Agriculture*. Available at: www.foei.org/resources/publications/publications-by-subject/food-sovereignty-publications/a-wolf-in-sheeps-clothing (Accessed 15 May 2016).

Friedmann, H. (2005). From Colonialism to Green Capitalism: Social Movements and the Emergence of Food Regimes. In: Buttel, F.H and McMichael, P. (eds), *New directions in the sociology of global development: Research in rural sociology and development: Volume 11*, Elsevier Press, Oxford, 227–264.

Friedmann, H. (2009). Discussion: Moving Food Regimes Forward: Reflections on the Symposium Essays. *Agriculture and Human Values* 26, 4, 335–344.

Friedmann, H. and McMichael, P. (1989). Agriculture and the State System: The Rise and Fall of National Agriculture. *Sociological Ruralis* 29, 2, 93–117.

Gardner, B. (2009). *Global Food Futures: Feeding the World in 2050*. Bloomsbury Academic, London and New York.

Garnett, T., Appleby, M.C., Balmford, A., Bateman, I.J., Benton, T.G., Bloomer, P., Burlingame, B., Dawkins, M., Dolan, L., Fraser, D., Herrero, M. Hoffmann, I., Smith, P., Thornton, K., Toulmin, C., Vermeulen, S.J. and Godfray, H.C.J. (2013). Sustainable intensification in agriculture: Premises and policies. *Science* 5, 33–34.

Gliessman, S. (2013). Agroecology: Growing the Roots of Resistance. *Agroecology and Sustainable Food Systems* 37, 1, 19–31.

Godfray, H.C.J. and Garnett, T. (2014). Food Security and Sustainable Intensification. *Philosophical Transactions of the Royal Society B* 369, 20120273.

Guthman, J. (2004). *Agrarian Dreams: The Paradox of Organic Farming in California*. University of California Press, Berkeley.

Hightower, J. (1978). *Hard Tomatoes, Hard Times*. Routledge, New York.

Holt Gimenez, E. and Shattuck, A. (2011). Food Crises, Food Regimes, and Food Movements: Rumblings of Reform or Tides of Transformation? *The Journal of Peasant Studies* 38, 1, 109–144.

IAASTD (2008). International Assessment of Agricultural Knowledge, Science and Technology for Development. Available at: www.fao.org/fileadmin/.../Agriculture_at_a_Crossroads_Global_Report_IAASTD.pdf (Accessed 22 May 2017).

IPES-Food (International Panel of Experts on Sustainable Food Systems) (2016). *From Uniformity to Diversity: A Paradigm Shift from Industrial Agriculture to Diversified Agroecological Systems*. Available at: www.ipes-food.org (Accessed 22 May 2016).

Kuyper, T.W. and Struik, P.C. (2014). Epilogue: Global Food Security, Rhetoric, and the Sustainable Intensification Debate. *Science Direct* 8, 71–79.

Levidow, L. (2015). European Transitions Towards a Corporate-Environmental Regime: Agroecological Incorporation or Contestation? *Journal of Rural Studies* 40, 76–89.

Lipietz, A. (1987). *Mirages and Miracles: The Crisis of Global Fordism*. Verso, London.

Magdoff, F., Foster, J.B. and Buttel, F.H. (eds) (2000). *Hungry for Profit: The Agribusiness Threat to Farmers, Food, and the Environment*. Monthly Review Press, New York.

Marsden, T. (2013). From Post-Productivism to Reflexive Governance: Contested Transitions in Securing More Sustainable Food Futures. *Journal of Rural Studies* 29, 123–134.

Marsden, T. and Morley. A. (eds) (2015). *Sustainable Food Systems: Building a New Paradigm*. Routledge, London.

McMichael, P. (2005). Global Development and the Corporate Food Regime. In: Buttel, F.H. and McMichael, P. (eds) *New Directions in the Sociology of Global Development, Research in Rural Sociology and Development: Volume 11*. Elsevier Press, Oxford, 265–300.

McMichael, P. (2009). A Food Regime Genealogy. *The Journal of Peasant Studies* 36, 1, 139–169.

McMichael, P. (2014). Rethinking the Land Grab Ontology. *Rural Sociology* 79, 34–55.

Otero, G. (2011). Neoliberal Globalization, NAFTA, and Migration: Mexico's Loss of Food and Labor Sovereignty. *Journal of Poverty* 15, 4, 384–402.

Pechlaner, G. and Otero, G. (2010). The Neoliberal Food Regime: Neoregulation of the New Division of Labor in North America. *Rural Sociology* 75, 2, 179–209.

Petersen, B. and Snapp, S. (2015). What is Sustainable Intensification? Views from Experts. *Land Use Policy* 46, 1–10.

Pretty, J. (1997). *The Sustainable Intensification of Agriculture*. Natural Resources Forum. Blackwell.

Rockstrom, J., Williams, J., Daily, G., Noble, A., Matthews, L.G., Wetterstrand, H., DeClerck, F., Shah, M., Steduto, P., de Graiture, C., Hatibu, N., Unver, O., Bird, J., Sibanda, S. and Smith, J. (2017). Sustainable Intensification of Agriculture and Human Prosperity and Global Sustainability. *Ambio* 46, 1, 4–17.

Scoones, I. (2016). The Politics of Sustainability and Development. *Annual Review of Environmental Resources* 41, 293–319.

Struik, P.C., Kuyper, T.W., Burssaard, L. and Leeuwis, C. (2014). Deconstructing and Unpacking the Scientific Controversies in Intensification and Sustainability: Why the Tensions in Concepts and Values. *Science* 8, 80–88.

Thompson, P. (2010a). *The Agrarian Vision: Sustainability and Environmental Ethics*. The University Press of Kentucky, Lexington.

Thompson, P. (2010b) *The Ethics of Intensification: Agricultural Development and Social Change*. Springer, New York.

Trauger, A. (ed.) (2015). *Food Sovereignty in International Context: Discourse, Politics and Practice of Place*. Routledge, London.

UNFAO (United Nations Food and Agricultural Organization) (2009). *Global Agriculture Towards 2050*. Paper presented to the High-level Expert Forum on 'How to Feed the World in 2050'. Rome, 12–13 October. Available from: www.fao.org (Accessed 11 March 2016).

Velten, S., Leventon, J., Jager, N. and Newig, J. (2015). What is Sustainable Agriculture? A Systematic Review. *Sustainability* 7, 7833–7865.

Wittman, H., Desmarais, A.A. and Wiebe, N. (eds) (2010). *Food Sovereignty: Reconnecting Food, Nature and Community*. Food First Books, Oakland.

Part II

Framing the contested discourse

2 Sustainable intensification

Agroecological appropriation or contestation?

Les Levidow

Introduction

Since at least the 1980s 'sustainable agriculture' has been a contested term, encompassing divergent aims and policy agendas (Buttel, 2006; Carolan, 2006; Clapp and Fuchs, 2009; Constance, 2010; Douglass, 1984; Kloppenburg, 1988; Lélé, 1991; Marletto *et al.*, 2016). In the past decade, an extra focus has been agri-intensification, especially around the concept 'sustainable intensification.' Such debate intersects with a contest over how agroecology relates to the dominant agro-food regime (Giraldo and Rosset, 2017; Levidow *et al.*, 2014).

This chapter analyses how sustainable intensification (henceforth SI) has been promoted in the European context, through the following questions: How do agri-intensification agendas appropriate agroecological practices for reinforcing the dominant agro-food system? Or else for contesting it? How do such issues arise in various policy arenas? Through what alliances and agendas?

As a basis to address those questions, the next section introduces theoretical concepts on the incumbent food regime – its agents, its reform agenda and its contestation. Later sections analyse the following: the global drivers and debate on SI; intensification priorities, means and trade-offs; European farmers' agendas and their conflicting role in policy arenas, especially for agri-research and subsidy criteria. The Conclusion returns to the above questions.

Food regime contested: theoretical concepts

'Sustainability' has a long, contested history, especially in agriculture. Conventional agriculture has often renewed its capacity for 'sustaining the unsustainable,' e.g. promoting GM crops as more sustainable than conventional agriculture (Buttel, 2005, 2006). Various techno-fixes, each promoting specific agri-futures, have been targets of controversy. To analyse the recent debates over SI, this section introduces several theoretical concepts: neoproductivism, a market-driven food regime (with a reform trend), and a corporate-environmental food regime.

Since the 2007–08 spike in food prices, global institutions have promoted greater agricultural productivity. With an imperative to double food production by 2050, this dominant agenda conflates people's needs, marketing strategies and 'demand' for much greater meat consumption (e.g. FAO, 2009a). Proponents have highlighted long-term difficulties: higher energy costs, competing land uses, GHG emissions, resource burdens and other environmental harms.

That agenda has recast agro-industrial productivism: an incipient neo-productivist paradigm faces the challenge to locate the environmental sustainability and resilience of national food-supply systems within current globalization patterns (Marsden, 2012: 307). Neoproductivism has been repositioned in different ways across national political systems (Almås and Campbell, 2012). Although it encompasses a cooperative version enhancing farmers' knowledge, the prevalent form is a competitive market-driven productivism (Burton and Wilson, 2012). The neoproductivist agenda has been widely articulated as 'sustainable intensification.' With its diverse meanings, this shift can be analysed through political economy theories of the agro-food system as a global regime.

A food regime has been theorized as a 'rule-governed structure of production and consumption of food.' Since the 1990s the dominant agro-food regime has been widely called 'free trade,' a misnomer for the GATT-WTO agreements overcoming or deterring trade barriers. In this market-driven system, agro-industrial methods maximize yield and generate surpluses, for which subsidy gains global export. This undermines productive capacities and less-intensive methods elsewhere, thus pushing farms everywhere to adopt intensification methods. In this dominant regime, 'agrofood corporations are the major agents attempting to regulate agrofood conditions, that is, to organize stable conditions of production and consumption which allow them to plan investment, sourcing of agricultural raw materials, and marketing' (Friedmann, 2003: 52). This regime rests on assumptions that international trade more efficiently accommodates food needs; this has become normalized, even 'as a key delivery mechanism for food system sustainability' (Clapp, 2017).

The dominant regime has faced global resistance from small-scale food producers. Launched by Via Campesina at the 1996 World Food Summit, 'food sovereignty' has since been taken up by numerous farmers' and civil society organizations, while also contesting the capital-intensive agri-modernization project. Over the past decade such groups have jointly linked food sovereignty with agroecology as a transformative agenda (Holt Gimenez and Shattuck, 2011). Such a role depends on linking the three main forms of agroecology – a scientific discipline, an agricultural practice and a social movement (Wezel *et al.*, 2015).

Partly in response to such resistance, mainstream institutions have been promoting changes such as more public finance for agri-development, third-party certification and product labelling to address weak sustainability, and higher-productivity cultivation methods for market competitiveness (e.g. UN, 2008). Together these changes have been theorized as a 'reformist' model, meant to integrate smallholders into global markets, often in the name of social inclusion and environmental sustainability. Facing global revolts against agbiotech, the reformist model accepts that addressing global hunger needs all possible solutions, while narrowing these to various techniques (Holt Gimenez and Shattuck, 2011: 121–123).

Meanwhile counter-hegemonic global food movements contest the corporate food regime and its productivist agenda. These movements embrace agroecology and community-based food systems but vary in political perspectives, which may co-exist in the same organization. The 'progressive' trend promotes alternatives to agro-industrial foods, implicitly accepting the dominant regime, while the 'radical' trend seeks to confront it explicitly through an agenda for food sovereignty. Each trend is heterogeneous and politically fluid. The current food crisis offers opportunities for strategic alliances between them, but also opportunities for the reformist agenda to split the progressive from the radical trend (Holt Gimenez and Shattuck, 2011: 133–134).

Indeed, for a long time, capitalism has had success in co-opting resistance, especially in the global North. Its socially articulated nation-states seek to maintain a complementarity between the role of the labour force as producers and consumers, in turn as a basis for legitimacy. Here the 'progressive' trend encompasses small-farm petty commodity producers and domestic agri-business facing threats to their local markets by globalized circuits of corporate agri-food capital (Tilzey, 2017: 23).

As the dominant regime responds to protest and new market opportunities, an overall trend has been theorized as a 'corporate-environmental food regime.' This shifts agro-industrial production methods towards reducing some harmful environmental effects, deepening commodity relations in agriculture, and fulfilling consumer demand for 'green' products. Such a new regime (or variant) potentially emerges from capitalizing such alternatives: 'a new round of accumulation appears to be emerging in the agrofood sector, based on selective appropriation of demands by environmental movements, and including issues pressed by fair trade, consumer health, and animal welfare activists' (Friedmann, 2005: 229, 249).

In this nascent regime, some new standards originate from alternatives to the corporate-industrial regime, e.g. organic food and functional foods. But most health, environmental and social problems cannot be reduced to consumer demand for novel products. So a nascent corporate-environmental food regime has been contested by the alternative food movements that it appropriates (Friedmann, 2005: 257).

Given those regime shifts and conflicts, what are possible outcomes? For regimes in general, alternatives have been theorized as niche innovations, which can either fit-and-conform to the incumbent regime or else stretch-and-transform it (Smith and Raven, 2012: 1030). Any arena is more conducive to one strategy or the other. The two strategies are 'exercised in contrasting arenas, with potentially very different outcomes in terms of form and function of the emerging socio-technical system, who holds control and what sustainability criteria are maintained' (Smith and Raven, 2012: 1033). For the European debate, next let us examine different priorities of intensification, and then how stakeholder groups promote them.

Sustainable intensification (SI): global drivers and debate

The 'sustainable intensification' (henceforth SI) concept originally addressed problems of African smallholders. Attempting to maintain livelihoods and local food supplies, they face pressures to over-exploit the soil and/or to take over more land for cultivation. Such efforts could worsen environmental degradation. To enhance the resource base and livelihoods together, knowledge-based agroecological methods should conserve soil and water, as well as manage nutrient flows and pests. Through SI as participatory improvement methods, 'yields are increased without adverse environmental impact and without the cultivation of more land' (Pretty, 1997).

Intensification was given different priorities a decade later, especially after the 2007 increase in global food prices, signalling a longer-term global threat of food insecurity. This has been attributed to various causes – climate change (environmental stresses including pests and water scarcity), land and water shortages, competition for biomass, growing market demand for biofuels and livestock products, etc. (Conway, 2009). Such 'causes' ideologically naturalize global market forces as objective imperatives which must be accommodated through higher yields. Towards remedies, this intensification agenda conflates higher productivity, yields, better livelihoods, global market competitiveness and food security. A similar agenda has been widely taken up by mainstream institutions, aiming to integrate some peasants into the neoliberal food regime (Escobar, 2011).

For this agenda, a key UN document was the *Comprehensive Framework for Action* (CFA). It recognizes that smallholder farmers are central to the long-term problems of hunger and poverty. Its agenda assumes that smallholders will benefit from 'expansion of agricultural marketing and processing enterprises which integrate smallholders into domestic and international food supply chains' (UN, 2008: 16, 28).

According to the CFA, the 'resilience and sustainability of agriculture under intensification' can be ensured through various techniques such as biotech and Conservation Agriculture. The latter is 'sometimes called agro-ecology because it combines agricultural practice and effective use of ecological knowledge and direct seeding into crop residues.' Conservation Agriculture features crop rotation as a means of energy efficiency and Integrated Pest Management. Public-Private Partnerships would help 'in enabling greater smallholder participation in market-oriented food production,' e.g. by ensuring that any problems are trackable and that produce is traceable through the whole chain back to the producers (UN, 2008: 26, 30).

How to reconcile higher yield with sustainability? For the UK's Royal Society, sustainable intensification (SI) production systems are 'knowledge-, technology-, natural capital- and land-intensive.' To avoid environmental damage, the 'intensity of use of non-renewable inputs must in the long term decrease' by substituting various techniques including agroecology and GM crops (Royal Society, 2009: 46), thus very broadly defining 'renewable' inputs.

This imperative is linked with globalized, volatile markets: 'Domestic patterns of food production and consumption have become interconnected in global markets' (Royal Society, 2009: 4). In those ways, SI agendas promote a 'toolkit' of various options for reconciling higher productivity with environmental sustainability, especially for a competitive advantage in commodity markets.

Exemplifying SI, Conservation Agriculture promotes no-till methods to conserve soil fertility and avoid erosion. Alternatives to tillage include treatments with total herbicides (e.g. Syngenta, 2013), sometimes with herbicide-tolerant crops, all in the name of sustainable intensification. Along those lines, some agri-input supply companies have rebranded their products as SI tools (Constance and Moseley, 2018; this volume).

The UN's Food and Agriculture Organization (FAO) has promoted SI for crop production. Its agenda 'allows countries to achieve sustainable increases in agricultural productivity through an ecosystem approach,' especially for going beyond subsistence agriculture. The FAO emphasizes wider market access, i.e. farmers competing on global markets:

> Increasing agricultural productivity through improved use of resources to achieve higher yields while promoting the sustainability of the farming systems and progressing from subsistence farming to market-oriented agriculture, supported by Conservation Agriculture (CA) and Integrated Plant Nutrient Management (IPNM) ...
>
> Strengthening livelihoods using the benefits of increased productivity and diversification within the value chain, including through providing the conditions for access to good agricultural practices and knowledge, quality seeds, post-harvest and agro-processing technologies, food safety systems, markets and credit.
>
> (FAO, n.d.)

This narrowly defines the objective as food products, while potentially aggravating farmers' dependence on input suppliers and market intermediaries.

Given that higher-yield methods often have negative impacts, SI agendas propose various technical remedies. There is a pervasive imperative to 'rebuild research and technology transfer capacity in developing countries in order to provide farmers with appropriate technologies,' through 'a rich toolkit of relevant, adoptable and adaptable ecosystem-based practices' (FAO, 2011; cf. FAO, 2009b: 31). Such a toolkit needs 'every possible solution, including agroecology and biotechnologies' (FAO, 2016a).

'Transfer' implies that Southern farmers adopt Northern innovations. Indeed, SI has been promoted as standard technology packages, generally inappropriate for diverse conditions. Any successful SI approach would need to understand relationships between smallholders, their organizations, other stakeholders and policies (Arora and Nijbroek, 2016).

From the perspective of farmer-civil society alliances, the SI agenda acknowledges the potential contribution from agroecological methods, yet these are readily subordinated to the dominant agro-food system. The SI agenda neglects issues of procedural and distributive justice, e.g. collective empowerment to define needs and how they are met; this could strengthen agroecosystem approaches (Loos *et al.*, 2014). Indeed, from the standpoint of food sovereignty, a global smallholder alliance warns against 'the threat of co-optation': agroecology is being appropriated 'as a set of production techniques that can conform to the industrial agriculture model,' and thus as 'a tool to legitimize, sustain and replicate the dominant model' (IFA, 2015: 22).

This threat was likewise identified by La Via Campesina.

> agroecology itself is under dispute by corporations, governments and the World Bank, with the scientists and intellectuals who knowingly or unwittingly work for them. This neoliberal attempt to co-opt agroecology can be seen in government 'organic agriculture' programs that promote monoculture-based organic exports for niche markets, and subsidize companies to produce organic inputs that are even more expensive than the agrotoxics whose costs led to the debt-trap so many rural families find themselves in.
>
> (LVC, 2013: 70)

The food sovereignty movement says likewise: 'In this scenario we can see how green capitalism has "discovered" agroecology as a way of incorporating peasant agriculture, its territories and agro-ecological practices into global circuits of accumulation' (Nyeleni, 2016: 3).

Thus SI agendas provoke debate on several issues. Is the aim to 'feed the world sustainably' or rather to 'help the world feed itself' in socially just ways? How to compare the productivity of conventional and agroecological methods? Is the aim simply to make the former less harmful? How to enhance and evaluate entire agroecosystems? (Bernard and Lux, 2017). Such questions arise in distinctive ways in Europe – the focus of subsequent sections.

European debates on agri-intensification: priorities and trade-offs

As a distinctive context, European agriculture has been the focus of debates over divergent aims, e.g. intensification versus extensification, land-sparing versus land sharing, and 'feeding the world' versus Europe feeding itself in more sustainable ways. In this context, expert reports provide entry points to analyse divergent forms and priorities of intensification.

Which intensification?

There has been a long-standing debate about how to intensify or ecologize agriculture, each agenda drawing on a different concept of ecology (Ollivier and Bellon, 2013). Agroecologists have promoted farm redesign around agroecosystems, whereby biodiversity provides various ecological services beyond producing food. Such services include: recycling nutrients, regulating microclimate and local hydrological processes, suppressing undesirable organisms and detoxifying noxious chemicals (Altieri, 1999). By contrast, an 'ecoagriculture' agenda has demarcated ecologically intensified areas from nature conservation areas, which are thereby spared from biodiversity loss.

Conflict between those two agendas erupted at the 2004 meeting of the International Union for the Conservation of Nature (IUCN). From an agroecological standpoint, the ecoagriculture agenda was attacked for a production model which 'hinders attempts to provide adequate food for a growing world population' (Altieri, 2004). In more subtle ways, an analogous conflict emerged later between agendas for land-sparing versus land-sharing (see later section).

To clarify those issues, a European multi-stakeholder initiative elaborated the new concept 'eco-functional intensification.' It means

> more efficient use of natural resources, improved nutrient recycling techniques and agro-ecological methods for enhancing diversity and the health of soils, crops and livestock. Such intensification builds on the knowledge of stakeholders using participatory methods … [It means] activating more knowledge and achieving a higher degree of organization per land unit. It intensifies the beneficial effects of ecosystem functions, including biodiversity, soil fertility and homeostasis.
>
> (Niggli *et al.*, 2008: 34)

Originating from the organic sector, this concept was eventually promoted as agroecological, with broader relevance for improving all agriculture (ARC2020 *et al.*, 2012; Bellon, 2016).

Eco-functional intensification emphasizes resource conservation and recycling, towards lower dependence on external inputs:

> Diversified land use can open up new possibilities for combining food production with biomass production and on-farm production of renewable energy from livestock manure, small biotopes, perennial crops and semi-natural non-cultivated areas. Semi-natural grasslands may be conserved and integrated in stockless farm operations by harvesting biomass for agro/bio-energy and recapturing nutrients from residual effluent for use as supplementary organic fertiliser on cultivated land.
>
> (Schmid *et al.*, 2009: 26)

For example, spatial crop diversification encompasses intercropping annual grain species, cultivar mixes, perennial grains, or forage species and forestry and annual crops. For greatest success, such methods would depend on changes in plant breeding, agricultural extension services and marketing (Jensen *et al.*, 2015).

In those ways, eco-functional intensification has overlaps with the earlier concept 'ecological intensification.' Initially it meant more precise, efficient use of external inputs for commodity crops (Cassman, 1999). Later it acquired agroecological meanings, i.e. mimicking and incorporating natural ecological interactions to intensify productivity. Thus the concept has meant 'intensification in the use of the natural functionalities that ecosystems offer' (Chevassus au Louis and Griffon, 2008).

Seen as a set of techniques, 'ecological intensification' has trade-offs between several aims – productivity versus resilience, biodiversity, ecosystem services, etc. But those aims have potential synergies, depending on agroecosystem design (Geertsema *et al.*, 2016). There is debate on the aims of agroecological practices – for merely incremental improvement or else transformational change (Brym and Reeve, 2016: 214).

The full potential depends on wider agroecosystems beyond the farm:

> New avenues for agronomy to strengthen agroecological intensification should go beyond the cultivated field or the mixture of species in a given landscape. They should explore desirable properties and mechanisms that operate at the scale of complex socio-ecological systems, i.e. that take into account sociological and ecological dynamics and interactions in agroecosystems.
>
> (Doré *et al.*, 2011: 203)

Ecological intensification emphasizes ecological processes and ecosystem services. These depend on stakeholder involvement to strengthen practitioners' knowledge of such processes (Wezel *et al.*, 2015).

Those diverse meanings and aims have entered debates on sustainable intensification (SI). According to participants at a UK multi-stakeholder workshop, SI encompasses various biotechnological, agroecological and other methods – meant to increase yield, while also lowering the burdens on land use and natural resources. The workshop report elaborated ways to link SI with an agroecosystem approach, central to agroecology (Garnett and Godfray, 2012). But the latter's advocates remain doubtful about its prospects within SI. From the Soil Association, promoting organic certification, one participant described his experience of SI agendas as follows:

> for unquestioning investment in developing new agricultural inputs instead of a greater focus on agroecology; for a less caring approach to animal welfare or biodiversity in the name of upping production; for a narrow definition of sustainability that leaves little room for fairness, health or changes in consumption practices. In effect, for agribusiness as usual with a light green tinge.
>
> (Tom Macmillan, in Garnett and Godfray, 2012: 41)

Indeed, SI easily becomes a proxy for techno-fixes. At a Royal Society meeting on sustainable intensification, all five speakers emphasized the imperative to increase yields as, in turn, an imperative for GM crops. As a commentator reported sarcastically: 'In their minds, only GM has the potential to solve all the world's food problems and be kinder to the environment and wildlife at the same time.' Instead 'we need to put much more of our effort into improving agriculture in ways we can be sure won't actually make food production less sustainable' (Young, 2013).

Techno-fixes also complement biodiversity protection through land-sparing rather than land-sharing. The former has been the dominant policy agenda, favoured by major interests such as the European Crop Protection Association and the European Landowners Organization. According to their joint briefing on agri-biodiversity, 'the loss of ecological heterogeneity as a consequence of agricultural intensification and landscape homogenization has had negative impacts on other bird species' (ELO & ECPA, 2009: 26). From their standpoint, however, modern agricultural technologies have raised yields and can do so more sustainably on existing agricultural land. Such improvements minimize demand for land, thus leaving more available for biodiversity conservation (ibid.). This implicitly supports land-sparing, rather than land-sharing through broader agroecosystems (EP, 2016: 7; Green *et al.*, 2005). Such policy conflicts underlie divergent forms and aims and intensification.

Relevance to that debate depends on different ways of framing 'productivity.' As a flexible output/input ratio, this can be interpreted in multiple ways as a basis for framing trade-offs between productivity versus biodiversity. Next let us examine two expert reports exploring trade-offs in a European agricultural context

Minimizing trade-offs through land-sparing?

A high-profile report on SI has come from the Rural Investment Support for Europe (RISE) Foundation, chaired by Franz Fischler, the former EU Commissioner for Agriculture (Buckwell *et al.*, 2014: 95). The study was funded partly by the European Landowners Organization (ibid: 95). Some affiliates seek to reserve non-cultivated land for lucrative recreational purposes such as game-hunting, consistent with a land-sparing policy (cf. ELO & ECPA, 2009 above).

According to the RISE report, Europe's main agricultural problem is land abandonment, alongside unsustainable intensification and rising food imports:

> Agricultural encroachment onto new lands is not the problem in the EU; indeed the reverse process of agricultural abandonment is more often of concern for environmental and social reasons. The critical EU issue is that the past intensification of agriculture is associated with pervasive undesirable environmental impacts in Europe. An additional concern is that agricultural imports into the EU are associated with environmental damage in the exporting countries.
>
> (Buckwell *et al.*, 2014: 7)

This problem-diagnosis informs the report's rationale for SI, namely: Further increments in global food output must come largely from higher yields on existing agricultural land, e.g. as a means to substitute for imports. Sustainability aspects need most attention in Europe, which anyway cannot feed the world and should reduce its dependence on imports. How?

Towards improvements, the key concept is defined as follows: 'Sustainable Intensification means simultaneously improving the productivity and environmental management of agricultural land…' Productivity is attributed different meanings: 'A sustainable intensification path could mean an increase in the output per hectare of environmental services of the farm or an increase in agricultural products per hectare; it does not only mean the latter' (Buckwell *et al.*, 2014: 7, 76). Broader than yield alone, productivity can encompass those public goods.

Agricultural strategies have various trade-offs, as depicted in Figure 2.1. Generally, intensification 'involves some reduction in environmental performance in exchange for increased food production, but staying in the sustainable quadrant. In the past, production choices have led to sacrifice of some environment for food output' (Buckwell *et al.*, 2014: 17). The latter appears where the black curve moves downwards, denoting loss of plant species.

From the large dot, arrows towards the black curve indicate familiar trade-offs, stereotypically known as intensification versus extensification. By contrast, vertical and horizontal arrows towards the biodiversity-yield frontier bound the

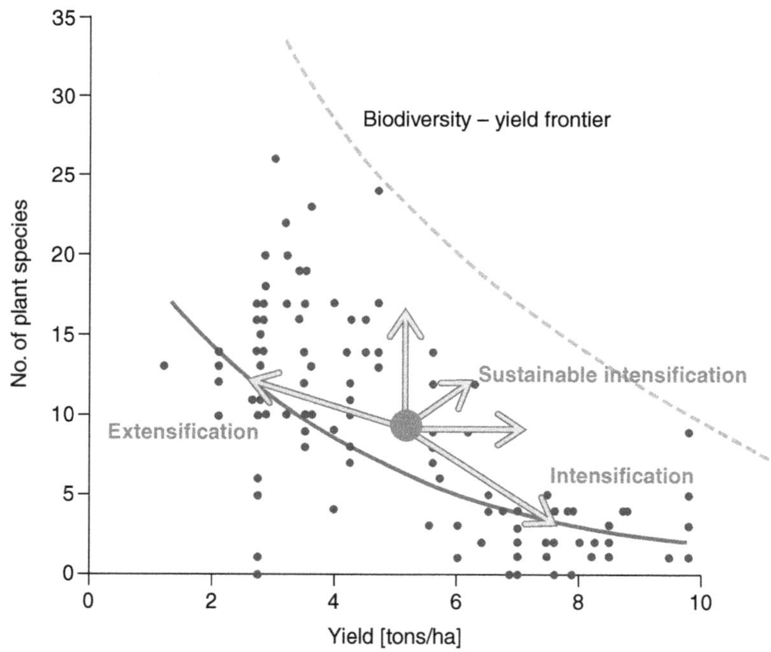

Figure 2.1 Sustainable intensification – biodiversity (based on Buckwell *et al.*, 2014: 70). Source: by permission of the RISE Foundation.

'sustainable quadrant.' This encompasses various methods which maintain or increase biodiversity, while also potentially increasing yield (Buckwell *et al.*, 2014: 70).

How to reconcile those aims? The report mentions six sustainable systems: agroecology, biodynamic, organic, integrated, precision farming and conservation agriculture. To reduce chemical inputs, for example, these inputs can be substituted through nutrient recycling (ibid: 65).

As a key message, protecting natural resources could mean withdrawing some land from cultivation: 'In this context the potential output loss from the further withdrawal of a few percentage points of land to provide biodiversity and water protection could be replaced by a relatively few years' productivity growth' (Buckwell *et al.*, 2014: 55). Consistent with a land-sparing strategy, the report emphasizes biodiversity as an environmental good – but not as agroecosystems, a crucial basis for agroecological methods to be effective through land-sharing.

For policy relevance, the report anticipates SI gaining impetus from agri-environmental and subsidy criteria of the EU's Common Agricultural Policy, especially its requirement for Ecological Focus Areas. However, that policy framework exemplifies conflicts around intensification (see penultimate section).

Creating public goods through agroecosystems

To clarify preferable forms of SI, UK state agencies commissioned a report from the Land Use Policy Group (LUPG), who used the opportunity to promote agroecological methods. Looking beyond yield alone, the report defines productivity in terms of lower resource inputs and public goods. The report notes trade-offs of several kinds:

> Productivity also implies efficiency with respect to resources used (and their related emissions), which may involve a trade-off between yields per hectare and, for example, fossil energy use and GHG emissions per kg of food produced
>
> (Lampkin *et al.*, 2015: 66)

As some farmers shift from agrochemical to agroecological methods, they replace external with internal inputs. There are still trade-offs between productivity versus biodiversity:

> In situations where there was a greater reduction in agrochemical use, and a greater uptake of practices such as reliance on biological nitrogen fixation and soil fertility-building phases in rotations, there was a trade-off involving reduced yields alongside an increased output of environmental goods and associated (non-provisioning) ecosystem services. Despite this, levels of efficiency (in terms of non-renewable resource use and emissions) were not necessarily any worse and often improved.
>
> (Lampkin *et al.*, 2015: 110)

This counts as advantageous by emphasizing environmental goods, and broadly defining efficiency in terms of resource usage rather than yield per unit land.

The report warns against single-technique fixes, which can cause problems. Specific practices in isolation (e.g. bio-substitutes for fertilizers and pesticides) can undermine crop protection and yields.

> Sustainable intensification, if focused mainly on producing more with less, represents only the first step on the way. While some initiatives may encourage input substitution, for example replacing harmful pesticides with less harmful alternatives, this does not imply implementation of an agroecological approach.
>
> (Lampkin *et al.*, 2015: 10)

Towards the latter approach, combining suitable practices can generate synergistic relationships. Substituting for chemical fertilizers, legumes fix nitrogen; they also support pollinators, as well as improving the nutritional and health value of forage crops for livestock (Lampkin *et al.*, 2015: 109).

A modest step, input substitution could stimulate a farmer to make greater changes towards agroecosystems: 'a system redesign approach based on ecological principles is considered more likely to get closer to a sustainable end point' (ibid). Such systems are 'knowledge-based rather than technology-intensive' (ibid: 108). Greatest benefits come from 'a whole-system redesign approach focused on the farm ecosystem,' dependent on biodiversity beyond cultivated fields (cf. Altieri, 1999).

If an SI agenda emphasizes greater provision of environmental goods and ecosystem services, then agroecology has a significant contribution. Likewise if productivity broadly encompasses public goods per unit resource input, rather than simply yield per unit land (see earlier). While the RISE report implicitly complements a strategy of land-sparing, the LUPG report advocates land-sharing (Buckwell *et al.*, 2014: 92). This resonates with a wider agroecological agenda:

> A major argument for wildlife-friendly farming and agroecological intensification is that crucial ecosystem services are provided by 'planned' and 'associated' biodiversity, whereas the land sparing concept implies that biodiversity in agroecosystems is functionally negligible.
>
> (Tscharntke *et al.*, 2012; ibid)

What are the obstacles? Market signals incentivize higher yield. Moreover, UK institutions serving agriculture 'are resistant to changes in their worldview, with a continuing strong focus on technology-driven production increases ...' (Lampkin *et al.*, 2015: 116). Indeed, this could describe how SI selectively appropriates aspects of agroecological methods. Next let us look at European policy agendas for agri-intensification and their subsidy incentives.

EU policy tensions around agri-intensification

In the European Union, farmers' alliances promote different forms of innovation, especially in agendas for agri-innovation and for agri-subsidy. Both relate likewise to different forms of intensification. Let us examine these differences.

Stakeholders' divergent agri-innovation agendas

Europe's conventional farmers are represented politically by COPA-COGECA, promoting a neo-mercantilist export-orientated strategy as the main basis to continue subsidies under the Common Agricultural Policy or CAP (Potter and Tilzey, 2005). According to this lobby, 'To be able to satisfy the demand for sustainable intensification it is essential to have better conditions for more efficient use of water, fertilizers and other resources at farm level' (COPA-COGECA, 2012: 7). Here SI becomes an output-input efficiency for global market competitiveness, thus setting criteria for preferable agronomic techniques. Along these lines, many European farmers have sought to become more competitive by reducing external inputs.

By contrast, widespread agroecological initiatives in the global South have inspired similar efforts in Europe, where the ecological aspects have been articulated with socio-political ones (Stassart *et al.*, 2012). According to a broad European alliance of farmers' and civil society groups: 'the solution lies in a high degree of self-sufficiency and food sovereignty at local, regional, national or continental level,' where people have 'the right to establish their own agriculture and food policy' (ARC2020 *et al.*, 2010). For the European affiliate of La Via Campesina, 'Agroecology as understood by social movements is complementary and inseparable from food sovereignty we want to build' (ECVC, 2013; also 2015).

In an EU policy context emphasizing innovation, mainly meaning capital-intensive technology, agroecology has been promoted as a different kind of innovative practice. It combines four types of innovation – know-how, organizational, social and technological (IFOAM EU Group *et al.*, 2012). This reverses the stereotype of experts transferring techniques to farmers. Agroecological methods have gained interest from conventional farmers facing a cost-price squeeze, potentially driving them off the land.

Agroecological perspectives have been incorporated into the EU's research priorities. When the Directorate-General for Research hosted expert foresight studies, these advocated agroecological approaches:

> Approaches that promise building blocks towards low-input high-output systems, integrate historical knowledge and agroecological principles that use nature's capacity and models nature's system flows, should receive the highest priority for funding.
>
> (SCAR FEG, 2011: 8)

Since 2008 the EU organics lobby has been making detailed proposals for agroecological research topics (Niggli, 2008), with support from COPA-COGECA's

Organics section. Many were adopted and funded in Framework Programme 7 (2007–14), eventually with the umbrella term 'eco-functional intensification.' This diverges from the productivist agenda of agri-industry (for detail, see Levidow *et al.*, 2014). Both agendas were extended into the successor programme, Horizon 2020 (2014–2020).

Meanwhile, within their agenda for social justice and sustainability, civil society organizations have jointly promoted agroecology. This is 'based on applying ecological concepts and principles to optimize interactions between plants, animals, humans and the environment while taking into consideration the social aspects that need to be addressed for a sustainable and fair food system' (GHA, 2017: 5). This cites the FAO's agroecology initiative, whose European conference declared, 'Agroecology principles should be formulated and used as the principle guideline to transform and improve the current food system' (FAO, 2016b: 3). Civil society groups shaped its agroecology initiative, by contrast with the prevalent FAO agenda for a productivist intensification.

As another policy arena, the European Commission set up European Innovation Partnerships (EIP) in several sectors. The one for Agricultural Productivity and Sustainability had a High Level Steering Board representing diverse stakeholders, especially the agri-supply industry, as well as several CSOs allied with small-scale farmers' groups. Disagreements arose about what forms of productivity would be environmentally sustainable. Agri-industry representatives emphasized greater yield with fewer external inputs; by contrast, the CSO-farmer network emphasized environmental sustainability and biodiversity through practices enhancing farmers' knowledge (cf. ARC2020 *et al.*, 2012).

EIP-Agri's Strategic Implementation Plan encompassed all those approaches, e.g. organic farming, low-external input systems, sustainable intensification, etc. 'The diversity of knowledge (local/traditional know-how and practices, common knowledge and expert knowledge) in the definition of research problems, the definition of people concerned, and in finding solutions should be valorised …' (EIP-A, 2013: 5, 9). Designed for practitioners, its Operational Groups have been facilitating farmers' joint knowledge-production with experts, including agroecological methods (TP Organics, 2017).

Greening the CAP: productivity versus biodiversity

By contrast with that modest success in research and innovation arenas, the farmer-CSO network has had greater difficulty influencing the EU's criteria for agricultural subsidy. European agriculture has attracted a long-standing debate on its future trajectory: either to intensify cultivation methods for greater yield and global economic competitiveness, or else to extensify methods through lower external inputs for higher-quality products and biodiversity conservation. The latter trajectory depends on agroecological methods, enhancing various public goods within and beyond agriculture (Schmid *et al.*, 2012).

Despite global pressures for trade liberalization, the European Union maintains significant subsidy aiding agri-production through the Common Agricultural

Policy (CAP) under the broad concept 'multifunctionality.' Underlying the CAP is a neo-mercantilist techno-modernist agenda seeking competitive advantage for commodity export; this agenda has shaped the CAP's pillar 1 Single Farm Payments, remunerating each farm according to its production units. In parallel pillar 2, the European Agricultural Fund for Rural Development (EAFRD), supports farmers' livelihoods in marginal, less productive areas; these have undergone economic pressures to become more market-orientated, e.g. through 'quality' products and/or agroecological methods aimed at affluent reflexive consumers (Potter and Tilzey, 2005). Tensions between those trajectories have pervaded the CAP.

In the run-up to the 2013–2020 CAP, numerous CSOs and farmers' groups formed ARC2020, an alliance bringing together progressive and radical trends of the alternative food movement. Resulting partly from their campaign, the European Commission proposed measures for 'greening' the subsidy criteria, namely: In the CAP's first pillar, Green Direct Payments must comprise at least 30 per cent of the national budget for direct farm payments. Farmers can be remunerated for three obligatory practices – maintenance of permanent grassland, ecological focus areas and crop diversification (DG Agriculture, 2013; EC, 2013). For each farm larger than 15 hectares of arable land, 5 per cent must be covered by Ecological Focus Areas (EFA). These are meant to bring environmental benefits, improve biodiversity and maintain attractive landscapes – such as through landscape features, buffer strips, afforested areas, fallow land, areas with nitrogen-fixing crops, etc.

However, the agro-industry lobby criticized the Commission's proposals for undermining farmers' economic competitiveness. COPA-COGECA successfully lobbied the European Parliament to weaken the requirements. Consequently the EFAs have flexible criteria allowing member states and farmers to bypass or even contradict the biodiversity aim. Agronomists warned that the EFAs had weak prospects for benefitting biodiversity (Pe'er, 2014).

Those early warnings were soon vindicated. As a wider context, catch crops are fast-growing crops grown between successive plantings of a main crop, sometimes helping to retain minerals. Some member states have favoured productive options such as catch crops and nitrogen-fixing crops, thus facilitating intensive cultivation methods rather than biodiversity. In many cases, such crops supplement chemical fertilizers rather than replace them (Lanker, 2016). Moreover, some farmers plough up semi-natural grassland, despite its importance for biodiversity (ARC 2020, 2016). Such harmful practices remain eligible within the flexible rules for subsidy.

Having advocated the reforms, an EU-wide NGO-farmer alliance then advised national campaigns how to shape and use the EU rules for a truly 'greening' agenda, especially through agroecological methods (ARC2020, 2013). They jointly published a guide for how 'the new CAP can help transition EU agriculture towards agroecological approaches,' linking these with food sovereignty. As the guide warned, however, their agenda conflicts with the EU's efforts at 'promoting and incentivising factory-style agriculture based on models of sustainable intensification or by simplifying the system at the expense of agroecological outcomes' (ARC2020, 2015: 1).

In those ways, the CAP encompasses divergent forms of intensification. Subsidy rewards bio-substitution methods for greater yield with lower input costs for farmers. As an implicit form of SI, these practices serve a neo-mercantilist agenda whereby agro-industrial farmers are meant to become more globally competitive. This agroecological appropriation reinforces the dominant food system. Thus a critical perspective on SI helps to illuminate conflicts over 'greening' European agriculture.

Conclusion: tensions of SI

Let us return to the original questions: How do agri-intensification agendas appropriate agroecological practices for reinforcing the dominant agro-food system? or else for contesting it? How do such issues arise in policy arenas, especially in Europe? through what alliances and agendas?

The term SI itself has undergone change and debate. After the 2007 global food crisis, mainstream institutions extended the original SI concept from small-scale peasant farmers to global relevance. SI was now meant to address multiple problems including food insecurity, environmental degradation and climate change, all within dominant assumptions about farmers competing on global markets through higher yield, conflated with 'productivity.' From earlier debates about sustainable agriculture, the question 'what to sustain?' has been displaced by the question 'what and how to intensify?'

Agroecological practices are niche innovations which can play various roles in relation to the incumbent agro-food regime. Such divergent roles correspond somewhat to 'stretch-transform' versus 'fit-conform' strategies (cf. Smith and Raven, 2012), as follows.

For at least a decade, agroecological practices have been promoted by alliances among farmers, civil society organizations (CSOs) and agronomists. Their agenda valorizes wider public goods, partly to enhance agroecological methods themselves, based on farmers' knowledge-exchange. Their agenda links agroecology with food sovereignty for contesting and transforming the dominant agro-food regime.

More recently some agroecological methods have been selectively appropriated by the dominant regime, especially through the umbrella term sustainable intensification (SI). This is meant to raise yields with less environmental harm through a broad 'toolkit' – including agroecological methods, GM crops, no-till (optionally with herbicides), etc. With better techniques from experts, farms would become more competitive within conventional agro-food markets, thus potentially legitimizing agri-industrial products as 'green.'

In its latter prevalent role, SI can be understood by linking three theoretical perspectives, namely: A neo-productivist paradigm seeks to reconcile productivist aims with resource conservation (Marsden, 2012). The neoliberal food regime has generated a reformist agenda seeking to incorporate smallholders who would otherwise be marginalized or expelled. This agenda acknowledges that all potential solutions are needed, while narrowing them to technical options (Holt Gimenez and Shattuck, 2011). And a nascent corporate-environmental food regime aims to alleviate environmental harm from agro-industrial systems and/or to establish distinctive 'green' production chains (Friedmann, 2005, 2009). Moreover, here those perspectives have illuminated how SI implicitly relates to European policy agendas, e.g. land sparing versus land sharing, research priorities, and agri-subsidy criteria.

Agricultural systems entail trade-offs between productivity versus sustainability, as a rationale to seek innovative methods which can minimize such trade-offs or even offer synergies. Although productivity can broadly include environmental services, in practice it is more narrowly seen as the yield of specific commodity crops, in turn attributed to specific techniques such as bio-input-substitutes. The latter approach complements a land-sparing strategy, i.e. raising yield on some land in order to keep other land outside cultivation, e.g. for biodiversity conservation. By contrast, an agroecosystems land-sharing strategy enhances synergies between biodiversity and productivity, understood more broadly. This divergence is manifest in two European expert reports analysed earlier.

Prevalent SI agendas involve only two components of agroecology – scientific knowledge and agronomic practices. As the third component, social movements are essential for agroecological experiments to build farmers' collective knowledge and build support for a transformative role (Wezel *et al.*, 2009). Combining all three components, a European CSO-farmer alliance has sought policy change favouring agroecological practices within a food sovereignty perspective.

For that CSO-farmer alliance, policy arenas have had different opportunities and outcomes. They have gained greater support for their agroecological agenda in EU research and innovation priorities. But they had only modest success in the agri-subsidy 'greening' agenda, originally meant to enhance biodiversity, crucial for agroecosystem transformation. After strong lobbying from the agri-industry lobby, subsidy payments remain available for high-yield 'productive options,' even if such methods supplement agrochemicals, thus fitting the dominant system.

Like other 'reformist' initiatives, SI potentially divides the progressive and radical trends of alternative food movements (cf. Holt Gimenez and Shattuck, 2011: 133–134). Yet instead the two trends have come closer together through joint CSO-farmer policy agendas, at least in Europe. From their standpoint, the prevalent SI agenda has become suspect for subordinating agroecological methods to agro-industrial systems and conventional supply chains.

What strategic implications? From a transformative perspective on the incumbent agro-food regime, SI may seem yet another example of conventional agriculture's capacity to sustain the unsustainable (cf. Buttel, 2005, 2006). Nevertheless SI offers entry points for sharpening debate on diverse aims and forms of productivity. Agri-intensification initiatives also may offer opportunities for CSO-farmer alliances to press for agroecosystem approaches based on farmers' knowledge. These can push practices beyond the dominant neoliberal regime, even beyond 'green' corporate-environmental alternatives. By recognizing different forms of intensification, alliances can develop better strategies for intervening in agri-policy arenas, regardless of whether SI is explicit.

Questions to explore

- How do some forms of agri-intensification become favoured or presumed by policy agendas?
- How do the issues around SI arise in various policy contexts, explicitly or implicitly?
- How do various stakeholder groups form alignments supporting or opposing specific forms of agri-intensification?
- How do those alignments strengthen or marginalize efforts to transform the dominant agri-food regime?

Acknowledgements

Versions of the argument here were presented at three events: the 'European Agroecological Practices: Action-research for a transformative role' workshop in Brussels, May 2016; the International Sociological Association (ISA) Vienna conference in July 2016; and the Agroecology Europe Forum, Lyon in October 2017. All elicited useful comments from participants. Likewise comments fom Lili Balogh, Stéphane Bellon, Doug Constance and Mark Tilzey.

References

Almås, R. and Campbell, H. (eds) (2012). *Rethinking Agricultural Policy Regimes: Food Security, Climate Change and the Future Resilience of Global Agriculture*. Emerald Group Publishing Limited, Bingley, UK.

Altieri, M.A. (1999). The ecological role of biodiversity in agroecosystems, *Agriculture, Ecosystems and Environment* 74(1): 19–31.

Altieri, M.A. (2004). Agroecology versus Ecoagriculture: balancing food production and biodiversity conservation in the midst of social inequity. *CEESP Occasional Papers* 3, agroeco.org/wp-content/uploads/2011/02/ECOAG.pdf

ARC2020 (2013). *CAP and Rural Development Policy Reform Deal for 2014–2020*, https://docs.google.com/file/d/0B0aA8VrCTsRAcHdkdllNYnU5ZG8/edit

ARC2020 (2015). *Transitioning Towards Agroecology: Using the CAP to Build New Food Systems*. Berlin: Agricultural and Rural Convention (ARC2020), IFOAM-EU and FoEE, www.arc2020.eu/wp-content/uploads/2015/02/arc2020-brochure-5-with-hyperlinks.pdf

ARC2020 (2016). CAP greening threatens ecologically precious land, www.arc2020.eu/2016/03/cap-greening-threatens-ecologically-precious-land/

ARC2020, IFOAM EU & TP Organics (2012). *Agro-Ecological Innovation Project: Progress and Recommendations*. Brussels: Agricultural & Rural Convention 2020, International Federation of Organic Agriculture Movements, Technology Platform Organics, http://agro-ecoinnovation.eu/toolbox/july-2012-workshop-materials/

Arora, S. and Nijbroek, R. (2016). How to move beyond technology in 'sustainable intensification,' STEPS Centre, Brighton, http://sro.sussex.ac.uk/65564/, http://steps-centre.org/blog/is-something-missing-from-sustainable-intensification/

Bellon, S. (2016). Contributions croisées de l'agriculture biologique à la transition agroécologique, *Innovations Agronomiques* 51: 121–138.

Bernard, B. and Lux, A. (2017). How to feed the world sustainably: an overview of the discourse on agroecology and sustainable intensification, *Regional Environmental Change* 17(5):1291–1301, DOI 10.1007/s10113–016–1027-y

Brym, Z.T. and Reeve, J.R. (2016). Agroecological principles from a bibliographic analysis of the term agroecology, Chapter 5 of E. Lichtfouse (ed.), *Sustainable Agricultural Reviews* 19: 203–231, www.springer.com/gp/book/9783319267760

Buckwell, A., Nordang Uhre, A., Williams, A., Polakova, J., Blum, W.E.H., Schiefer, J., Lair, G., Heissenhuber, A., Schiebl, P., Kramer, C., Haber, W. (2014). *The Sustainable Intensification of European Agriculture*. Brussels: RISE Foundation.

Burton, R. and Wilson, G. (2012). The rejuvenation of productivist agriculture: the case for 'cooperative neo-productivism,' in Reidar Almås, Hugh Campbell (ed.) *Rethinking Agricultural Policy Regimes: Food Security, Climate Change and the Future Resilience of Global Agriculture* (*Research in Rural Sociology and Development*, Volume 18): 51–72. Bingley BD16 1WA, UK: Emerald Group Publishing Limited.

Buttel, F.H. (2005). The environmental and post-environmental politics of genetically modified crops and food, *Environmental Politics* 14: 309–323.

Buttel, F.H. (2006). Sustaining the unsustainable: agro-food systems and the environment in the modern world, in Cloke, P., Marsden, T., Mooney, P.H. (eds), *Handbook of Rural Studies*, pp. 212–229. London: Sage Publications.

Carolan, M. (2006). Social change and the adoption and adaptation of knowledge claims: Whose truth do you trust in regard to sustainable agriculture? *Agriculture and Human Values* 23(3): 325–339.

Cassman, K.G. (1999). Ecological intensification of cereal production systems: Yield potential, soil quality, and precision agriculture, *PNAS* 96(11): 5952–5959, www.pnas.org/content/96/11/5952.full

Chevassus au Louis, B. and Griffon, M. (2008). La nouvelle modernité: une agriculture productive à haute valeur ecologique. *Déméter: Économie et Stratégies Agricoles* 14: 7–48, http://clubdemeter.com/pdf/ledemeter/2008/la_nouvelle_modernite_une_agriculture_productive_a_haute_valeur_ecologique.pdf

Clapp, J. (2017). The trade-ification of the food sustainability agenda, *Journal of Peasant Studies* 44(2): 335–353.

Clapp, J. and Fuchs, D. (eds) (2009). *Corporate Power and Global Agrifood Governance*. Cambridge, MA: MIT Press.

Constance, D. (2010). Sustainable agriculture in the United States: a critical examination of a contested process, *Sustainability* 2: 48–72, www.mdpi.com/2071-1050/2/1/48/pdf, DOI:10.3390/su2010048

Constance, D. and Moseley, A. (2018). Unpacking sustainable intensification: discourses from agribusiness, in this volume.

Conway, G. (2009). Food security and agricultural development: the challenges for R&D. Presentation to CGIAR Science Forum, www.scienceforum2009.nl/Portals/11/Conway-pres.pdf

COPA-COGECA (2012). Research and Innovation as the driving force for sustainable, productive and competitive European agriculture, RES(12)5658:4. Brussels: Committee of Professional Agricultural Organisations (COPA), www.copa-cogeca.be

DG Agriculture (2013). *Overview of CAP Reform 2014e2020*. http://ec.europa.eu/agriculture/policy-perspectives/policy-briefs/05_en.pdf

Doré, T., Makowski, D., Malézieux, E., Munier-Jolain, N., Tchamitchian, M., Tittonell, P. (2011). Facing up to the paradigm of ecological intensification in agronomy: revisiting methods, concepts and knowledge, *European Journal of Agronomy* 34: 197–210.

Douglass, G.K. (ed.) (1984). *Agricultural Sustainability in a Changing Word Order*. Boulder, CO: Westview.

EC (2013). Regulation (EU) no 1307/2013 of the European Parliament and of the Council of 17 December 2013 establishing rules for direct payments to farmers under support schemes within the framework of the common agricultural policy. OJL 347/608, 608e670.

ECVC (2013). Proposal of position text on agroecology. Brussels: European Coordination Via Campesina (ECVC), www.eurovia.org/

ECVC (2015). Highlights from the public debate on Agroecology in the EU. Brussels: European Coordination Via Campesina (ECVC), www.eurovia.org/spip.php?article1299

EIP-A (2013). *Strategic Implementation Plan*: European Innovation Partnership, Agricultural Productivity and Sustainability (EIP-A), Directorate-General for Agriculture and Rural Development, http://ec.europa.eu/agriculture/eip/pdf/strategic-implementation-plan_en.pdf

ELO-ECPA (2009). *Biodiversity and Agriculture*. Brussels: European Crop Protection Association and the European Landowners Organisation, www.elo.org/UserFiles/File/42445%20Agriculture%20Biodiversity.pdf

EP (2016). *Biodiversity and Agriculture*. European Parliament, Research Service, www.europarl.europa.eu/RegData/etudes/BRIE/2016/583842/EPRS_BRI(2016)583842_EN.pdf

Escobar, A. (2011). *Encountering Development: The making and unmaking of the third world*. Princeton, NJ: Princeton University Press.

FAO (n.d.). Sustainable Crop Production Intensification (SCPI). Rome: Food and Agriculture Organisation, www.fao.org/agriculture/crops/thematic-sitemap/theme/spi/scpi-home/framework/sustainable-intensification-in-fao/en/

FAO (2009a). *Global Agriculture Towards 2050*. Paper presented to High-level Expert Forum on 'How to feed the world in 2050,' 12–13 October. Rome: Food and Agriculture Organization, www.fao.org

FAO (2009b). *Strategic Framework 2010–2019*.

FAO (2011). *Save and Grow: A policymaker's guide to the sustainable intensification of smallholder crop production*, www.fao.org/3/ea20b790-4a84-57fc-aa98-a367a012ddb8/i2215e.pd

FAO (2016a). *Key messages for agricultural biotechnologies symposium*. Rome: FAO, www.fao.org/3/a-bc613e.pdf

FAO (2016b). *Recommendations from the participants*. Regional Symposium on Agroecology for Sustainable Agriculture and Food Systems in Europe and Central Asia, 23–25 November 2016 – Budapest, www.agroecology-europe.org/wp-content/uploads/2017/01/Recommendations-Regional-Symposium-Europe-and-Central-Asia.pdf

Friedmann, H. (2003). The political economy of food: a global crisis, *New Left Review* 197: 29–57, www.neaculture.it/The%20Political%20Economy%20of%20Food.pdf

Friedmann, H. (2005). From colonialism to green capitalism: Social movements and the emergence of food regimes, in *New Directions in the Sociology of International Development: Research in rural sociology and developments*, vol. 11, F.H. Buttel, and P.D. McMichael (eds): 227–264. Amsterdam: Elsevier, www.emeraldinsight.com/doi/abs/10.1016/S1057-1922%2805%2911009-9

Friedmann, H. (2009). Discussion: Moving Food Regimes forward: Reflections on the Symposium Essays. *Agriculture and Human Values* 26, 4: 335–344.

Garnett, T. and Godfray C. (2012). *Sustainable Intensification in Agriculture. Navigating a course through competing food system priorities*, Food Climate Research Network and the Oxford Martin Programme on the Future of Food, University of Oxford, UK. Funded by the UK Government's Foresight Programme as part of its follow up activities to the *Future of Food and Farming* Project. Plus commentaries.

Garnett, T., Appleby, M.C., Balmford, A., Bateman, I. J., Benton, T.G., Bloomer, P., Burlingame, B., Dawkins, M., Dolan, L., Fraser, D., Herrero, M. Hoffmann, I., Smith, P., Thornton, K., Toulmin, C., Vermeulen, S.J., and Godfray, H.C.J. (2013). Sustainable Intensification in Agriculture: Premises and Policies. *Science* 5: 33–34.

Geertsema, W., Rossing, W., Landis, D.A. (2016). Actionable knowledge for ecological intensification of agriculture, *Frontiers in Ecology and the Environment* 14(4): 209–216

GHA (2017). *For Peace, People and Planet: A Civil Society perspective on the next EU Research Framework Programme (FP9)*. Brussels: Global Health Advocates and numerous other organisations.

Giraldo, O.F. and Rosset, P.M. (2017). Agroecology as a territory in dispute: between institutionality and social movements, *Journal of Peasant Studies*, http://dx.doi.org/10.1080/03066150.2017.1353496

Green, R.E., Cornell, S.J., Jo'rn, P.W., Scharlemann, P.W. and Balmford, A. (2005). Farming and the fate of wild nature, *Science* 307: 550−555f.

Holt Giménez, E and Shattuck, A. (2011). Food crises, food regimes and food movements: rumblings of reform or tides of transformation? *Journal of Peasant Studies* 38(1): 109–144.

IFA (2015). Declaration of the International Forum for Agroecology, Nyeleni Centre, Mali, 24–27, www.foodsovereignty.org/forum-agroecology-nyeleni-2015

IFOAM EU Group, ARC 2020, TP Organics (2012). *Agro-ecology: Ten examples of successful innovation in agriculture*, Brussels, http://agro-ecoinnovation.eu/wp-content/uploads/2012/11/Eco_Innovation_broch_24pages_ENG_lr.pdf

Jensen, E.S., Bedoussac, L., Carlsson, C., Journet, E-P., Justes, E., Hauggaard-Nielsen, H. (2015). Enhancing yields in organic crop production by eco-functional intensification, *Sustainable Agriculture Research* 4(3): 42–50, 10.5539/sar.v4n3p42

Kloppenburg, J. (1988). *First the Seed: The Political Economy of Plant Biotechnology, 1492–2000*. NY & Cambridge: Cambridge University Press.

Lampkin, N.H., Pearce, B.D., Leake, A.R., Creissen, H., Gerrard, C.L., Girling, R., Lloyd, S., Padel, S., Smith, J., Smith, L.G., Vieweger, A., Wolfe, M.S. (2015). *The Role of Agroecology in Sustainable Intensification*. Report for the Land Use Policy Group (LUPG). Organic Research Centre, Elm Farm and Game & Wildlife Conservation Trust.

Lanker, S. (2016). Lighter shade of green – CAP fails in Germany & beyond, www.arc2020.eu/2016/04/a-lighter-shade-of-green-cap-fails-in-germany-beyond/, https://slakner.wordpress.com/2016/04/13/greening-of-direct-payments-first-preliminary-figures-on-the-eu-level/

Lélé, S.M. (1991). Sustainable development: a critical review, *World Development* 19(6): 607–621.

Levidow, L., Pimbert, M., Vanloqueren, G. (2014). Agroecological research: conforming — or transforming the dominant agro-food regime? *Agroecology and Sustainable Food Systems* 38(10): 1127–1155.

Loos, J., Abson, D.J., Chappell, M.J., Hanspach, J., Mikulcak, F., Tichit, M. (2014). Putting meaning back into 'sustainable intensification,' *Frontiers in Ecology and Environment* 12(6): 356–361.

LVC (2013). *From Maputo to Jakarta: 5 years of agroecology in La Vía Campesina*. Jakarta: La Vía Campesina, www.landaction.org/IMG/pdf/De_Maputo_a_Yakarta_EN-web.pdf

Marletto, G., Franceschini, S., Ortolani, C., Sillig, C. (2016). Towards sustainable food: a contested transition, Mapping Sustainability Transitions: 23–38. SpringerBriefs in Business book series.

Marsden, T. (2012). Towards the new productivism: sustaining the unsustainable in a new period of regulatory instability, in Gert Spaargaren, Peter Oosterveer, Anne Loeber (eds), *Food Practices in Transition: Changing Food Consumption, Retail and Production in the Age of Reflexive Modernity*: 291–311. London: Routledge.

Niggli, U. (ed.) (2008). *Vision for an Organic Food and Farming Research Agenda to 2025*, Brussels: IFOAM- EU Group, www.organic-research.org/index.html, http://orgprints.org/13439/

Nyéléni (2016). Agroecology at a crossroads – between institutionalization and social movements, Nyéléni Newsletter 28, https://viacampesina.org/en/wp-content/uploads/sites/2/2016/12/2016-12-14 Nyeleni_Newsletter_Num_28_EN.pdf

Ollivier, G. and Bellon, S. (2013). Dynamiques des agricultures écologisées dans les communautés scientifiques internationales: une rupture paradigmatique à rebondissements, *Nature, Sciences et Sociétés* 21(2): 166–181, www.nss-journal.org/articles/nss/abs/2013/02/nss130093/nss130093.html

Pe'er, G. (2014). EU agricultural reform fails on biodiversity, *Science* 344(6188): 1090–1092.

Potter, C. and Tilzey, M. (2005). Agricultural policy discourses in the European post-Fordist transition: neoliberalism, neomercantilism and multifunctionality, *Progress in Human Geography* 29(5): 581–600.

Pretty J. (1997). The sustainable intensification of agriculture, *Natural Resources Forum* 21(4): 247–256.

Royal Society (2009). *Reaping the Benefits: Science and the sustainable intensification of global agriculture*, London.

SCAR FEG (2011). 3rd SCAR foresight exercise: Sustainable Food Consumption and Production in a Resource-Constrained World, Brussels: Standing Committee on Agricultural Research, Foresight Expert Group, available from http://ec.europa. eu/research/agriculture/scar/pdf/scar_feg3_final_report_01_02_2011.pdf.

Schmid, O., Padel, S., Levidow, L. (2012). The bio-economy concept and knowledge base in a public goods and farmer perspective, *Bio-based and Applied Economics (BAE)* 1(1): 47–63, http://orgprints.org/20942/, www.fupress.net/index.php/bae/article/view/10770

Schmid, O., Padel, S., Halberg, N., Huber, M., Darnhofer, I., Micheloni, C., Koopmans, C., Bugel, S., Stopes, C., Willer, H., Schluter, M., Cuoco, E. (2009). *Strategic Research Agenda for Organic Food and Farming*, Brussels: Technology Platform Organics, www.organic-research.org/index.html

Smith, A. and Raven, R. (2012). What is protective space? Reconsidering niches in transitions to sustainability, *Research Policy* 41(6): 1025–1036.

Stassart, P.M., Baret P., Grégoire J-C., Hance T., Mormont M., Reheul D., Stilmant D., Vanloqueren G., Visser M. (2012). L'agroécologie: trajectoire et potential pour une transition vers des systèmes alimentaires durables, in D. Van Dam, J. Nizet, M. Streith et P.M. Stassart (eds), *Agroécologie entre pratiques et sciences sociales*: 25–51. Dijon: Educagri Editions, www.philagri.net/wp-content/uploads/2013/05/Agroecologie-Stassart-Baret-et-al.-GIRAF-version-web-2.2.pdf

Syngenta (2013). *Syngenta Contribution to Sustainable Agriculture*, Basel: Syngenta, www4.syngenta.com/~/media/Files/S/Syngenta/documents/syngenta-sustainable-agri-brochure.pdf

Tilzey, M. (2017). Reintegrating economy, society, and environment for cooperative futures: Polanyi, Marx, and food sovereignty, *Journal of Rural Studies* 53: 317–334, www.sciencedirect.com/science/article/pii/S0743016716306970?via%3Dihub

TP Organics (2017). *Innovating for Organics: Organics in EIP-AGRI Operational Groups*. Brussels: Technology Platform Organics.

Tscharntke, T., Clough, Y., Wanger, T.C., Jackson, L., Motzke, I., Perfecto, I., Vandermeer, J., Whitbread, A. (2012). Global food security, biodiversity conservation and the future of agricultural intensification, *Biological Conservation* 151: 53–59.

UN (2008). *Comprehensive Framework for Action* (CFA), NY: UN High-Level Task Force on the Global Food Crisis, www.un.org/ga/president/62/letters/cfa160708.pdf

Wezel, A., Soboksa, G., McClelland, S., Delespesse, F., Boissau, A. (2015). The blurred boundaries of ecological, sustainable, and agroecological intensification: a review, *Agronomy for Sustainable Development* 35(4): 1283–1295, DOI 10.1007/s13593-015-0333-y

Young, R. (2013). Sustainable intensification. Bristol, UK: Sustainable Food Trust, http://sustainablefoodtrust.org/articles/sustainable-intensification/

3 Sustainable intensification as a sociotechnical imaginary

Paul B. Thompson

Sustainable intensification has arguably emerged as the mainstream agricultural scientist's vision for agricultural production systems of the twenty-first century. This chapter will sketch the contours of this vision by recounting some possible origin stories for it. Succinctly, the advocates of sustainable intensification accept the imperative to increase the productivity of agriculture (this is what intensification means). They also recognize imperatives to limit the ecological footprint of agriculture while ameliorating some of the more dramatic environmental consequences of the industrial production techniques of the twentieth century (that is what sustainable means). Thus one possible origin story for sustainable intensification lies in a particular interpretation of the challenges that the agricultural sector must meet during the twenty-first century on a global basis. Related to this is a story that recounts a particular philosophy of agriculture that is driven by specific ethical commitments. Before discussing these narratives, I will tell a more conventional story about how I personally came to be acquainted with sustainable intensification – an acquaintance that predates the emergence of widely circulated scientific manifestos in the early 2010s. However, it will prove helpful to undertake some preliminary expectorations prior to these exercises in storytelling.

Before talking about what the advocates of sustainable intensification have in mind for future food systems, some clarification of vision is in order. That is, what, in general, *is* a vision for agriculture such that participants in mainstream agricultural science could have one? It is at least arguable that many scientific disciplines operate without any social or cognitive construct that could be characterized as a 'vision' at all. As such, the chapter is organized as follows. First, a discussion of future food systems as socio-technical imaginaries – a theoretical construct derived from science studies that substantially refines the notion of a vision for agriculture. Second, a memoire of my own encounter with this socio-technical imaginary dating back to the beginnings of the twenty-first century. The third section of the chapter provides a more analytic discussion of some specific challenges that global agriculture is expected to encounter during the coming decades, and argues that sustainable intensification has been imagined as a way that traditional applied biology disciplines in agricultural research institutions can meet them. The fourth section draws upon this analysis but stresses a normative conception of agriculture's services to society along with a diagnosis of how the agricultural technologies of the last century failed these norms in key respects. The chapter concludes with a few critical remarks on sustainable intensification.

Socio-technical imaginaries

The concept of socio-technical imaginaries must have its origins in the philosophy of Cornelius Castoriadis, a French intellectual who, like more celebrated figures of the 1970s, struggled with the *ménage à trois* of structural linguistics, Marxism and Freudian psychoanalysis that dominated European social theory in the decades following World War II. Castoriadis introduced the social imaginary in his book *The Imaginary Institution of Society*. It was a deeply ingrained and largely unconscious set of constructs that, like a Hegelian *Zeitgeist*, shaped a given generation's understanding of reality and hence determined (for them) the limits of practical possibility. Castoriadis stressed the way that these constructs attained causal efficacy as performatives, as utterly contingent (and hence capable of being different) but nevertheless attaining what Michel Foucault was calling 'power' in virtue of being acted out and acted upon by people who did not think to question them (Castoriadis, 1987). Although for Castoriadis the social imaginary was pervasive, for later theorists such as Charles Taylor it has been pluralized: several (possibly many) distinct social-imaginaries emerge and compete with one another (Taylor, 2004).

The notion of a social imaginary has been developed by Sheila Jasanoff along with her students and collaborators working in science studies, who, like Taylor, have emphasized the role of science in fixing modern conceptualizations of the real. Going beyond Taylor, Jasanoff sees science-led technological innovation as a material performance that realizes contingent possibilities, shaping the potential for future action. The socio-technical imaginary is offered as a theoretical construct for organizing both historical and sociological analysis of technological innovation in a given sector or region. The construct is intended to stress both the co-production of technical means and social institutions, and also the sense in which the innovation and implementation of a socio-technical system is temporally structured. Jasanoff is thus drawing both on the post structural social theory of Castoriadis and Taylor, and on an older and more modest idea that came from the literature in organizational development. The socio-technical system concept was coined by scholars who recognized that the work-flow process in advanced manufacturing was a function of both engineering design and institutions such as work rules, organizational hierarchy and customary practices within a work force. The term implies that analysis of work-flow process requires a systems model that includes both technical and social elements (Cooper and Foster, 1971).

Writing with Sang-Hyun Kim, Jasanoff argues that changes in socio-technical systems occurring over time can be mapped in terms of a plan or anticipatory sequence. Key actors coordinate their activity according to this shared vision in order to move the system from one state to another. In the cases Jasanoff and her students have studied, operationalization of change involves not only transformation of social practices, but also invention and development of tools and techniques that do not even exist at the time when the imagined futures originate in the minds of actors. It is, in fact, misleading to characterize this coordinating schema as a plan because effecting the transformations is known to require significant innovation in both technical means and social organization. Jasanoff and Kim thus refer to the schema that allows agents to cooperatively enact the transformative operations as an 'imaginary.' It is a cognitive artefact – a mental picture – shared in varying degrees (and with varying degrees of overlap) by actors who engage in constructive and co-productive socio-technical innovations (Jasanoff and Kim, 2009; 2013).

In some studies on socio-technical imaginaries, the artifact might accurately be characterized as an evolving scenario that is held in mind by a visionary individual. Cecil Rhodes, for example, brought a fairly typical bourgeois attitude to the extractive industry of Southern Africa in the 1870s: he saw mining as an opportunity for profit. Yet in realizing this opportunity, he engaged in both social and technical innovations through networking, labour relations and, on the technical side, the development of open pit mining and physical infrastructure for the work force and the creation of a rail transport system, (Storey, 2015). This should not be taken to imply that Rhodes envisioned this totality at the outset. Nevertheless, in William Kelleher Storey's analysis the creation of a mining industry in Southern Africa required close coordination between the property rights, labour rules and governmental authorities that were being created and the as-yet-to-be-innovated technical means for converting unmined diamonds into a pecuniary asset. In Southern Africa, the trace of a schematic and evolving plan for this coordination arose in the mind of a single individual.

In other studies, the socio-technical imaginary is both less specific and more widely shared. Clark Miller's work on 'securitization' of contemporary politics stresses the role of science in identifying such wide-ranging phenomena as economic development and climate change as 'threats' that can be registered by governing bodies and mobilized politically in support of policy initiatives. The theme has permeated all levels of state administration and beyond to emerging institutions for global governance. While politics might be thought of as a process for achieving solidarities in support of diverse human ends, science has come to be in service to politics by articulating the consonance of personal and state security (Miller, 2015). While Storey's characterization of sociotechnical imaginaries in Southern Africa emphasized the mind of a single individual, Miller's approach to global governance traces key discursive tropes that link science and policy across a wide variety of distinct social networks. Here the sociotechnical imaginary enables power-relations supporting multiple actors with diverse interests. The sociotechnical imaginary of technologically enforced security spreads as a meme throughout a variety of contexts.

Jasanoff's students have drawn upon a wide range of research methodologies in developing the concept of socio-technical imaginaries, including archival research, analysis of contemporary documents and texts, interviews and participant-observation in research development activities. Kim uses archival sources to trace South Korea's resistance to imports of U.S. beef in the wake of the bovine spongiform encephalitis (BSE) crisis, and interprets these sources as an instance of neoliberal political discourse (Kim, 2015). Nancy Chen relies on planning documents and informant interviews to construct her account of how Chinese molecular biologists have usurped the traditional assumption that innovation will be farmer led, replacing it with a sociotechnical imaginary that is expert led, consigning farmers to roles with no substantial leadership or intellectual contribution (Chen, 2015). Elta Smith relies primarily on open-ended informant interviews to characterize the emergence of a socio-technical imaginary around the development of golden rice (Smith, 2015). The socio-technical imaginary construct is thus not dependent on a particular social science research method. It is an ideal type that allows the analyst to explain and predict innovation and co-production of science and society by organizing the action of individuals and groups around vaguely shared goals and expectations concerning the prospects for change in the status quo. This methodological pluralism is also evident in the analysis that follows.

Sustainable intensification: early textual sources

Given this range of specification, sustainable intensification exemplifies a sociotechnical imaginary that is less the product of a visionary individual than it is a widely shared understanding of the purposes and possible capabilities for agricultural production systems, as well as the obstacles to realization of those capabilities that can be anticipated over a time horizon of two to five decades. Jules Pretty offered an early characterization of sustainable intensification in a *National Resources Forum* article published in 1997. Pretty situated his analysis between what he characterized as 'business as usual optimism' – the view that mainstream agricultural science could continue to deliver impressive gains in yields – and 'environmental pessimism,' the prediction of utter environmental collapse in global food systems. Pretty's view of sustainable intensification accepted the need to constrain global agricultural production within its current ecological foot print, but he also argued that low-input production technologies could meet food needs, provided that agricultural science was coupled tightly to farmer-led research and innovation programmes (Pretty, 1997).

Other authors were also using the phrase 'sustainable intensification' in the late 1990s. Dan Clay, Tom Rearden and Jaakko Kangasniemi (1998) use the concept to describe a series of conservation investments being made by Rwandan smallholders. Writing with three other co-authors, Reardon uses the phrase to examine the impact of agricultural policy changes on farm-level investment in conservation practices (Reardon *et al.* 1999). Indeed, Reardon utilized a number of related phrases to describe the focus of his work on agricultural development on the African continent, including 'sustainable agricultural productivity,' and simply 'sustainability issues.' This work had a specifically

African focus and an emphasis on farm-level strategies for conserving soil and water resources. It was implicitly focused on smallholder production, but Reardon's articles do not articulate an ethical or ideological commitment to either low input approaches or smallholder participation in the research planning process.

Timothy Reeves uses the concept in a 1999 editorial that appeared in *Nature Biotechnology*. Reeves was then Director General of the International Maize and Wheat Improvement Center (CIMMYT) in Mexico, the international agricultural research centre that was most closely associated with the work of Norman Borlaug. Reeves cites Borlaug frequently in urging the need for continued agricultural research in order to keep up with the pace of global food demand. He argues specifically for research on key plant pests, rather than general approaches to increase yield potential, and advocates biotechnology as a promising approach for breakthroughs in these areas (Reeves, 1999). Although Reeves emphasizes that new agricultural technologies should strengthen the economic viability of small farmers, he does not echo Pretty's stress on low-input approaches or on research methods that rely on strong input from small farmers.

Significantly, none of these authors offer stipulative definitions for sustainable intensification, nor do they suggest that the term is being coined specifically for the ideas being presented in these early papers. This indicates that the phrase 'sustainable intensification' was, if not familiar, at least in use among scholars contributing to the literature on international agricultural development. Other sources from the 1990s suggest that the phrase may have emerged from an elision of 'sustainable livelihoods,' and 'agricultural intensification.' By the 1990s, the phrase 'sustainable livelihoods' was being used to emphasize developmental initiatives that emphasize the stabilization and enhancement of rural household income. It was a successor to development theories that stress 'putting people first' (Cernea, 1985) and targeting interventions that could be incorporated into the production systems of smallholders having little access to credit and capital. Agricultural intensification was a commitment to the strategy of increasing the efficiency of production processes, as opposed to 'extensification,' or increasing total food production simply by bringing new land into production. White papers exploring the links between sustainable livelihoods and agricultural intensification were being circulated contemporaneously with the appearance of the papers that were introducing the phrase sustainable intensification (Carswell, 1997; Scoones, 1998). By the year 2000, a white paper by Scott Swinton began by referring to 'the debate over sustainable intensification' (Swinton, 2000) further indicating that this turn of phrase was being used frequently, at least in international development.

Given this literature from the 1990s, it is clear that the phrase 'sustainable intensification' was being used in development circles. However, it is not clear that there is enough coherence among these textual sources to warrant the claim that it had emerged as a 'socio-technical imaginary.' While for Pretty – a researcher long associated with low-input and sustainable production strategies – the phrase did indeed embrace an alternative to mainstream 'Green Revolution' style agricultural development, for Reeves it was being used to advocate an updated version of the Green Revolution approach. Reardon was using the phrase specifically for conservation oriented initiatives in Africa, but they were largely consistent with conservationist strategies that have been studied within agricultural research at least since the Dust Bowl years of the 1930s. Given that an Internet search turns up no references to sustainable intensification prior to 1990, it seems that this phrase had emerged in the wake of the World Commission on Environment and Development report of 1987, which had stressed 'sustainable development,' and that a number of development specialists had merged longstanding themes and theories into a thrust that would eventually come to be identified as 'sustainable intensification.'

Sustainable intensification: a memoir

My personal encounter with sustainable intensification began in one of the dining rooms adjacent to the cafeteria at the Food and Agricultural Organization (FAO) headquarters building in Rome. The year was late 1999 or early 2000. FAO was in the midst of an agricultural ethics initiative that consisted primarily in the development of white papers framing and reflecting on issues in the development and application of new food production, processing and distribution technologies. An initial document had established that ethics could be characterized simply as an explicit articulation and defence of how food production *ought* to be configured, and had illustrated how controversies over chemical use and the relative winners and losers from yield enhancing technological innovations reflect contrasting answers to this normative question. Other papers were still in development (Stannard, 2008). Peter Kenmore and Clive Stannard were engaging me in conversation about the need for an ethics white paper that would perform two functions. There was a felt need to issue a white paper that displayed firm cognizance of reigning ideas in ethical theory, and to sketch in broad strokes how competing approaches in theoretical ethics would dictate alternative ways of conceptualizing right and wrong with respect to food systems. There was also a need to produce a white paper on sustainable intensification. The substance of our conversation was whether both of these objectives could be combined in a single effort.

This was a conversation, not a semi-structured interview. My recollection of it should be interpreted neither as a definitive portrait of the mindset at FAO, nor as an accurate reflection of my conversationalists' own views. It was, nonetheless, the basis for my own indoctrination into sustainable intensification, and my memory of it can thus serve as one data point among many for tracing the evolution of sustainable intensification as a sociotechnical imaginary. What follows in this section of the chapter is a blend of reminiscence and reconstruction. It is my testimony as to what sustainable intensification appeared to mean within the FAO, roughly between 2000 and 2005.

Stannard and Kenmore described an FAO that was divided over the record of previous attempts to transfer science-based and Western-style technology to agricultural sectors that were still dominated by smallholders who were still dedicating a significant portion of their efforts towards meeting their own subsistence needs. It was, in effect, a difference of opinion on the Green Revolution and whether or not investment in technologies that increase the efficient conversion of land, labour and energy inputs into harvestable commodities should continue. At the bottom of it all was an impressive body of social science research showing how the introduction of higher-yielding dwarf varieties in Mexico and Southeast Asia had been accompanied by dramatic social transformations. Green Revolution varieties of wheat and rice were more responsive to fertilizers, but the infrastructure for supplying seeds and fertilizer was poorly developed. Farmer adoption of new varieties was uneven and proved risky for smallholders with virtually no margin for error. Even slightly better capitalized farmers could bear these risks and began to realize significant gains. This fuelled economic growth in the agricultural sector, but at the expense of the poor and with significant turmoil in local property and governance institutions, (Griffin, 1979). More fertilizer use generally led to more monoculture and to the introduction of toxic chemicals for controlling insects and other plant pests. In addition to eutrophication from fertilization itself, these chemicals – generally unregulated in less-developed economies – caused pollution of ground and surface waters and had significant harmful impacts on farmworker health and safety, (Dahlberg, 1979).

By the early years of the twenty-first century, there were two points of view on the Green Revolution. On one side, many biologically trained scientists had simply concluded that the social costs were more than offset by the substantial gains in total farm commodity production. The argument was often put in Malthusian terms: whatever the social costs, industrialization of the farm sector was at least producing enough edible commodities to meet a constantly growing global demand, (Trewavas, 2002). The leading advocate of this perspective was, of course, Norman Borlaug himself, winner of the Nobel peace prize for his work on higher-yielding crop varieties. By the time of my conversation at FAO, Borlaug had begun to make the argument for continued efficiency-oriented innovation in connection with gene transfer and biotechnology (Borlaug, 2000). This perspective apparently had a fairly robust constituency at FAO, especially among those trained in the biologically-based agricultural sciences and among a cadre of economists who saw growth in the agricultural sector as a key prod to a more general economic expansion in the less industrially developed economies of Africa, Asia and Latin America.

The opposing perspective was held by those who stressed the unequal distribution of harmful impacts, which were disproportionately being borne by the poor, women and where relevant, by ethnic minorities. Vandana Shiva had resuscitated the earlier criticisms of the Green Revolution, framing them as threats to environmental justice and re-interpreting them as forms of gender oppression, (Shiva, 1991). Although Shiva's failure to acknowledge the earlier round of critiques that had been generated from within the community of scholars and activists working on agricultural development limited the influence of her critique at FAO itself, her notoriety brought new critics of technology-led development strategies into the debate. Inside the FAO, the opposition to Borlaug-style variety development and technology transfer emphasized less intrusive activities that were designed above all else to protect and preserve the social institutions on which the poorest of the poor depend. This had translated into a great deal of scepticism about biotechnology, but more broadly into a posture of opposing all food production technologies that were premised on increasing biological efficiencies.

In the minds of Stannard and Kenmore (or at least the way I remember it), sustainable intensification was being imagined as a way to break the deadlock between those who venerated Borlaug and the Green Revolution, and those who demonized this period in the history of agricultural development. Sustainable intensification was thus a third way. Sustainable intensification was an explicitly values-based approach to agricultural development. On the one hand, it acknowledged that total food production *would* need to increase, and that increasing food production by allowing existing smallholder approaches to expand geographically was unacceptable. Conversion of tropical rainforests to farming or livestock production would harm biodiversity and contribute to climate change by destroying carbon sinks. On the other hand, smallholder lives and livelihoods should not be treated as a 'cost' that might be offset by benefits either in the form of increased food production or in terms of environmental goals like biodiversity preservation. Recognizing both of these values, as opposed to only one of them, narrowed the set of acceptable strategies for agricultural development programmes. Yet it was also a way to envision how the two schools of thought on the Green Revolution's legacy might converge on a strategy for the future.

Given this framing, sustainable intensification was a sociotechnical imaginary in the following sense. First, it articulated parameters for both technical innovation and the social institutions governing land use, commodity production and infrastructure development. Although all these things needed to change, the change that was needed is specified in terms of twin goals: constraining the ecological footprint of agriculture while respecting the integrity and livelihoods of smallholders. Second, although my recollection is sketchy on this point, it *did* envision certain existing agricultural technologies as models for the type of innovation that would be consistent with these goals. Biological control of plant pests, for example, was seen as a means more consonant with the limited financial means of smallholders than chemical and genetic methods that required both purchased inputs and a regime for protecting intellectual property. Broadly speaking, Pretty's adaptation of the low-input, sustainable agriculture (LISA) model was deemed to be more attuned to the socioeconomic sustainability of farmers using traditional methods. Finally, sustainable intensification was being promoted as a forward-looking vision that could shape and steer innovations in policy as well as tools and techniques. Indeed, within the context of FAO it was envisioned as an alternative to then dominant visions that were stymied by seemingly irreconcilable differences.

The ethics of sustainable intensification

Subsequent to my conversation, I worked with Stannard, Kenmore and others at FAO to produce a white paper entitled *The Ethics of Sustainable Intensification* (FAO, 2004). My role in this project was to help FAO staff in articulating how key values being discussed in connection with agricultural development strategies would be supported (or not) by recognized literatures in philosophical ethics. The white paper illustrates how increases in efficiency of production are supportable on utilitarian social choice grounds, while neo-Kantian rights theories would constrain initiatives that achieve otherwise compelling goals at the expense of individuals. In short, to pursue social goods in a manner that accepts loss of life, destruction of a household or the sacrifice of dignity simply as a 'cost' is to treat the individuals so affected simply as a means to an end, violating the master principle of morality articulated by Immanuel Kant (1724–1804). The white paper also identified the importance of virtues in the evaluation of technical means, arguing that highly disruptive technical changes can undercut the solidarities and communal identity processes that are necessary for reinforcing norms of character such as friendship, fidelity, piety, citizenship and even courage, (FAO, 2004).

After the paper was completed, Michigan State University and the Wye agricultural campus of Imperial College, London organized a small conference to further explore the ethical implications of sustainable intensification. Andrew Dorward's summary report emphasizes that ethics has a complex meaning in the context of agricultural development. First, it is an acknowledgement of the values held by farmers and others who live at the site where development is intended to occur. Second, ethics is (or at least should be) a component of the development policy planning process, where planners, government officials and development specialists review and discuss the rationale for configuring agricultural development initiatives in accordance with one set of goals and constraints, or another. These plans and policies are, in turn, the causal impetus for changes in the distribution of costs, benefits, and the power to control the course of future events. These redistributions have ethical resonance that is felt by affected parties, and *this* sense of policy ethics should be distinguished from the process-role for ethics in the actual planning process. Given this redistributive impact, there is then a *further* sense in which the use of technology – now conditioned by the new institutional structure – can have ethical significance. Dorward argues that explicit attention to these multiple roles for ethics would strengthen the planning and implementation of agricultural development, but his summary of the meeting does not explicitly address sustainable intensification as a framework (Dorward, 2008). Again based on my recollection of the meeting, Dorward's analysis is an accurate one: it was ethics, rather than sustainable intensification as such, that was the main focus of conversation.

However, several papers presented at the meeting or prepared subsequently do address sustainable intensification more directly. John Perkins and Rachel Jamison provided a historical overview of the values driving agricultural research. They note a focus of increasing biological yields that is rather insensitive to the impact of resulting production systems on the lifestyle or economic viability of farm households or on rural communities. Within industrial economies, farm households accepted the adverse impacts of technical change in exchange for public policies that strengthened land tenure through legal property rights, but this political bargain was becoming less and less tenable as the environmental costs were becoming harder and harder to ignore. Perkins and Jamison conclude by emphasizing the environmental challenges implied by sustainable intensification (Perkins and Jamison, 2008). Commenting on their analysis, Michiel Korthals argues that Perkins and Jamison understate the extent to which agricultural scientists' battle against hunger also entailed a campaign against any and all alternatives. Echoing the neo-Kantian themes in the FAO white paper, Korthals says that fairness and respect for the autonomy of both farmers and food consumers have been sacrificed by an agricultural development strategy in which economic incentives and market processes have been the underlying drivers for change (Korthals, 2008).

Luis Camacho underlines this theme with a critique that addresses sustainable intensification from the perspective of development ethics. Camacho rejects the notion that agricultural development can be seen as distinct from larger ideological forces that shaped all development policy, especially during the last quarter of the twentieth century. He notes that development theory came to be dominated by a neoliberal philosophy of governance which specified the role of governments and international agencies (such as CYMMET or FAO). Under this philosophy, policy should be constructed so that all producers of goods and services are placed into competition with one another, while the workings of a supposedly neutral market will determine which will survive in this competitive environment. Without being too specific about the likely policies that would emerge under a regime of sustainable intensification, Camacho worries that without a development ethics perspective that directly challenges the neoliberal perspective, sustainable intensification would simply replicate the competitive structures that still dominate international agreements and institutions (Camacho, 2008).

The papers by Dorward, Perkins and Jamison, Korthals and Camacho were all published in a volume that I edited, (Thompson, 2008). In the interests of concision, I will report that other contributors have followed Dorward, Camacho and Korthals in stating that the chief virtue of sustainable intensification lay in the way that procedures for ethical reflection and debate were being incorporated within this sociotechnical imaginary. They also went on to explore additional ways in which a full discussion of intensification would be much more complex than the original FAO white paper might have been taken to imply. Clare Palmer discussed how thrusts in environmental ethics might be applied to more clearly specify the criteria that might be applied in evaluating whether any agricultural production system is sustainable (Palmer, 2008). David Fraser adapted one of the other FAO white papers on animal agriculture to argue that impacts on livestock species themselves should be included in any ethical analysis of sustainable intensification, (Fraser, 2008). Succinctly, sustainable intensification was being imagined as a sociotechnical imaginary that was consistent with the FAO vision described in the previous section, but with the important amendment that explicit attention to ethics was to be an important component in the social institutionalization of the sustainable intensification ideal.

Sustainable intensification: back to the future

The 2008 volume on ethics is situated near the end of a hiatus in the textual record for sustainable intensification, and the attempt to incorporate ethics into this imaginary has not drawn much attention from subsequent scholars. In October of 2009 The Royal Academy released a report entitled *Reaping the benefits: Science and the sustainable intensification of agriculture*. The report took a narrowly biological approach to its analysis, arguing that population growth, climate change and economic development would each place pressure on the global food supply, while acknowledging that high input agricultural technology developed during the heyday of the Green Revolution had sometimes caused adverse impacts on water, soils and biodiversity (Royal Society, 2009). The report was silent on social displacement associated with earlier intensification strategies, as well as ignoring the negative socio-economic consequences experienced by women, ethnic minorities and other vulnerable groups. This was followed in 2011 by a government document referenced colloquially as 'the Foresight report.' This study from the UK Office of Science echoed the 2009 message of the Royal Academy and called for increases in research on technologies that could promote sustainable intensification. It is, without question, these two documents, rather than any of the previous literature, which have shaped the subsequent discourse on sustainable intensification.

This literature has grown quite large, especially in the United Kingdom. In 2012 the Food Ethics Council of the UK devoted an issue of its newsletter to a range of viewpoints. A contribution from Tim Benton argues for increasing total production along with efficiency as a way to relieve pressure on the agricultural resource base (Benton, 2012), offering virtually no deviation from the mantra that guided agricultural research throughout the course of the twentieth century. Vicki Hird, from Friends of the Earth, expresses extreme scepticism, describing sustainable intensification as 'business as usual,' from mainstream agribusiness (Hird, 2012). Other contributors to the issue might be said to fall somewhere in between, with most acknowledging that simply increasing the biological efficiency of agricultural production was only one piece of a more complex picture. None of these authors indicate cognizance of a prior literature on sustainable intensification. Several express the opinion that the idea had been invented by the government in the Foresight report.

An American literature was also accumulating. In 2011 the phrase 'sustainable intensification' was used prominently in a paper published in the prestigious and influential *Proceedings of the National Academy of Science* (PNAS). The paper by ecologists David Tilman, Christian Balzer, Jason Hill and Belinda

Befort emphasized agricultural contributions to greenhouse gas emissions. The authors situated their concerns about continued increase in the use of nitrogen fertilizers and other chemical inputs within a context of global environmental justice by arguing that the industrialized North should not be permitted to impose the costs of adjusting agricultural production systems on comparatively poor farmers in less developed nations (Tilman *et al.*, 2011). Although literature on sustainable production systems is discussed, the paper does not cite any of the previous literature on sustainable intensification that is discussed above. Furthermore, the argument presented by Tilman, Balzer, Hill and Befort is largely disconnected from an agricultural development context, and simply does not take up the Green Revolution debate that had shaped thinking within FAO a decade earlier.

By 2013, there appeared to be a significant effort to mount enthusiasm for a new round of investment in agricultural research under the banner of sustainable intensification. A two-page article with many signing co-authors appeared in the Policy Brief section of *Science*. Entitled *Sustainable intensification in agriculture: Premises and Policies*, the article identified four principles that the authors were endorsing under the heading of sustainable intensification:

1 The need to increase production.
2 Meeting this need through higher yields because increasing the area of land in agriculture carries major environmental costs.
3 Emphasizing environmental sustainability as well as productivity.
4 Sustainable intensification is a goal that is not specified in terms of how it should be achieved.

The authors went on to clarify the meaning of the fourth principle by writing,

> The merits of diverse approaches—conventional, 'high-tech,' agro-ecological, or organic—should be rigorously tested and assessed, taking bio-physical and social contexts into account. Building the social and natural science evidence base to allow formulation of context-dependent SI strategies is a research priority.
>
> (Garnett *et al.*, 2013: 33)

Although the article in fact mentions many of the concerns noted within *The Ethics of Intensification*, including animal welfare, the idea of specific attention to ethics in reconciling multiple desiderata is not mentioned.

A 2014 article in the *Philosophical Transactions of the Royal Society* by Charles Godfray and Tara Garnett (the last and lead authors on the 2013 paper in *Science*) spells out the sociotechnical imaginary in slightly more detail. Although Godfray and Garnett begin their analysis by noting the complex and multi-dimensional nature of food systems, they move immediately to a focus on increasing global food supplies. The article includes more detailed discussion of environmental constraints on both 'extensification' (e.g. the strategy of bringing new land into food production) and past approaches to increasing yields through technological innovations. They note especially many of the environmental impacts that have been discussed above, and also reiterate Fraser's emphasis on the well-being of livestock confined within intensive production systems, (Godfray and Garnett, 2014). Notably, the authors do not discuss aspects of food security that are unrelated to total global food supply, and their analysis also omits concerns about fairness, due process and the potential for adverse impacts on smallholder livelihoods that were prominent in the FAO white paper. In many respects, the sociotechnical imaginary being portrayed in the literature from this decade reverts to the vision being articulated in the 1990s. It emphasizes social and technological innovations that economize on the utilization of scarce natural resources such as land, water and soil fertility, and that achieve efficiencies by reducing the production of harmful externalities, including environmental toxins and net increases in greenhouse gas emissions. The addition of animal welfare is a significant modification, but ethics does not appear as a component in this imaginary. The execution of this vision is described as if it could be realized entirely by well-funded technical elites.

A 2017 article presents a meta-analysis of articles discussing sustainable intensification. The authors of this study report massive growth in the use of this expression since 2010, and included over 500 publications in the study. They cite usage by the USDA in connection with the Obama administration's 'Feed the Future Initiative' and they note references to sustainable intensification in materials released by Monsanto as well as the peer-reviewed literature coming out of the UK. The study examines both desired goals articulated for sustainable intensification and specific indicators for sustainability that are proposed. The authors conclude that sustainable intensification does indeed indicate a shift from productivist agricultural research strategies of the twentieth century, largely in virtue of increased attention to indicators of biodiversity and ecosystem functionality that were entirely absent. However they also note that measures for positive contributions of agricultural production continue to be defined in terms of yield. Socio-economic indicators are not prominent (Mahon, Crute, Simmons and Islam, 2017).

The exception to this trend is an entry with Jules Pretty, again, as the lead author in 2011 (and also a member of the team that produced the Foresight report). Following up on the imaginary he had articulated in his 1997 article, Pretty continues to stress the environmental dimensions of sustainability by discussing low-input/low-impact production methods associated with alternative agriculture. Pretty and his co-authors review several projects where such methods were introduced into African production systems, arguing that they do in fact increase yields without also causing harmful environmental impacts or substantially reducing the future availability of resources. They do not discuss ethics, but they do situate their report on sustainable intensification within the context of past debates over the Green Revolution and the need to incorporate protections for fairness and involvement of smallholders into the planning and execution of sustainable intensification. The paper includes an extensive section discussing needed innovations in social infrastructure that build trust and confidence, and they list empowerment and education of women as one of the social elements that they include in their sociotechnical imaginary (Pretty *et al.*, 2011).

Conclusion: a contested imaginary

It would be entirely reasonable to question whether *any* kind of coherent sociotechnical imaginary emerges out of the review completed above. There are key points of agreement across all sources, however. Agricultural systems of the future will need technological innovations, and these innovations cannot achieve increases in total food production in ways that expand on the ecological footprint of agriculture. These limitations stress the importance of biodiversity and carbon sinks. Beyond this, different authors stress different tools and techniques, ranging from conventional techniques of soil and water conservation to innovations in low-input production, on the one hand, or biotechnology, on the other. Authors also differ in terms of whether they see the sacrifice of the autonomy, culture and perhaps even the very lives of poor or marginalized groups in pursuit of agricultural efficiencies as a theme that is worth mentioning in specifying their vision. Only the philosophers are willing to argue that planned and explicit consideration of these issues should be a built-in component of sustainable intensification.

Given these differences, one might conclude that there really *is* no coherence exhibited amongst these varied projects. Indeed, the range of proposals and plans being discussed in the published literature would have been expanded considerably had we reviewed everyone who discusses simply the sustainability of food systems, or mentions some aspect of sustainable agriculture. This suggests that even in the cases reviewed, sustainable intensification is less a thing having definite contours than it is a happenstance phrase, conjured at random, and equivalent to many other ways in which agricultural scientists and policy analysts might think about the sustainability of future food systems. As I have noted elsewhere, there has been a longstanding tendency for many agricultural scientists to simply equate sustainability with the food system's ability to generate a quantity of food *capable* of feeding the world population. Whether or not the food system actually *does* feed everyone is not deemed relevant (Thompson, 2010).

Although my own testimony and remembrance contradicts this interpretation, it is difficult to deny that Vicki Hird appears to be correct in associating sustainable intensification with 'business as usual.' At least for a moment and within a segment of the agricultural development specialist community, sustainable intensification was envisioned as a 'thing,' a roughly imagined future in which certain kinds of technological innovation would go forward in tandem with certain kinds of social change. They would not merely avoid but would in fact correct both social and environmental insults that had been associated with the yield enhancing approach of the Green Revolution. They would be as mindful of fairness and the individual rights of smallholders as they would be of biodiversity and the need to mitigate the processes driving climate change. This imaginary was augmented in a few cases by a recognized need to incorporate critical thinking and engaged discussion of how ethical imperatives were implicated within processes of technical change, and to develop procedures for respecting the rights of affected parties.

Rather than emerging as a socio-technical imaginary that challenges the standing order, mobilizing technical innovation to perform an alternate reality, sustainable intensification is starting to look more like what Castoriadis had in mind when he analysed *the* social imaginary. Largely unreflective routines of agricultural R&D continue, while those who present an alternative vision are simply ignored. The imagination of agricultural scientists is constrained by the actors' inability to recognize contingency and envision different research paths having practical exigency. They remain oblivious to their own participation in those practices that lend an aura of inevitability to the trajectory of technical change. The point of linking ethics to intensification was to break down the group-think that allows the agricultural establishment to just keep doing what it has always been doing for the most part, while looking for band-aids that cover up the bad outcomes.

As I have argued elsewhere, the mindset of agricultural science remains hidebound by the inability to conduct an ethical conversation in a public forum. They may have adopted the discredited philosophy of science advocated by mid-twentieth century logical positivists, which holds discussion of norms and values to be inherently unscientific. Alternatively, they have accepted a simplistic version of economic utilitarianism which holds that farmer adoption of a technology is proof of its ethical superiority (Thompson, 2017). This is less the kind of visionary socio-technical imaginary discussed by Jasanoff and her students than it is a limitation in the agricultural research community's ability to imagine alternatives. It is, in short, the socio-technical imaginary of 'There is no alternative.'

References

Benton, T. (2012). 'Supply and demand: Increasing production and efficiency sustainably,' *Food Ethics: The Magazine of the Food Ethics Council* 7(3): 10–11.

Borlaug, N. (2000). Ending world hunger: The promise of biotechnology and the threat of antiscience zealotry, *Plant Physiology* 122: 487–490.

Camacho, L. (2008). Agriculture intensification from the perspective of development ethics, in *The Ethics of Intensification: Agricultural Development and Cultural Change*, P. B. Thompson (ed.): 97–110, Springer, Dordrecht, NL.

Carswell, G. (1997). Agricultural intensification and rural livelihoods: A 'think piece,' IDS Working Paper 64, Accessed 20 March 2017 at https://opendocs.ids.ac.uk/opendocs/bitstream/handle/123456789/3368/Wp64.pdf?sequence=1

Castoriadis, C. (1987). *The Imaginary Institution of Society*. K. Blamey (tr.), The MIT Press, Cambridge, MA

Cernea, M. M. (ed.), (1985). *Putting people first: Sociological variables in rural development*. Oxford University Press, Oxford, UK.

Chen, N. N. (2015). Consuming biotechnology: Genetically modified rice in China, in *Dreamscapes of Modernity: Sociotechnical Imaginaries and the Fabrication of Power*, S. Jasanoff and S-H. Kim (eds): 219–232, University of Chicago Press, Chicago.

Clay, D., Reardon, T. and Kangasniemi, J. (1998). Sustainable intensification in the highland tropics: Rwandan farmers' investments in land conservation and soil fertility, *Economic Development and Cultural Change* 46: 351–377.

Cooper, R. and Foster, M. (1971). Sociotechical systems. *American Psychologist*, 26: 467–474.

Dahlberg, K. A. (1979). *Beyond the Green Revolution: The Ecology and Politics of Global Agricultural Development*. Plenum Press, New York.

Dorward, A. (2008). Ethics in agricultural change: Questions and proposals, in *The Ethics of Intensification: Agricultural Development and Cultural Change*, P. B. Thompson (ed.): 204–212, Springer, Dordrecht, NL.

FAO (Food and Agricultural Organization of the United Nations) (2004). The ethics of sustainable intensification, FAO, Rome, Accessed 22 March 2017 at www.fao.org/docrep/007/j0902e/j0902e00.HTM

Fraser, D. (2008). Animal welfare and the intensification of animal production, in *The Ethics of Intensification: Agricultural Development and Cultural Change*, P. B. Thompson (ed.): 167–190, Springer, Dordrecht, NL.

Garnett, T., Appleby, M. C., Balmford, A., Bateman, I. J., Benton, T. G., Bloomer, P., Burlingame, B., Dawkins, M., Dolan, L., Fraser, D., Herrero, M., Hoffmann, I., Smith, P., Thornton, P. K. Toulmin, C. Vermeulen, S. J. and Godfray, H. C. J. (2013). Sustainable Intensification in Agriculture: Premises and Policies, *Science* 341: 33–34.

Godfray H. and Garnett T. (2014). Food security and sustainable intensification. Phil. Trans. R. Soc. B 369: 20120273. http://dx.doi.org/10.1098/rstb.2012.0273, http://rstb.royalsocietypublishing.org/content/369/1639/20120273.full.pdf, http://rstb.royalsocietypublishing.org/content/369/1639/20120273.short.

Griffin, K. (1979). *The Political Economy of Agrarian Change: An Essay on the Green Revolution*. 2nd edn Macmillan, London.

Hird, V. (2012). The politics of plenty, *Food Ethics: The Magazine of the Food Ethics Council* 7(3): 19–21.

Jasanoff, S. and Kim, S-H. (2009). Containing the atom: Sociotechnical imaginaries and nuclear regulation in the U.S. and South Korea, *Minerva* 47: 119–146.

Jasanoff, S. and Kim, S-H. (2013). Sociotechnical imaginaries and national energy policies, *Science as Culture* 22: 189–196.

Kim, S-H. (2015). Social movements and contested sociotechnical imaginaries in South Korea, in *Dreamscapes of Modernity: Sociotechnical Imaginaries and the Fabrication of Power*, S. Jasanoff and S-H. Kim (eds): 152–173, University of Chicago Press, Chicago.

Korthals, M. (2008). Two battles in the history of agriculture: Against hunger and against alternatives. Comments on John Perkins' and Rachel Jamison's 'History, Ethics and Intensification in Agriculture,' in *The Ethics of Intensification: Agricultural Development and Cultural Change*, P. B. Thompson (ed.): 91–96, Springer, Dordrecht, NL.

Mahon, N., Crute, I., Simmons, E. and Islam, M. M. (2017). Sustainable intensification–'oxymoron' or 'third-way'? A systematic review. *Ecological Indicators*, 74: 73–97.

Miller, C. A. (2015). Globalizing security: Science and the transformation of contemporary political imagination, in *Dreamscapes of Modernity: Sociotechnical Imaginaries and the Fabrication of Power*, S. Jasanoff and S-H. Kim (eds): 277–299, University of Chicago Press, Chicago.

Palmer, C. (2008). Environmental ethics and agricultural intensification, in *The Ethics of Intensification: Agricultural Development and Cultural Change*, P. B. Thompson (ed.): 131–148, Springer, Dordrecht, NL.

Perkins, J. H. and Jamison, R. (2008). History, ethics and intensification in agriculture, in *The Ethics of Intensification: Agricultural Development and Cultural Change*, P. B. Thompson (Ed): 59–84, Springer, Dordrecht, NL.

Pretty, J. N. (1997). The sustainable intensification of agriculture, *Natural Resources Forum* 21: 247–256.

Pretty, J. N., Toulmin, C. and Williams, S. (2011). Sustainable intensification in African agriculture, *International Journal of Sustainable Agriculture* 9: 5–24.

Reardon, T., Barrett, C., Kelly, V. and Savadogo, K. (1999). Policy reforms and sustainable agricultural intensification in Africa, *Development Policy Review* 17: 375–395.

Reeves, T. V. (1999). Intensification for the nine billion. *Nature Biotechnology* 17: BV33–BV34.

Royal Society (2009). *Reaping the Benefits: Science and the Sustainable Intensification of Global Agriculture*. RS Policy document 11/09. The Royal Society, London.

Scoones, I. (1998). Sustainable rural livelihoods: A framework for analysis, IDS Working Paper 72, Accessed 20 March 2017 at https://opendocs.ids.ac.uk/opendocs/bitstream/handle/123456789/3390/Wp72.pdf?sequence=1.

Shiva, V. (1991). *The Violence of the Green Revolution: Third World Agriculture, Ecology and Politics*. Zed Books, London.

Smith, E. (2015). Corporate imaginaries of biotechnology and global governance: Syngenta, golden rice and corporate social responsibility. In *Dreamscapes of Modernity: Sociotechnical Imaginaries and the Fabrication of Power*, S. Jasanoff and S-H. Kim (eds): 254–276, University of Chicago Press, Chicago.

Stannard C. (2008). Doing Ethics in Food and Agriculture. In: Thompson P. B. (ed.) *The Ethics of Intensification*. The International Library of Environmental, Agricultural and Food Ethics, vol. 16. Springer, Dordrecht.

Storey, W. K. (2015). Cecil Rhodes and the making of a sociotechnical imaginary for South Africa. In *Dreamscapes of Modernity: Sociotechnical Imaginaries and the Fabrication of Power*, S. Jasanoff and S-H. Kim (eds): 34–55, University of Chicago Press, Chicago.

Swinton, S. M. (2000). More social capital, less erosion: Evidence from Peru's Altiplano. Department of Agricultural Economics, Michigan State University, East Lansing. Accessed 20 March 2017 at www.biw.kuleuven.be/aee/clo/idessa_files/Swinton2000.pdf

Taylor, C. (2004). *Modern Social Imaginaries*. Duke University Press, Durham, NC.

Thompson, P. B. (ed.) (2008). *The Ethics of Intensification: Agricultural Development and Cultural Change*, Springer, Dordrecht, NL.

Thompson, P. B. (2010). *The Agrarian Vision: Sustainability and Environmental Ethics*. The University Press of Kentucky, Lexington.

Thompson, P. B. (2017). *The Spirit of the Soil: Agriculture and Environmental Ethics*. 2nd edn. Routledge, New York.

Tilman, D., Balzer, C., Hill, J. and Belfort, B. L. (2011). Global food demand and the sustainable intensification of agriculture, *Proceedings of the National Academy of Science of the United States* 108: 20260–20264.

Trewavas, A. (2002). Malthus foiled again and again. *Nature* 418: 668–670.

UK Office of Science (2011). *Foresight, The Future of Food and Farming: Challenges and Choices for Global Sustainability*. The Government Office for Science, London.

4 Agrifood discourses and feeding the world

Unpacking sustainable intensification

Douglas H. Constance and Athena Moseley

Agrifood studies and feeding the world

The industrial agrifood system is in the midst of a legitimation crisis regarding its negative ecological, economic, and social externalities (Magdoff *et al.*, 2000; Constance *et al.*, 2014; Campbell *et al.*, 2017). This crisis has accelerated steadily over the past 40 years and has now reached a tipping point based on the generalized realization that the current agrifood system is the major contributor to global climate change. To complicate the scenario, by the year 2050 nine billion people will need to be fed without expanding the agricultural land base, while reducing negative environmental impacts such as greenhouse gas emissions (UNFAO 2009; Gardner, 2009; Campbell *et al.*, 2017). In response to this challenge, two visions of agriculture have emerged as the better path forward: (1) sustainable intensification, and (2) agroecology (Levidow, 2015). These two transition paths are the outcome of a long history of competing visions and contested discourses regarding the preferred model of the agrifood system. This chapter examines the historical progression and current manifestation of this critical discourse. The chapter begins with some context on the historical tension between the 'agrarian' and 'industrial' visions of US agriculture. Then it presents the various dimensions of the crisis of industrial agriculture leading to the 'food security' versus 'food sovereignty' discourse. Finally, it ends with a discussion and analysis of sustainable intensification as the new paradigm for agriculture.

Agrifood studies: grounding the discourse[1]

The agrifood system is the outcome of thousands of years of agricultural 'intensification' and 'extensification' (Lenksi, 1970; Thompson, 2010a). Extensification refers to the process of bringing more lands into cultivation; intensification refers to using technical and cultural innovations to increase yield on those lands. Increased yield creates food surplus, which allows population growth and the division of labour, which generates social stratification and conflict over control of the surplus. The control of food surplus is the original source of power by humans over other humans.

After two million years of slow change living as hunter-gatherers, around 12,000 years before the present (YBP) humans domesticated a narrow genetic range of plants and animals which created food surpluses and thereby released many of us from dependence on daily hunting and gathering. Climate and geography combined favourably in some regions to support the development of horticultural, fishing, and pastoral societies. Around 6,000 YBP the first cities emerged, followed by complex, centralized agricultural dynasties about 4,000 YBP. The agricultural revolutions combined the traction and manure fertilizer of domesticated animals with the improved digging stick (hoe, then plough) and domesticated plants. The purposeful manipulation of the conducive environments allowed settlements, which led to a new organizational form – the city, which needed a bureaucratic staff to manage it. The ruling class and its officers rationalized urban life with theology as the source of truth and systematically extracted wealth from rural peasants and slaves to support the urban elites and nascent armies. The state managed the war machine to build empire and extract wealth with force, as militaristic societies replaced peaceful societies. Agricultural surplus generated social hierarchy, a complex division of labour, and systemic rural to urban migration. The synergistic interplay between food provisioning technology, food surplus, population growth, and social organization progressed. Civilization is possible because of agricultural surplus.

Neolithic humans became food producers using subsistence technology such as plant and animal domestication and the digging stick. Per acre yield of grain increased slowly and steadily as early farmers improved their land, labour, breeds, and food technologies. Through conquest and empire, extensification spread the model over new lands and intensification increased productivity. While agriculture developed simultaneously in various parts of the world, the model that emerged in the Fertile Crescent region of the Middle East was diffused to Europe by the Roman Empire, where it was rationalized in Great Britain in the 1600s with the adoption of rotations and formalized agricultural research to stave off the Malthusian crisis. Colonialism spread the model based on a systematic transfer of wealth from the 'New World' to the 'Old World.' Three hundred years ago the industrial revolution began to replace animal power with machines, generating even more food surplus, which made possible rapid population growth and urbanization. The enclosure movement in Great Britain and Europe privatized collective lands, the state created a national market for agriculture, and the railroads linked these markets. Trade and colonialism diffused industrial agriculture around the world.

Scientific agriculture: the paradigm shift

The political, economic, and scientific revolutions of the eighteenth and nineteenth centuries set the stage for wide-scale transformation of society (Polanyi, 1944). Modern thought reduced agriculture to its component parts, which were manipulated to serve humanity. Utilitarian philosophies and Protestant theologies provided a rationale for self-maximizing behaviour as the basis of capitalist society. Social Darwinism justified colonialism and laissez-faire economics dominated politics. This paradigm shift in human thought and organization transformed society and agriculture. The modernization of agriculture is a key dimension of this transformation. The rational application of science and industry to agriculture increased technological innovations exponentially. The new organizational forms called democracy, capitalism, industrialism, and Protestantism provided the cultural innovations. Food surplus increased exponentially.

Industrialization supported agricultural mechanization, which combined with Mendelian seed breeding science and soil science (fertilizers, pesticides, irrigation), enhanced plant and animal yields. Machines provided traction and chemicals provided fertility and pest-control. Economies of scale led to larger fields, specialization, monoculture and steady rural to urban migration. Science was employed to maximize food production and avoid Malthusian famine. In the 1800s universities and government bureaux emerged in Europe, dedicated to scientific agricultural intensification. This modern system was spread globally, especially to the settler states (Perkins and Jamison, 2010).

The US model: the discourses in the land grant universities

During the Civil War the US government took several actions to modernize agriculture (Danbom, 1979): the United States Department of Agriculture (USDA) was created; the Land Grant University system (LGUs) was created; the Homestead Act of 1862 was passed; immigration policies provided industrial workers and prospective farmers; the transcontinental railroad was subsidized; and the Native Americans were subdued. The actions, policies, and programmes accelerated extensification and intensification.

The early LGU system consisted of two competing discourses regarding the preferred model of agriculture and the role of agricultural sciences (Thompson, 2010b): the agrarian ethic and the industrial ethic. The agrarian ethic, referred to as Jeffersonian Agrarianism, sees agriculture as a virtuous social structure with particular cultural norms that enhance quality of life (Berry, 1978). This view was supported by rural sociologists and institutional agricultural economists in LGUs and USDA. The industrial ethic views agriculture as another commodity sector whose role is to efficiently produce food with positivist science. Any negative externalities could be corrected with better technology. This path was supported by natural (soil, animal, plant) scientists, neo-classical agricultural economists, and urban elites.

Populist unrest, the New Deal, and World War II

In the late 1800s and early 1900s populist farmers called on the US government to protect them from the power of the agricultural trusts (grain processing and meat packing). The government responded with laws such as the Packers and Stockyards Act (1921) and the Capper-Volstead Act (1922) to give farmers some protection from predatory monopoly power. In 1929 the Country Life

Commission was created by political and industrial elites to counter the success of the populist farmers (Danbom, 1979; Gilbert, 2015). Contrary to the populist views that saw rural poverty as a result of agribusiness power and economic predation, the Commission argued that rural social problems were due to technological backwardness. The government and the LGUs mobilized to modernize rural America, with special focus on the Cooperative Extension Service. The Country Life Commission and LGU system worked together to create the intensification technologies (machines, genetics, chemicals, confinement animal production) and extend those innovations to new lands. The American Farm Bureau Federation (AFBF) grew out of LGU Cooperative Extension activities as rural counties created farmer education 'bureaus' to teach and support modern farming. A major role for rural social scientists was to conduct research to enhance the adoption and diffusion of modern innovations.

The New Deal institutionalized farm programmes to protect and preserve family-farms, while at the same time supporting extensification and intensification. The Agricultural Adjustment Act and Commodity Credit Corporation in 1933 provided market protections for farmers. The Bureau of Agricultural Economics and Farm Service Agency oversaw progressive programmes to improve the lives of tenant farmers, black farmers, and agricultural workers. Both views of agriculture existed in the LGU system and USDA bureaus during this time, the more industrial, business-oriented view and the more agrarian, community-oriented view. During World War II a coalition led by the AFBF purged the progressive elements in the USDA. The industrial ethos focusing on adoption and diffusion of productivist agriculture dominated state policies and programmes for the next 40 years (Gilbert, 2015).

After World War II, and especially during the Cold War, the productivist approach dominated the LGUs. Productivism combined the mechanization of the industrial revolution, selective breeding and hybrid seeds, and chemical pesticides and fertilizers to maximize yield per acre. The resulting food surplus was employed as a weapon to counter the spread of communism. This adoption-diffusion model of agricultural modernization, based on technological improvements adopted by modern, innovative farmers, was spread to the world through the Green Revolution as 'packages' of agricultural intensification. The US diet based on that model of agriculture was spread through the world through food aid programmes such as Public Law 480: Food for Peace (Buttel and Newby, 1980; McMichael, 1996).

GMOs

Genetically modified organisms (GMOs) are the newest biotechnology of agricultural intensification (Falkner 2009; Williams 2009). GMO technology began in 1973 with the development of a recombinant DNA technique, which enables foreign genes to be inserted into organisms to enhance desired traits or suppress undesired traits. A US Supreme Court ruling in 1980 allowed intellectual property rights for living organisms, which prompted rapid investment and growth of the industry. The first commercial GMOs in the United States were canola, corn, cotton, soybeans, and tomatoes in the 1990s. Life science companies merged with seed and pharmaceutical firms to form a single integrated production process. Further mergers and acquisitions created an oligopolistic structure with a few dominant firms (Howard 2016).

GMO supporters emphasize economic efficiency, environmental sustainability, and food security, while opponents point out the environmental, health, and community risks (Shiva, 1992; Kuntz, 2014). GMO advocates view it as the new paradigm for agriculture that will sustainably feed the world through intensive monoculture. Detractors counter that GMOs are an insufficiently tested technology with environmental and human risks such as genetic pollution and ecosystem degradation, which supports monoculture, corporate concentration, and 'sustaining the unsustainable' (Buttel, 2006). This view supports agroecological systems for local and regional consumption as the preferred agricultural paradigm.

The legitimation crisis: productivism challenged

In the 1970s the productivist model was criticized as a system whereby the USDA and the LGUs were coopted by agribusiness (Hightower, 1973; Buttel and Newby, 1980). In the LGUs a new 'Sociology of Agriculture' emerged to counter the dominant productivist paradigm. It departed from the structural functional frameworks and positivist methodologies that underpinned the adoption/diffusion paradigm to embrace a critical view focusing on the negative relationship between the industrialization of agriculture and quality of life in rural areas (Constance, 2014). Simultaneously, environmental sociology frameworks exposed the extractive and unsustainable ecology of conventional, productivist agriculture (Buttel, 1987). US agriculture was part of a global agrifood system dominated by powerful TNCs employing global sourcing to construct global commodity chains that benefitted corporate shareholders over agricultural producers (Bonanno *et al.*, 1994; McMichael, 1994; Goodman and Watts, 1997). At the international level the value-neutral modernization framework was challenged by the value-laden dependista framework focusing on neocolonialism, whereby the Global North continued to exploit the Global South through corporate domination (McMichael, 1996).

Agrifood studies: chronology of the discourse

Four key questions inform the critical Sociology of Agrifood literature; The Environmental Question, The Agrarian Question, The Food Question, and The Emancipatory Question (Constance, 2008; Constance *et al.*, 2014). The questions progress chronologically but overlap as crises persist. Environmental degradation is the first crisis to reach critical mass, generating social movement resistance and government regulation. The Environmental Question asks, 'What is the relationship between industrial agriculture and the quality of the rural environment?' The answer is that productivist, chemical-intensive, GMO-based monoculture has a negative relationship with the environment, which it extracts wealth from and externalizes costs to in the forms of soil, water, air, and species degradation. The fact that industrial agriculture is the primary contributor to global warming provides evidence that extensification and intensification based on the productivist paradigm is ecologically unsustainable.

The Agrarian Question asks, 'What is the relationship between industrial agriculture and the quality of life for farmers and rural communities?' The answer is that as industrial agriculture increases, the quality of life in rural areas decreases as wealth is transferred from rural communities to agrifood TNCs and their shareholders (Goldschmidt, 1947; Lobao and Stofferahn, 2008). Agrifood globalization in the form of the WTO-coordinated Corporate Food Regime enhances corporate rights over human rights (Magdoff *et al.*, 2000; Fold and Pritchard, 2005; McMichael, 2005; Burch and Lawrence, 2007; Bonanno and Constance, 2008). In response to the legitimation crisis of the 'Food from Nowhere' regime, a broad-based alternative agriculture social movement is creating a 'Food from Somewhere' system (Campbell, 2009; see The Food Question below). The 'Food from Somewhere' agrifood system re-embeds agro-ecological agriculture in local and regional relationships, often referred to as 'Civic Agriculture' and/or 'Fair Trade' (Lyson, 2004; Raynolds *et al.*, 2007).

The Food Question asks, 'What is the relationship between the conventional agrifood system and quality of food we eat?' The answer is that the dominant food system is characterized by an industrial diet of unhealthy 'pseudo-food' commodity chains centred on fats, sugars, starches, salt, and empty carbohydrates (Dixon 2009; Winson, 2013), which generates burgeoning public health crises, inhumane treatment of food animals, and hazardous work lives for farmworkers and processing plant workers. The 'Quality Turn' captured the growing interest in the multi-dimensional aspects of quality in production, processing and consumption (Goodman, 2003; Wright and Middendorf, 2008).

The three previous questions point out barriers to the creation of a just, equitable, and sustainable agrifood system. The Emancipatory Question focuses directly on the ethical dimensions of agrifood systems. The Emancipatory Question asks, 'What is the relationship between the conventional agrifood system and social justice?' The answer is that the Corporate Food Regime privileges the market over civil society, which marginalizes the civil rights of the majority of the people and beings on the planet (McMichael, 2005; Holt Gimenez and Shattuck, 2011; Gertel and Sippel, 2014). The Corporate Food Regime is grounded in neoliberal philosophies and neoconservative politics, and is coordinated by global governance organizations such as the World Trade Organization (WTO), International Monetary Fund, World Bank, North American Free Trade Agreement, and the European Union. This 'Food from Nowhere' regime is implemented by nation-states who compromise their sovereignty to attract corporate investment.

The Emancipatory Question asks specifically, 'What would a fair, just, equitable, and sustainable agrifood system look like?' Social justice is the crucial fault line in agrifood studies (Allen, 2008). How can we transform the agrifood system to be socially, economically, and environmentally just and sustainable for all, including universal human rights, including food animal rights? Does such a system include a right to safe and nutritious food for all? Does it include peasant and farmer rights to access to the land, to the means of production, to keeping their cultures? Does it include social justice for undocumented farm and food workers? The food security versus food sovereignty discourse captures well The Emancipatory Question.

The 'modern' (though still much informed by Malthus) food security discourse begins in the 1940s when the United Nations Food and Agricultural Organization (FAO) was created to establish global food security. Although the FAO embraced the scientific modernization of world agriculture (extensification and intensification), it also included the UN's Universal Declaration of Human Rights, which maintained that food was an essential right of life rather than a commodity. The Cold War subverted FAO multilateralism as the United States employed bilateral food aid to counter the spread of communism. The FAO vision of food as a right was formally replaced in 1986 when the World Bank redefined food security as the ability to buy food. In 1994 the WTO institutionalized the global free trade regime and this market vision of food security, whereby countries grow and trade agrifood products based on comparative advantage and people buy these foods instead of grow them. The WTO's Agreement on Agriculture in 2008 furthered this vision by defining the 'new agriculture' as a system of global entrepreneurial farmers employing 'sustainable intensification' practices linked to agrifood TNCs in flexible arrangements governed by sustainability standards (Ingram *et al.*, 2010; McMichael, 2014).

In contrast, the food sovereignty movement posits a counter frame to food security approaches. Created by La Via Campesina, a global, broad-based, peasant-centred, social movement committed to social justice and human rights, this view from the global South challenges the Corporate Food Regime through protests where it denies the validity of the WTO-sanctioned food security framework based on free trade, corporate intellectual property rights, and land grabs. Instead, La Via builds coalitions to create agrifood self-sufficiency through land reform, indigenous knowledge and agro-ecological principles (Desmarais, 2007; Rosset, 2008; Wittman *et al.*, 2010; Fairbairn, 2012). This perspective proposes to heal the global metabolic rift of industrial agriculture through repossession and regionalization of agrifood systems, between farmers, farm animals, and eaters.

The food sovereignty discourse is grounded in an alternative conception of modernity (Desmarais, 2007; McMichael, 2014). Food security includes a land commodification ontology that assumes that the problem of food supply can be solved through ecological modernization and sustainable intensification, a high-tech repackaging and greening of the modernist adoption and diffusion approaches of the productivist paradigm. This bio-capitalist 'Second Green Revolution' links entrepreneurial global farmers practicing sustainable intensification to agrifood TNC constructed global value chains governed by the WTO free-trade regime. Friedmann (2005) calls this system of green consumers linked to 'greener' companies the Corporate Environmental Food Regime.

In contrast, the land sovereignty ontology views land through a multifunctional lens rather than the commodity lens. Food sovereignty embraces a triple-bottom line, full-cost accounting approach that internalizes the environmental externalities and embraces a rights-based rather than market-centred framework, where rights are defined in collective terms rather than through the liberal conception of individual rights (McMichael, 2014). This ontology requires a repossession of the land in the face of the continuing enclosures based on accumulation through dispossession. The intellectual property rights/copy right framework advanced by the WTO is countered by the copy-left and open-source framework of La Via Campesina. Domestic agrifood production is a better path to food security than global commodity chains (de Schutter, 2008). Moderate and smaller scale agro-ecological farming is more resilient to climate shocks. The battle between La Via Campesina and the GMO seed TNCs over seed sovereignty is a crucial example of the ontological fracture (Kloppenburg, 2010).

The food security and food sovereignty frames proceed from non-reconcilable ontological differences (McMichael, 2014). The food security discourse separates the social and physical sciences and casts traditional agriculturalists as primitive laggards whereas the food sovereignty frame values interdisciplinary approaches, honours indigenous knowledge, and pursues social justice (Rivera-Ferre, 2012). The food security approach lacks a social justice component, which is a central feature of the food sovereignty perspective (Guthman, 2008; Fairbairn, 2012).

Agrifood studies: the two paths forward

The tension between the food security and the food sovereignty visions aligns directly with the two proposed transition paths to a sustainable global agrifood future. The food security path is based on neo-productivist, high-tech solutions using all available tools and technologies, including intellectual property and GMOs, as the new paradigm to meet the challenge of feeding the world with sustainable intensification (Almås and Campbell, 2012; Marsden, 2013; Levidow, 2015). The food sovereignty path is based on agroecology and a social justice framework. The food security path is patterned on consequentialist philosophy grounded in utilitarian assumptions about agrifood science and rurality. The greater good for the most people outweighs the negative impacts on the few. The agroecology path employs rights-based rhetoric grounded in deontological assumptions to support its social justice agenda (Thompson, 2010a). The food security path includes incremental, green reforms to the existing system, while the food sovereignty path pushes for transformative change to the system (Holt Gimenez and Shattuck, 2011; Constance *et al.*, 2014). The agroecologists warn that sustainable intensification is an oxymoron at least, and more probably a 'wolf in sheep's clothing.' The neo-productivists promise their hi-tech green solution can feed the world, while the low-tech agroecology approach cannot.

The chronology of sustainable intensification

The term 'sustainable intensification' (SI) originates in the 1990s as an agro-ecological programme designed to increase food production (intensification) in developing countries – in particular in Africa – without bringing more marginal and/or pristine land into production (extensification), while at the same time reducing negative environmental externalities on the existing cultivated lands (see Pretty, 1997; Levidow, 2015). To increase community food supplies and livelihoods, appropriate technologies informed by indigenous, knowledge-based, agro-ecological methods would increase yields, conserve soil and water, and carefully manage nutrients and pests. These technologies are employed to encourage local processes of innovation and adaptation whereby the byproducts of each cycle should become the inputs to another, reducing external inputs and environmental impact. In this context SI is a culturally-sensitive and lower-tech alternative to high-tech Green Revolution approaches that have proven unrealistic and/or problematic for much of the developing world.

In 1996 the UNFAO-sponsored World Food Summit concluded that a new round of intensification was needed to feed the world, but also to avoid the negative environmental consequences of industrial agriculture à la the Green Revolution. While there are numerous practical and ethical challenges involved to intensify sustainably, it is the responsibility of national governments to support and facilitate the task (UNFAO, 1996). In 2006 the World Bank defined SI as a combination of production practices such as Integrated Pest Management, Conservation Farming, Low External Input and Sustainable Agriculture, Organic Agriculture, Precision Agriculture and Diversification (Worldbank, 2006).

After the world food crisis in 2007, the United Nations (2008) incorporated SI into the discourse on global food security. SI would be the approach to move small-holder peasants past subsistence production by linking them to improved marketing channels and national and international supply chains. To accomplish SI, a combination of biotech and Conservation Agriculture would be employed. While Conservation Agriculture includes agro-ecological approaches, it also employs no-till methods reliant on herbicide resistant crops – GMOs. Technologies developed in the global North would be part of the 'tool kit' transferred to producers in the global South to maximize production per acre while conserving soil and water resources. In light of the mounting evidence of industrial agriculture's correlation with global climate change, SI became the model for all of agriculture – the new paradigm to feed the world sustainably (UNFAO, 2009).

The Royal Society of the United Kingdom (2009) repeated the SI agenda to reduce reliance on non-renewable inputs through increased adoption of agro-ecology and GM techniques to increase yields without adverse environmental impacts and without the cultivation of more land. Because global agrifood markets are increasingly volatile in the wake of the 2007 food crises, SI provides a toolkit to reconcile increased productivity with environmental sustainability and thereby provide competitive advantage to ecological producers in commodity markets. The UK Environment, Food and Rural Affairs Committee (EFRA 2011) stated the need to enable EU farmers to achieve 'sustainable intensification' to help mitigate the imminent crises of 2050. In 2012 the UK Commission on Sustainable Agriculture and Climate Change, documented the potential of SI as 'the most promising means of simultaneously increasing food production while achieving land-based mitigation [of greenhouse gases]...' (Beddington *et al.* 2012: 28). In 2013 a group of researchers from universities, global organizations, government organizations and the UN redefined SI. The new definition broadens the scope to ecosystem services, biodiversity and human health (Macmillan, 2013).

SI is now widely adopted as a concept and guiding principle of agrifood production by national and international organizations, such as the 'Feed the Future' programme of the USDA, The Consultative Group of International Agricultural Research (CGIAR), the UNFAO, the Montpellier Panel and the Sustainable Development Solutions Network, as well as agrifood TNCs and international donor organizations such as the Gates Foundation (Tittonell, 2014). Agrifood GMO TNCs are increasingly embracing the food security discourse from a sustainable intensification perspective, especially regarding increasing production in developing countries (Constance *et al.*, 2016). For example, Monsanto's Sustainability Report (2014: 17) defines sustainable agriculture as

> growing the right amount of food on less land using resources in an efficient manner that preserves natural ecosystems, addresses water scarcity and climate change, improves farmer livelihoods and benefits society. This approach is often referred to as sustainable intensification.

Dr Robb Fraley, Monsanto Chief Technology Officer, adds that 'sustainable intensification is key to meeting food security needs for our growing planet while also reducing agriculture's impact on the environment' (Monsanto 2014: 20). Monsanto's (2013: 5) ambition is to 'help improve lives around the world by procuring goods and services in ways that conserve and protect natural resources and are ethically and socially responsible.' To accomplish this, Monsanto will 'continue to innovate with products and services that help farmers grow food sustainably' (Monsanto, 2014: 14).

The other major GMO firms have similar pronouncements on their websites regarding food security and sustainable intensification (Constance *et al.*, 2016). For example, DuPont reports that it is 'addressing food security with science-based innovations' (2015: 1) to increase yields and produce 'more safe, healthy food to the world's growing population' (2016: 1). Dow (2016a) defines sustainable intensification as addressing food security through maximizing food production in limited space, increasing the availability of healthy foods, reducing food waste and utilizing advances in crop protection chemistry, formulations, and traits. The comprehensive strategy will boost 'agriculture productivity to maximum sustainable levels to keep pace with the growing needs of our world's rapidly expanding population' (Dow 2016b: 1). Bayer Crop Science (2016a: 1) reports that in the face of climate change, population growth, and limited natural resources 'what is urgently needed is a significant and sustainable increase in agricultural production worldwide.' 'Realistically, the only way forward is to produce much more food from what land we have left in a more efficient, sustainable way' and to do that 'crop protection products are crucial' for 'climate-adapted agriculture' (Bayer Crop Science 2016b: 1). Finally, Syngenta (2016a) proposes to 'bring greater food security in an environmentally sustainable way to an increasingly populous world by creating a worldwide step-change in farm productivity.' To do so, 'the chemistry at the heart of Syngenta crop protection products and seed treatments are vital to sustainable agriculture' (Syngenta 2016b: a).

SI is criticized for being too focused on intensifying production rather than minimizing ecological externalities and social justice disparities, thereby subordinating agroecology to the dominant industrial model (Garnett *et al.*, 2013). Petersen and Snapp (2015) find that the term is not well defined or understood among researchers. It 'has become a buzzword' that allows people to put 'a positive spin' on solutions that are truly unsustainable (Nink, 2015). For civil society critics, SI is not a modest greening of industrial agriculture, but rather a radical rethinking of the agrifood system to not only reduce environmental externalities, but more importantly enhance animal welfare, human nutrition and sustainable rural development. For many agricultural researchers, while SI is necessary in the face of climate change, population growth, and ecological constraints, it is fraught with conceptual and programmatic inconsistencies, which tend to privilege agricultural intensification over ecological sustainability (Kuyper and Struik, 2014; Struik *et al.*, 2014; Petersen and Snapp, 2015). The politics of the possible tends to push SI towards an incremental greening of the dominant system and away from its transformative agenda for Africa 20 years ago.

Conclusions

This chapter has provided an historical overview of the contested discourses regarding the preferred model of agriculture. The legitimation crisis of industrial agriculture, culminating most recently in the realization that the dominant, productivist farming system is the major contributor to global climate change, has created a tipping point in the discourse. The modern manifestation of the agrarian versus industrial ethos for agriculture is the agroecology versus sustainable intensification positions on agrifood transitions. SI is the counter proposal by the industrial camp to delegitimize the agroecologists' critique as incapable of feeding the world. In fact, SI incorporates and coopts agroecological principles in its programme, but then adds Conservation Agriculture and GMOs as part of the entire technological toolkit to both feed the world and save the planet. Proceeding from the food security versus food sovereignty discourse, the La Via Campesina approach to agroecology proposes a radical transformation of the agrifood system which embraces a rights-based philosophy for humans, animals, and the planet, while the SI approach remains grounded in reformist, technological fixes and utilitarian ethics which continue to privilege corporations and their shareholders (Holt Gimenez and Shattuck, 2011).

Further research is needed on how the SI programme proceeds and to what degree it can hold close to its original grounding in agroecology. The drift in SI over the past 20 years does not bode well for a deep grounding. It may well indicate a trend in agroecology similar to what Guthman (2008) notes in organics, a 'deep' agroecology following the La Via food sovereignty model and an agroecology 'light' aligned with SI and food security approaches orchestrated by the GMO TNCs, governments, some NGOs, and major international donor organizations. As the venerated rural sociologist Fred Buttel (2006) warned us, we should be not surprised at the ability of the supporters of industrial agriculture to sustain the unsustainable. The contested discourse over the better model of agriculture is well over 100 years old. The competing SI versus agroecology agrifood transition pathways continue the discourse for the foreseeable future.

Notes

1 See Constance (2018) for a more detailed sociological examination of the history of agriculture.

References

Allen, P. (2008). Mining for Justice in the Food System: Perceptions, Practices, and Possibilities. *Agriculture and Human Values* 25: 157–161.

Almås, R. and Campbell, H. (eds) (2012). *Rethinking Agricultural Policy Regimes: Food Security, Climate Change and the Future Resilience of Global Agriculture*. Emerald Group Publishing Limited, Bingley, UK.

Bayer Crop Science (2016a). *Bayer: A Science for Better Life. Commitment to Food Security*. Available from: www.cropscience.bayer.com/en/Commitment/New-Revolution-in-Agriculture.aspx (Accessed 4 May 2016).

Bayer Crop Science (2016b). *Bayer: A Science for Better Life. Commitment to Food Security*. Available from: www.cropscience.bayer.com/en/Commitment/Food-security.aspx (Accessed 4 May 2016).

Beddington, J., Asaduzzaman, M., Clark, M., Fernandez, A., Guillou, M., Jahn, M., Erda, L., Mamo, T., Van Bo, N., Nobre, C. A., Scholes, R., Sharma, R. and Wakhungu, J. (2012). *Achieving food security in the face of climate change: Final report from the Commission on Sustainable Agriculture and Climate Change*. CGIAR Research Program on Climate Change, Agriculture and Food Security (CCAFS). Copenhagen, Denmark. Available from: www.ccafs.cgiar.org/commission (Accessed 8 May 2017).

Berry, W. (1978). *The Unsettling of America: Culture and Agriculture*. Sierra Club Books, San Francisco.

Bonanno, A. and Constance, D. H. (2008). *Stories of Globalization: Transnational Corporations, Resistance, and the Sate*. University Park, PA, The Pennsylvania State University Press.

Bonanno, A., Busch, L., Friedland, W. H., Gouviea, L. and Mingione, E. (eds) (1994). *From Columbus to ConAgra: The Globalization of Agriculture and Food*. University Press of Kansas, Lawrence, KS.

Burch, D. and Lawrence, G. (2007). *Supermarkets and Agri-food Supply Chains: Transformations in the Production and Consumption of Food*. Edward Elgar, Cheltenham, UK.

Buttel, F. H. (2006). Sustaining the Unsustainable: Agro-food Systems and the Environment in the Modern World. In: Cloke, P., Marsden, T. and Mooney, P. H. (eds) *Handbook of Rural Studies*. Sage Publication, London: 212–229.

Buttel, F. H. and Newby, H. (1980). *The Rural Sociology of Advanced Societies: Critical Perspectives*. Allenheld, Osmun, Montclair, NJ.

Campbell, H. (2009). Breaking New Ground in Food Regime Theory: Corporate 'Environmentalism,' Ecological Feedbacks, and the 'Food from Somewhere' Regime. *Agriculture and Human Values* 26(4): 309–319.

Campbell, B. M., Beare, D. J., Bennett, E. M., Hall-Spencer, J. H., Ingram, J. S. I., Jaramillo, F., Ortiz, R., Ramankutty, N., Sayer, J. A., and Shindell, D. (2017). Agriculture production as a major driver of the Earth system exceeding planetary boundaries. *Ecology and Society* 22, 44: 8.

Constance, D. H. (2008). The Emancipatory Question: The Next Step in the Sociology of Agrifood Studies. *Agriculture and Human Values* 25: 151–155.

Constance, D. H. (2014). Rural Sociology. In: Van Alfen, N. K. (ed.), *Encyclopedia of Agricultural and Food Systems, Vol. 5*. Elsevier, San Diego, CA: 62–74.

Constance, D. H. (2018). Industrial Agriculture. (Forthcoming) in Konefal, J. and Hatanaka, M. (eds), *Twenty Lessons in the Sociology of Food and Agriculture*. New York: Oxford University Press.

Constance, D. H., Konefal, J., and Grant, K. (2016). *Unpacking Sustainable Intensification: Discourses from Agribusiness*. Paper presented at the International Sociological Associations meetings. Vienna, Austria. 12 July.

Constance, D. H., Renard, M. C., and Rivera-Ferre, M. (eds) (2014). *Alternative Agrifood Movements: Patterns of Convergence and Divergence*. Emerald Group Publishing Ltd, Bingley, UK.

Danbom, D. B. (1979). *The Resisted Revolution: Urban America and the Industrialization of Agriculture: 1900–1930*. Iowa State University Press, Ames, IA.

de Schutter, O. (2008). *Promotion and Protection of All Human Rights: Civil, Political, Economic, Social and Cultural Rights, Including the Right to Development*. United Nations, New York.

Desmarais, A. A. (2007) *La Via Campesina: Globalization and the Power of Peasants*. Pluto Press, London.

Dixon, J. (2009). From the imperial to empty calorie: How nutrition relations underpin food regime transitions. *Agriculture and Human Values*, 26: 321–333.

Dow (2016a). *Science and Sustainability: Food and Agriculture*. Available from: www.dow.com/en-us/science-and-sustainability/global-challenges/global-challenges-food-agriculture (Accessed 6 May 2016).

Dow (2016b). *Dow Agrosciences*. Available from: www.dowagro.com/en-US (Accessed 6 May 2016).

DuPont (2015). *DuPont Food Security Goals: 2015*. Available from: http://foodsecurity.dupont.com/wp-content/uploads/2015/10/DuPont_FoodSecurity_Goals_2015.pdf (Accessed 4 May 2016).

DuPont (2016). *DuPont: Food Security*. Available from: http://foodsecurity.dupont.com/ (Accessed 4 May 2016).

EFRA. (2011). *The Common Agricultural Policy after 2013. Environment, Food and Rural Affairs Committee: Fifth Report 2010–11* Vol. 1. Available from: www.publications. parliam/ant.uk/pa/cm201011/cmselect/cmenvfru/671/67102.htm (Accessed 7 May 2017).

Fairbairn, M. (2012). Framing Transformation: The Counter-hegemonic Potential of Food Sovereignty in the US Context. *Agriculture and Human Values* 29: 217–230.

Falkner, R. (2009). The Troubled Birth of the "Biotech Century": Global Corporate Power and its Limits. In J. Clapp and D. Fuchs (eds), *Corporate Power and Global Agri-food Governance*: 226–251. MIT Pres, Cambridge.

Fold, N. and Pritchard, B. (eds) (2005). *Cross Continental Food Chains*. Routledge, New York.

Friedmann, H. (2005). From Colonialism to Green Capitalism: Social Movements and the Emergence of Food Regimes. In Buttel, F. H. and McMichael, P. (eds), *New Direction in the Sociology of Global Development: Research in Rural Sociology and Development. Vol. 11*: 227–264. Elsevier, Oxford, UK.

Gardner, B. (2009). *Global Food Futures: Feeding the World in 2050*. Bloomsbury Academic, London and New York.

Garnett, T., Appleby, M. C., Balmford, A., Bateman, I. J., Benton, T. G., Bloomer, P., Burlingame, B., Dawkins, M., Dolan, L., Fraser, D., Herrero, M. Hoffmann, I., Smith, P., Thornton, K., Toulmin, C., Vermeulen, S. J., and Godfray, H. C. J. (2013). Sustainable Intensification in Agriculture: Premises and Policies. *Science* 5: 33–34.

Gertel, J. and Sippel, S. R. (eds) (2014). *Seasonal Workers in Mediterranean Agriculture: The Social Costs of Eating Fresh*. Earthscan, Routledge, London and New York.

Gilbert, J. (2015). *Planning Democracy: Agrarian Intellectualism and the Intended New Deal*. Yale University Press, New Haven and London.

Goldschmidt, W., 1947. *As You Sow*. The Free Press, Glencoe.

Goodman, D. (2003). The Quality 'Turn' and Alternative Food Practices: Reflections and Agenda. *Journal of Rural Studies* 19: 1–7.

Goodman, D. and Watts, M. (eds) (1997). *Globalising Food: Agrarian Questions and Global Restructuring*. Routledge, New York.

Guthman, J. (2008). Neoliberalism and the Making of Food Politics in California. *Geoforum* 39: 1171–1183.

Hightower, J. (1973). *Hard Tomatoes, Hard Times*. Schenkman Publishing Co., Cambridge, MA.

Holt Gimenez, E. and Shattuck, A. (2011). Food Crises, Food Regimes, and Food Movements: Rumblings of Reform or Tides of Transformation. *The Journal of Peasant Studies* 38: 109–144.

Howard, P. H. (2016). *Concentration and Power in the Food System: Who Controls What We Eat?* Bloomsbury, New York and London.

Ingram, J., Erikson, I. and Liverman, D. (eds) (2010). *Food Security and Global Environmental Change*. Earthscan, London.

Kloppenburg, J. (2010). Impeding Dispossession, Enabling Repossession: Biological Open Sourcing and the Recovery of Seed Sovereignty, *Journal of Agrarian Change* 110: 367–388.

Kuntz M. (2014). Is It Possible to Overcome the GMO Controversy? Some Elements for a Philosophical Perspective. In: Ricroch A., Chopra S. and Fleischer S. (eds) *Plant Biotechnology*. Springer, Cham: 107–111.

Kuyper, T. W. and Struik, P. C. (2014). Epilogue: Global Food Security, Rhetoric, and the Sustainable Intensification Debate. *Science Direct* 8: 71–79.

Lenski, G. (1970). *Human Societies: A Macrolevel Introduction to Sociology*. McGraw Hill, New York.

Levidow, L. (2015). European Transitions Towards a Corporate-Environmental Regime: Agroecological Incorporation or Contestation? *Journal of Rural Studies* 40: 76–89.

Lobao, L. and Stofferahn, C. (2008). The Community Effects of Industrialized Farming: Social Science Research and Challenges to Corporate Farming Laws. *Agriculture and Human Values*. 25: 219–240.

Lyson, T. A. (2004). *Civic Agriculture: Reconnecting Farm, Food, and Community*. Tufts University Press, Medford, MA.

Macmillan, S. (2013). Growing more food using fewer natural resources: Pipe dream or the 'only' development pathway possible? ILRI News. 8 July. Available from: www.ilri.org/ilrinews/index.php/archives/11409 (Accessed 3 March 2017).

Magdoff, F., Foster, J. B. and Buttel, F. (eds) (2000). *Hungry for Profit: The Agribusiness Threat to Farmers, Food, and the Environment*. Monthly Review Press, New York.

Marsden, T. (2013). From Post-Productivism to Reflexive Governance: Contested Transitions in Securing More Sustainable Food Futures. *Journal of Rural Studies* 29: 123–134.

McMichael, P. (ed.) (1994). *The Global Restructuring of Agro-food Systems*. Cornell University Press, Ithaca, NY.

McMichael, P. (1996). *Development and Social Change*. Pine Forge Press, Thousand Oaks, CA.

McMichael, P. (2005). Global Development and the Corporate Food Regime.' In: Buttel, F. H. and McMichael, P. (eds) *New Directions in the Sociology of Global Development, Research in Rural Sociology and Development: Volume 11*. Elsevier Press, Oxford: 265–300.

McMichael, P. (2014). Rethinking the Land Grab Ontology. *Rural Sociology* 79: 34–55.

Monsanto (2013). *Fall 2013: Progress Report*. Available from: https://s3.amazonaws.com/SustainabilityCSR/CSR_FallProgressReport.pdf (Accessed 3 April 2016).

Monsanto (2014). *From the Inside Out: Monsanto 2014 Sustainability Report*. Available From: www.monsanto.com/…/csr_reports/monsanto-2014-sustainability-report.pdf (Accessed 3 April 2016).

Nink, E. (2015). Harvesting the Research: Ecological Intensification Can Feed the World. Food Tank. Available from: http://foodtank.com/news/2015/08/harvesting-the-research-ecological-intensification-can-feed-the-world (Accessed on 4 April 2016).

Perkins, J. and Jamieson, R. (2010). History, ethics, and intensification in agriculture. In P. Thompson (ed.), *The Ethics of Intensification*: 59–83. Springer, Netherlands.

Petersen, B. and Snapp, S. (2015). What is Sustainable Intensification? Views from Experts. *Land Use Policy* 46: 1–10.

Polanyi, K. (1944). *The Great Transformation: The Political and Economic Origins of Our Time*. Beacon Press, Boston.

Pretty, J. (1997). *The Sustainable Intensification of Agriculture*. Natural Resources Forum. Blackwell.

Raynolds, L. T., Murray, D. and Wilkinson, J. (eds) (2007). *Fair trade: The Challenges of Confronting Globalization*. Routledge, New York.

Rivera-Ferre, M. G. (2012). Framing of Agri-food Research Affects of the Analysis of Food Security: The Critical Role of the Social Sciences. *International Journal of Sociology of Agriculture and Food*, 19, 2: 162–175.

Rosset, P. (2008). Food Sovereignty and the Contemporary Food Crisis. *Development*, 51(4): 460–463.

Royal Society (2009). *Reaping the Benefits: Science and the Sustainable Intensification of Global Agriculture*. London.

Shiva, V. (1992). *The Violence of the Green Revolution: Third World Agriculture, Ecology, and Politics*. Zed Books, New York.

Struik, P. C., Kuyper, T. W., Burssaard, L. and Leeuwis, C. (2014). Deconstructing and Unpacking the Scientific Controversies in Intensification and Sustainability: Why the Tensions in Concepts and Values. *Science* 8: 80–88.

Syngenta (2016a). *About Syngenta*. Available at: www.syngenta.com/global/corporate/en/about-syngenta/Pages/Strategy.aspx (Accessed 9 May 2016).

Syngenta (2016b). *Syngenta: Products and Innovation*. Available at: www.syngenta.com/global/corporate/en/products-and-innovation/research-development/Pages/collaboration.aspx (Accessed 9 May 2016).

Thompson, P. (2010a). *The Ethics of Intensification: Agricultural Development and Social Change*. Springer, New York.

Thompson, P. (2010b). *The Agrarian Vision: Sustainability and Environmental Ethics*. University of Kentucky Press, Lexington, KY.

Tittonell, P. (2014). Ecological Intensification in Agriculture: Sustainable by Nature. *Science Direct* 8: 53–61.

UNFAO (1996). *World Food Summit: Rome Declaration on World Food Security*. 13–17 November. Available from: www.fao.org/docrep/003/w3613e/w3613e00.htm (Accessed 3 March 2016).

UNFAO (United Nations Food and Agricultural Organization) (2009) *Global Agriculture Towards 2050*. Paper presented to the High-level Expert Forum on "How to Feed the World in 2050. Rome, 12–13 October. Available from: www.fao.org (Accessed 11 March 2016).

United Nations (2008). *High-Level Task Force on the Global Food Crisis. Comprehensive Framework for Action*. Available from: www.un/org/ga/president/62/letters/cfa160708.pdf (Accessed 3 March 2016).

Williams, M. (2009). Feeding the World: Transnational Corporations and the Promotion of Genetically Modified Food. In J. Clapp and D. Fuchs (eds), *Corporate Power and Global Agrifood Governance*: 155–186. MIT Press, Cambridge.

Winson, A. (2013). *The Industrial Diet: The Degradation of Food and the Struggle for Healthy Eating*. NYU Press, New York.

Wittman, H. K., Desmarais, A. and Wiebe, N. (eds) (2010). *Food Sovereignty: Reconnecting Food, Nature and Community*. Food First Books, Oakland, CA.

World Bank (2006). *Module 4: Investments in Sustainable Agriculture Intensification*. Available from: http://siteresources.worldbank.org/EXTAGISOU/Resources/Module4_Web.pdf (Accessed 3 March 2016).

Wright, W. and Middendorf, G. (eds) (2008). *The Fight Over Food: Producers, Consumers, and Activists Challenge the Global Food System*. The Pennsylvania State University Press, University Park, PA.

5 Sustainability as the civil commons

Laying the groundwork for sustainable agriculture

Jennifer Sumner

Sustainability is a vague and contested term that is commonly used, but seldom understood. As competing factions jockey to control its meaning, it is crucial that it not be co-opted to support the continuation of corporate-controlled, chemical-based, monoculture agriculture. While some may dismiss the term entirely as too difficult or diffuse to define, we need to develop a clear meaning so we can avoid co-optation and ascertain whether our decisions and actions are making us more or less sustainable.

One understanding of the term that provides the basis for a counter-hegemonic definition proposes that sustainability is closely allied with the civil commons (Sumner, 2007), which McMurtry (1998: 376) defines as 'the organized, unified, and community-funded capacity of universally accessible resources of society to protect and to enable the lives of its members as an end in itself.' Public education systems, universal healthcare programmes, minimum wage laws, fair trade agreements, civil rights legislation and environmental protocols are all examples of the civil commons in action. Understanding sustainability in this way cuts through the vagueness associated with many other uses of the term and grounds it in basic life needs. As such, it can have a universal connotation with particular applications, thus allowing for a general understanding that can be applied in a myriad of ways, as well as measured and compared. This grounded understanding of sustainability also precludes transnational corporations from co-opting the term and using it as a form of sales promotion for their priced commodities.

This chapter aims to put forward a clear meaning for sustainability and develop an unambiguous term that will challenge the hegemonic discourses of conventional agriculture and point the way towards a more sustainable future. It will begin by laying out an alternative meaning for sustainability based on the civil commons. It will then apply this understanding to compound terms such as sustainable agriculture and sustainable food systems, and provide a means for establishing standards and metrics that become not only 'recipes for reality' (Busch, 2011) but also recipes for a reality that already exists – examples of the civil commons have been around for millennia – albeit in increasingly besieged form due to the ongoing corporate pressures for enclosure.

The civil commons

When we are faced with such anomalies as the Dow Jones Sustainability Index and Monsanto's 2015 Sustainability Report, it is clear that the concept of sustainability has moved a long way from its roots in nature. This movement has been fairly rapid, from its first use in the English language in 1972 through the Brundtland Report's use of sustainable development to today's feel-good catchword (Sumner, 2007). And while many have called upon the term, few have clarified it, leading to confusion about its meaning and leaving it open to co-optation.

One understanding has been put forward that links sustainability to building the civil commons (Sumner, 2007), a concept developed by philosopher and value theorist John McMurtry. According to McMurtry (2013), the civil commons is any co-operative human construct that protects and/or enables universal access to life goods. In other words, the civil commons is co-operative, not competitive, in its mode of engagement. It is a human construct, not a naturally occurring phenomenon, and so must be built by human agency. It protects through rules and regulations, and it enables through the provision of opportunities. The civil commons involves universal access, not paid access, and it provides life goods, or means of life, such as clean water, shelter, education, healthcare, open spaces and a safe workplace. For McMurtry (1998: 370), the civil commons is 'the middle term between life and more comprehensive life' because it makes the basic resources of life available to all its members.

The conceptualization of the civil commons is based in value theory. According to McMurtry (1998; 2002), two master codes of value underlie the long economic war expressed by history. From slave revolts through struggles to preserve common land to the Occupy Movement, value wars between the powerful and the disenfranchized have marked our path through time. On the one hand, he posits the life code of value, which reproduces or increases life by providing life goods, or means of life. In the life code of value, life is the regulating objective of thought and action, and a higher quality of life is always better by definition, regardless of the money that can be made. McMurtry (1998) symbolizes the life code of value as the following sequence:

Life → means of life → more life

On the other hand, the money code of value increases money through such means as the sale of commodities or speculation in the stock market. In essence, it is the transformation of money into more money. In the money code of value, money is the regulating objective of thought and action, and a larger quantity of money is always better by definition, whatever happens to life. McMurtry (1998) symbolizes the money code of value as the following sequence:

Money → commodity for sale → more money

As McMurtry (2002) notes, in opposition to the dominant money values embedded in global capitalism, not one civil commons institution or practice is developed or financed to generate profit for private investors. Instead, he argues,

> The nature of the civil commons can be expressed as follows: *It is society's organized and community-funded capacity of universally accessible resources to provide for the life preservation and growth of society's members and their environmental life-host*. The civil commons is, in other words, what people ensure together as a society to protect and further life, as distinct from money aggregates.
>
> (McMurtry, 1998: 24)

To differentiate the civil commons from traditional commons, McMurtry (1998) points out

> I have introduced the concept of 'civil commons' to distinguish it from the traditional 'commons' – the shared natural lands upon which an agricultural village economy depends. I mean by the civil commons both the traditional commons and all other universally accessible goods of life that protect or enable the lives of society's members ... the concept of the civil commons subsumes both the traditional commons and the built commons of universally accessible social goods evolved by public sectors since the Industrial Revolution and, in particular, since the end of World War II.
>
> (Ibid: 399)

The co-operative human construct of the civil commons that protects and/or enables universal access to environmental, social and economic life goods may seem a long way from nature and too anthropocentric to be associated with sustainability. However, sustainability would not even be an issue without human exploitation of the environment and of each other. The problems we face every day – climate destabilization, environmental degradation, an exponentially growing gap between rich and poor, endless war and the increasing financialization of every aspect of human and environmental life – have been created by humans and must be solved by humans. The causes of unsustainability emerged over millennia, and in recent decades many can be traced to the corporate food regime (Holt Giménez and Shattuck, 2011). The concept of the civil commons presents us with a solution to these problems and a path to sustainability.

Enclosure of the civil commons

This solution, however, has been under siege for centuries by those who operate from the money code of value and seek to privatize it. The modern legitimizing discourse for this ongoing enclosure is the 'tragedy of the commons,' first introduced by Garret Hardin (1968) when he argued that the commons would not work as a concept because of human greed. While assuming the self-maximizing principle of neoclassical economics, he advocated that 'the tragedy of the commons as a food basket is averted by private property, or something formally like it' (1245). McKibben (2007, 199) has countered that 'The "tragedy of the commons" really reflected what happened when hyper-individualism came into contact with older, more community-oriented ideas about the land.' This illustrates the history of the commons as a source of shared public wealth that provides for basic human needs by collective regulation. The shared public wealth of all forms of the civil commons allows people to subsist during difficult times and to live more comfortably during easier times.

The most famous enclosures took place in England during the Industrial Revolution. Driven by wealthy landowners who wanted to turn the commons into private sheep farms so they could profit from the international wool trade, the enclosure of the commons has been described by Polanyi (2001, 37) as 'a revolution of the rich against the poor.' He described how the enclosures had a powerfully unsettling effect:

> The war on cottages, the absorption of cottage gardens and grounds, the confiscation of the rights in the common deprived cottage industry of its two mainstays: family earnings and agricultural background. As long as domestic industry was supplemented by the facilities and amenities of a garden plot, a scrap of land, or grazing rights, the dependence of the laborer on money earnings was not absolute; the potato plot or 'stubbing geese,' a cow or even an ass in the common made all the difference; and family earnings acted as a kind of unemployment insurance. The rationalization of agriculture inevitably uprooted the laborer and undermined his social security.
>
> (Ibid: 96)

Without the means to feed, house and otherwise care for themselves, thousands were forced to migrate to the cities. In a scenario reminiscent of today's displaced rural people in the global market economy, they formed a desperate mass of starving humanity living in urban slums, with the lucky few who found work in the new factories forced to endure brutalizing conditions.

Enclosures have been ongoing, but particularly prevalent with the ascendency of neoliberalism, a theory of political-economic practices that proposes that human well-being can best be advanced by the maximization of entrepreneurial freedoms, which are supported by an institutional framework characterized by private property rights, individual liberty, free markets and free trade (Harvey, 2006). Under neoliberalism, the pressures of enclosure have been intense on such civil commons institutions as public healthcare, public education, public libraries, public parks and public knowledge, with negative implications for the majority of humans and for the environment. In terms of food, McMichael (2013: 9) points out the effect of enclosures:

> The food regime is premised on forms of enclosure across time and space. This dimension is critical because enclosure alters ecological relations: substituting world-extractive for local-extractive processes, thereby foreclosing local futures for a capitalist future driven by variable and unstable markets, rather than enduring socio-ecological needs.

To reverse this trend, we need to understand the importance of the civil commons and work to restore its place in our lives. One of the ways of doing this is to recognize its relationship to sustainability.

Sustainability as the civil commons

A term with great potential, the civil commons can be used as the basis for developing an understanding of sustainability. In keeping with Thompson's (2010) three broad, overlapping categories of sustainability – economic, ecological and sociological – Sumner (2007) proposes that sustainability involves building the civil commons. Examples of the economic aspects of the civil commons include minimum wage laws, legislation requiring equal pay for work of equal value, mothers' allowance programmes, fair trade agreements and guaranteed annual income schemes. Examples of the ecological aspects of the civil commons include the Kyoto agreement, the Montreal Protocol, greenbelt legislation, national and provincial/state parks, and bylaws banning the cosmetic use of pesticides. Examples of the sociological aspects of the civil commons include human rights legislation (including women's, children's, civil, indigenous and gay rights), the Geneva Convention, workplace safety laws, national childcare/eldercare programmes, public education systems and public libraries.

From this basic understanding of sustainability, we can posit that the more a society builds the civil commons, the more sustainable it becomes. In contrast, the more a society encloses, defunds, privatizes, vilifies, bans or destroys the civil commons, the less sustainable it becomes (Sumner and Sanders, 2016). Austerity programmes, flat tax laws, the privatization of the public sector, and international trade agreements represent just some of the means to enclose the civil commons and erode the universal provision of life goods. To reverse this unsustainable trend, we need to recognize the importance of the civil commons to our lives and then work towards preserving it in the name of sustainability.

Understanding sustainability as building the civil commons provides a fresh perspective for some well-known compound terms, such as sustainable development, in which sustainability morphs from a noun to an adjective. However, unless there is a clear understanding of the mother concept of sustainability, any offspring of this term will be just as vague as the root term. Shearman (1990) illuminates this issue when he observes that using *sustainability* as a modifier in compound terms such as *sustainable development* changes the way we come to understand the second half of those terms. In this way, *sustainable* is used not only as an adjective, but also as a contradiction. For Shearman, *sustainability* as a modifier implies that the status quo is inconsistent with the facts. If not, then terms like *sustainable development* would be redundant, because development would already be sustainable. Using Shearman's observation, McMurtry's (1998) value theory and Sumner's (2007) definition, sustainable development would entail development projects that operate out of the life code of value and build the civil commons (like community health care clinics or public education programmes), not private entrepreneurism or schemes to bring peasants to the global market.

Following this approach, we can examine terms like sustainable agriculture and sustainable food systems. Sustainable agriculture would entail agriculture that operates out of the life code of value and builds the civil commons. Agriculture that increases exports for entrepreneurs, enriches transnational corporations, functions as a sales site for GMOs or degrades the land, water and air would reflect the money code of value, not build the civil commons and thus not qualify as sustainable agriculture. But agriculture that preserves the soil, involves community/national control of food production or contributes to nutritious school lunch programmes would reflect the life code of value, provide universal access to life goods, build the civil commons and thus qualify.

Similarly, sustainable food systems would entail food systems that reflect life values and build the civil commons. Food systems that create obesogenic foodscapes, pollute the environment or operate for a profit would not qualify as sustainable food systems. But non-profit food systems that ensure everyone is fed nutritious food, within the ecological limits of the planet, would qualify as sustainable (Sumner, 2016).

Moving sustainability towards the civil commons

Over a quarter of a century ago, Stuart Hill (1984: 1) lamented that 'there is something seriously wrong with a society that requires one to argue for sustainability.' Sadly, we find ourselves in this predicament. Industrial agriculture is an evolutionary dead end (Sumner and Llewelyn, 2011) and as its legitimation crisis reaches critical mass, proponents are grasping at sustainability to justify their unsustainable path. For this reason, we need to clarify the meaning in a way that cannot be further compromised.

To begin this process, we can look to Pretty (1997: 249), who maintains that 'in any discussion of sustainability, it is important to clarify what is being sustained, for how long, for whose benefit and at what cost, over what area and measured by what criteria.' Following Pretty, we can understand that it is the civil commons that is being sustained: built, maintained and extended by co-operative human agency. It has been in existence for millennia and will continue indefinitely into the future. The civil commons has universal benefits, not just benefits for a select few. The cost of the civil commons is borne by the collective, sometimes through volunteer labour (e.g. barn raisings, neighbourhood care teams) but often through taxes (e.g. healthcare and education systems). It spans multiple scales, from local projects through national programmes to international covenants. And the criterion for measurement is the universal access to life goods.

Morgan and Morley (2014) add to the discussion when they point out that procuring food from a range of sustainable sources means buying local, organic or fairly traded products from both near and far. For these authors, this highlights the need to frame sustainability capaciously, so it can embrace multiple values, not just a single attribute. They go on to emphasize that the values of public health, social justice and ecological integrity represent the core values of a 'capacious sustainable development paradigm' (ibid: 94). These core values can be understood as life values and the civil commons as a capacious conceptual space to house them.

From a similar perspective, Rundgren (2016: 16) puts forward a related set of life values to undergird a capacious understanding of sustainability when he proposes that

> The rethinking of food as a right, of farming as a management system of the planet and the food system as a commons will lead us to develop new institutions that complement the roles of the market and the state

In particular, Rundgren's (2016) proposal of the food system as a commons opens the door for a capacious conceptualization of sustainability as the civil commons, in all its forms.

Pretty (1997) contends that precise and absolute definitions of sustainability (or its derivatives) are impossible, thus ensuring that a concept such as sustainable agriculture would not be 'a simple model or package to be imposed.' Indeed, the civil commons would look different in a rural community compared to an urban one, in a fishing village rather than a mining town or an indigenous community. The unifying factor would be the provision of life goods through the vehicle of the civil commons, but the expression of that provision would vary, depending on community needs.

The strengths of understanding sustainability in terms of the civil commons are many. It is an already-existing (although recently named) phenomenon, not just an untested idea. The civil commons has a collective orientation, which is more effective than individual approaches for the provision of life goods. It is underwritten by values that people recognize and can identify with. The civil commons provides life goods, not luxuries, for everyone, which few people would deny others, particularly children. It is capacious enough to encompass the three broad, overlapping categories of sustainability – economic, ecological and sociological – while not being so capacious as to be meaningless. And it is compatible with different economic models – it just emphasizes that life goods are not market commodities and so cannot be exploited for profit.

An example of the problem of not defining sustainability: sustainable intensification

Linking sustainability to the civil commons provides a clear and capacious understanding of sustainability. The need for such an understanding is illustrated by the concept of sustainable intensification, which has been hijacked by the corporate food regime. To begin, we look to Lohmann's (1990: 82) article entitled 'Whose common future?' published more than 25 years ago, in which he warned: 'Never underestimate the ability of modern élites to work out ways of coming through a crisis with their power intact.' In the face of the impending environmental crisis, he saw the more progressive global élites organizing themselves for the political management of the crisis by avoiding analysis of its causes, by using vague code words like 'security' to rally other members of the élite, by tailoring solutions not to the problems but to the interests of those who created them, by identifying the executors of the solution with the existing power structure, and by co-opting the NGOs to add credibility to their initiatives.

Today, we are experiencing déjà vu. As people around the world begin to seriously question industrial agriculture and the corporate food regime, transnational corporations are rushing to the sustainability banner to save their vulnerable enterprises. The neologism 'sustainable intensification' exemplifies this issue. The term was coined by Jules Pretty (1997), based on the perspective that substantial growth was possible in hitherto unimproved or degraded areas while simultaneously protecting or regenerating natural resources. Working from a broad perspective, he emphasized regenerative or low-input agriculture, farmer participation in all stages of technology development and extension, a combination of human capacity and ingenuity as well as biological and physical processes, an integrated suite of methods and technologies, and increased diversity of enterprises within farms complemented by increased links and flows among them. In effect, 'as the natural processes increasingly substitute for external inputs, gradually, the impact on the environment is reduced' (ibid: 248).

From this auspicious beginning, the term began to change, as carefully documented by Constance *et al.* (2016). In 2006, the World Bank (2006: 1) reduced sustainable intensification to production practices, 'such as Integrated Pest Management, conservation farming, low external input and sustainable agriculture, organic agriculture, precision agriculture and diversification.' In 2009, the Royal Society published a report that highlighted yields, describing sustainable intensification as a system where 'yields are increased without adverse environmental impacts and without the cultivation of more land.' By 2013 (Macmillan, 2013: 2), it had a new definition that included 'ecosystem services, biodiversity and human health.' Little wonder that in 2015, Petersen and Snapp (2015: 4) found that the term was not well defined or understood among researchers. They argued that the term 'has become a buzzword' that allows people to put 'a positive spin' on solutions that are truly unsustainable.

Constance *et al.* (2016) then go on to illustrate how large multinationals have seized upon the term and harnessed it to give a positive spin to their commercial campaigns: Monsanto, DuPont, Bayer Crop Science, Dow and Syngenta have all appropriated sustainable intensification to burnish their brands. Following Lohmann (1990), we can see how these corporations work to come through the current crisis in food and agriculture with their power intact by avoiding analysis of its causes (and their participation in the creation of the crisis), by using vague code words such as 'food security' and 'innovation,' by tailoring solutions to sales of their products, by identifying themselves as part of the solution, and by co-opting organizations like the Bill and Melinda Gates Foundation, a shareholder of Monsanto (Tittonell, 2014), to add credibility to their initiatives. As Tittonell (2014: 54) observes, 'Grass-root organisations and environmental movements around the world are weary of the term sustainable intensification which they often see as a window-dressing, green-washing strategy to justify any form of intensification.'

The case of sustainable intensification pinpoints the danger of the co-optation of sustainability to support the continuation of corporate-controlled, chemical-based, monoculture agriculture. And yet, if we had a clear and capacious understanding, we could see the possibilities in this new term. Using the meaning developed by Sumner (2007), we can understand sustainable intensification as intensification that builds the civil commons, such as more intensive use of organic agriculture or agro-ecology, not of private, for-profit commodities such as chemical fertilizers, pesticides and GMOs.

Conclusion

No reasonable person is opposed to the idea of sustainability (Pretty, 1997); yet its lack of a clear and capacious definition dilutes its effectiveness as a rallying point for much-needed change. As a result, sustainability has become susceptible to co-optation by market forces, as is evidenced in the corporate food regime's use of the term 'sustainable intensification.'

Rundgren (2016) maintains that the capitalist market economy appeals to many human traits but not the ones we need to make a great transition. It is built on, and encourages, constant competition, endless growth, increasing consumption and ongoing exploitation of humans and the environment. For Rundgren, 'The values that are implicit and explicit in the capitalist market economy are not the ones we need to face up to the challenges ahead' (ibid: 349). The values of the capitalist market economy can be understood as money values, which play out in decisions and actions taken every day around the world, particularly in the corporate food regime. For example, Rundgren points out that our food and farming systems are managed by market signals, not by signals from ecosystems or society. 'The system is simply not geared towards stewardship of the planet and living beings but to the maximization of marketable output and profit' (ibid: 15).

Transforming the corporate food regime will require 'a major shift from societies embedded in the market economy to economies and markets that are embedded in society' (Holt Giménez and Shattuck, 2011: 136–137). Part of that major shift will entail both reversing the ongoing enclosure of the civil commons and establishing the civil commons as the measure of sustainability. This combines a robust concept – the civil commons – with actually existing and longstanding practice by humans around the world, thus providing transformative impetus to the term sustainability and a bulwark against co-optation by transnational corporations.

That said, there are a number of challenges associated with understanding sustainability as the civil commons. To begin with, it is a complex term, not a neat soundbite. In this age of short attention spans, this complexity could create gaps in comprehension and uptake. There may also be difficulty in deciding between the provision of competing life goods (for example, choosing whether to use scarce water resources for human consumption or food production). As a result, criteria will need to be developed and there may be disagreements around who gets to carry this out. In addition, standards and metrics will need to be established, and this will cause tensions because some groups are inevitably left out in any standardization process. And, finally, the civil commons automatically engenders opposition from those who profit from the sale of life goods, and those with an ideological affinity for this perspective. This must be anticipated and planned for.

In spite of these challenges, understanding sustainability as the civil commons offers a clear, operationalizable definition of a contested term that has a basis in actual practice, is highly amenable to theorizing, provides a critique of conventional understandings of sustainability and indicates a way out of the crisis of unsustainability. As such, it confirms that a sustainable future is collective and lays the groundwork for clearer understandings of vital compound terms such as sustainable agriculture.

References

Busch, L. (2011). *Standards: Recipes for Reality*, MIT Press, Cambridge MA.

Constance, D. H., Konefal, J. and Moseley, A. (2016). 'The empire strikes back: Sustainable intensification in discourse and practice.' Paper presented at the annual Agriculture, Food and Human Values Conference, Toronto, 24 June.

Hardin, G. (1968). 'The tragedy of the commons,' *Science*, vol. 162: 1243–1248.

Harvey, D. (2006). 'Neo-Liberalism as Creative Destruction,' *Geografiska Annaler*, vol. 88B, no. 2: 145–158.

Hill, S. (1984). 'Redesigning the food system for sustainability.' Ecological Agriculture Projects, McGill University. Accessed 27 August 2010 at http://eap.mcgill.ca/publications/eap23.htm.

Holt Giménez, E. and Shattuck, A. (2011). 'Food crises, food regimes and food movements: Rumblings of reform or tides of transformation,' *The Journal of Peasant Studies* vol. 38, no. 1: 109–144.

Lohmann, L. (1990). 'Whose common future?' *The Ecologist*, vol. 20, no. 3: 82–84.

Macmillan, S. (2013). Growing more food using fewer natural resources: Pipe dream or the 'only' development pathway possible? *ILIR News*, 8 July. Accessed 15 April 2018 at www.ilri.org/ilrinews/index.php/archives/11409.

McKibben, B. (2007). *Deep Economy: The Wealth of Communities and the Durable Future*, Holt Paperbacks, New York.

McMichael, P. (2013). *Food Regimes and Agrarian Questions*, Fernwood Publishing, Halifax.

McMurtry, J. (1998). *Unequal Freedoms: The Global Market as an Ethical System*, Garamond, Toronto.

McMurtry, J. (2002). *Value Wars: The Global Market Versus the Life Economy*, Pluto Press, London.

McMurtry, J. (2013). *The Cancer Stage of Capitalism: From Crisis to Cure*, Pluto Press, London.

Monsanto (2015). *Growing Better Together: Monsanto's 2015 Sustainability Report.* Accessed 15 April 2018 at www.monsantoglobal.com/sustainability/documents/monsanto-2015-sustainability-report.pdf.

Morgan, K. and Morley, A. (2014). 'The public plate: Harnessing the power of purchase,' in T. Marsden and A. Morley (eds) *Sustainable Food Systems: Building a New Paradigm*, Routledge, New York.

Petersen, B. and Snapp, S. (2015). What is Sustainable Intensification? Views from Experts. *Land Use Policy* 46: 1–10.

Polanyi, K. (2001). *The Great Transformation*, Beacon Press, Boston.

Pretty, J. (1997). 'The sustainable intensification of agriculture.' *Natural Resources Forum*, vol. 21, no. 4: 247–256.

ROBECOSAM (2017). *DJSI Annual Review 2017.* Accessed 15 April 2018 at www.robecosam.com/en/sustainability-insights/about-sustainability/corporate-sustainability-assessment/ review.jsp.

Royal Society (2009). *Reaping the Benefits: Science and the sustainable intensification of global agriculture*, London: Royal Society.

Rundgren, G. (2016). *Global Eating Disorder*, Regeneration Publisher, Uppsala, Sweden.

Shearman, R. (1990). 'The meaning and ethics of sustainability,' *Environmental Management*, vol. 14, no. 1: 1–8.

Sumner, J. (2007). *Sustainability and the Civil Commons: Rural Communities in the Age of Globalization*, University of Toronto Press, Toronto.

Sumner, J. (2016). 'Conceptualizing sustainable food systems,' in M. Koç, J. Sumner and A. Winson (eds) *Critical Perspectives in Food Studies*, 2nd edn, Oxford University Press, Toronto.

Sumner, J. and Llewelyn, S. (2011). 'Organic solutions? Gender and organic farming in the age of industrial agriculture,' *Capitalism, Nature, Socialism*, vol. 22, no. 1: 100–118.

Sumner, J. and Sanders, C. (2016). 'Taking the next steps toward environmental sustainability: implementing the official plan on Pelee Island,' in L. K. Hallström, M. A. Beckie, G. T. Hvenegaard and K. Mündel (eds) *Sustainability Planning and Collaboration in Rural Canada: Taking the Next Steps*, University of Alberta Press, Edmonton.

Thompson, P. B. (2010). *The Agrarian Vision: Sustainability and Environmental Ethics*, The University Press of Kentucky, Lexington, Kentucky.

Tittonell, P. (2014). 'Ecological intensification of agriculture – sustainable by nature.' *Current Opinions in Environmental Sustainability*, vol. 8: 53–61.

World Bank (2006). *Module 4: Investments in Sustainable Agriculture Intensification.* Accessed 3 March 2016 at http://siteresources.worldbank.org/EXTAGISOU/Resources/Module4_Web.pdf.

Part III

Contested discourse in theory and practice

6 Zero Hunger discourse

Neoliberal, progressive, reformist or radical?

Kiah Smith

Introduction

The phrase *Zero Hunger* has come to indicate both food policy framework and a broad sustainability goal connecting food production with nutrition, poverty reduction and livelihoods. As a rallying call for addressing global food insecurity, Zero Hunger is supported by the UN's Secretary General, the Food and Agriculture Organization (FAO) and the Committee on World Food Security, based on multilateral agreement that 'a profound change of the global food and agriculture system is needed to nourish today's 795 million hungry and the additional two billion people expected by 2050.'[1] It is increasingly recognized that the current global food regime creates and reproduces unevenly distributed social, environmental and economic inequalities, and that the vulnerabilities contributing to, and arising from, food insecurity are deeply structural. The global currency of this discourse is evidenced by the framing of Sustainable Development Goal #2 – *End hunger, achieve food security and improved nutrition and promote sustainable agriculture*. While inequality is undeniably at the heart of these concerns, the ways in which we understand the relationships between hunger, justice and food regime transformation are diverse.

This chapter explores the multiple, and often contradictory, ways that food security, food justice and other 'food-specific' sustainability discourses are mobilized and contested under the banner of Zero Hunger. Historically, Zero Hunger (*Fome Zero*) emerged in Brazil in 2001 as a policy to achieve improved agricultural production, poverty reduction and food security. It was state-centric with high involvement of civic actors, redistribution via smallholder farmer cooperatives, social protection programmes and an emphasis on agroecology and justice. Success in Brazil led to the programme being taken up in other Latin American countries, reflecting decades-long regional movements to promote ecological farming and poverty eradication supported by the state. In 2012, world leaders at the UN Conference on Sustainable Development (Rio+20) reaffirmed the right of everyone to have access to safe and nutritious food, consistent with the right to adequate food and the fundamental right of everyone to be free from hunger. The UN Secretary-General's *Zero Hunger Challenge* was launched shortly thereafter, with specific targets for food and nutrition security, gaining wide support from many governments and other entities. *Zero Hunger West Africa* followed shortly after, launched in March 2014 by the FAO and the Economic Committee of West African States (ECOWAS). This programme is a collaboration between state and non-state actors in 14 West African countries aiming to promote sustainable agricultural investment alongside social protection. Then, 2015 saw the presentation of the global Sustainable Development Goals (SDGs), of which *SDG#2: Zero Hunger* highlights the interconnections between sustainable agriculture, empowering small farmers, promoting gender equality, ending rural poverty, ensuring healthy lifestyles, tackling climate change, and other issues addressed within the set of 17 SDGs in the post-2015 Development Agenda (UN 2015a).

In this chapter, I assess *Zero Hunger* not in terms of its achievements or outcomes as a policy platform, but from a perspective that considers the ways that emergent values, ideas and meanings about sustainable and just food systems have been constructed and mobilized at various times and scales. The first section outlines an analytical framework built on the works of Holt Gimenez and Shattuck (2011), Clapp and Dauvergne (2011) and the broader sociological literature on environmental discourses. Their typologies provide a tool for analysing the construction of food security, food and nutrition security, food justice and food sovereignty along a continuum between market/reformist and green/radical discourses. I ask: *How do multiple sustainability discourses inform the framing of Zero Hunger at national, regional and global levels? Which discourses dominate, how have they gained political traction, and with what outcomes for the framing of Zero Hunger approaches?* Next, I use secondary documents and a discourse analysis approach to investigate the framing of Zero Hunger at different levels: (a) national – Brazil; (b) regional – West Africa; and (c) global – SDG#2. Finally, I conclude by considering the implications that this discursive plurality might have for radical reform of the global food regime via Zero Hunger policies.

Analytical framework

Sustainability discourses in the realm of food

Following the world food and financial crises in 2008, critiques of the causes and consequences of global hunger have increasingly identified neoliberalism, corporatization and financialization (among other political economic conditions) as structural drivers of unsustainable food production and consumption patterns, with negative impacts for human and ecological wellbeing (see Lang, 2010; McMichael, 2008; Rosin *et al.*, 2012; Smith, 2016). Productivism and the 'calorie-ization' of food security have emphasized increasing agricultural output via Green Revolution technologies; 'neoliberalization' has prioritized trade liberalization and market integration; and 'empty calorie-ization' has promoted processed foods over foods that contribute to societal wellbeing (Carolan, 2013). These are defining tendencies of the current corporate food regime (see McMichael, 2005, 2009a, 2009b; Friedmann, 2005), and align with the hegemonic, neoliberal priorities of the ruling corporate class (Holt Gimenez and Shattuck, 2011).

Transnational food movements have responded with sustainability discourses that emphasize a rights-based approach to hunger governance and reform. These movements challenge modernization theory's understanding of development as linear and inequality as relative, and emphasize people's entitlements to basic rights (such as food, water, etc) (UN General Assembly, 2013; Pritchard *et al.*, 2014; Weber, 2014). Redistributive land (and water) reform, the right to food, smallholder empowerment, agroecological food production, alternative food networks, quality and authenticity, are focal points for an emergent social activism that is today driving the 'pre-figurative politics' aiming to challenge the limitations of the corporate food regime (Goodman, DuPuis and Goodman, 2014). Their critiques reflect a belief that 'our food system does not meet the fundamental criteria of social justice such as freedom from want, freedom from oppression, and access to equal opportunity' (Allen, 2008: 158). The discourses attached to hunger, food security, sustainability and justice, matter. As Dryzek (2013: 9) explains:

> a discourse is a shared way of apprehending the world, enabling those who subscribe to it to interpret information and put it together into coherent stories or accounts ... [Discourses] construct meanings and relationships, help define common sense and legitimate knowledge ... coordinate actions ... embody power and are bound with political practices – some interests are advanced, others suppressed.

Food security, food and nutrition security, the right to food, food sovereignty and food justice are key concepts framing how food/hunger is addressed within an increasingly contested global food regime. While literature on each of these concepts is too vast to include here, they can be summarized as:

- *Food security* is officially defined by the FAO as existing 'when all people at all times have physical, social and economic access to sufficient, safe and nutritious food to meet their dietary needs and food preferences for an active, healthy life' (FAO, 1996: 1). The four pillars of food security are availability, access, utilization and stability, and nutrition is central (FAO, 2016).
- *Food and nutrition security* connects food and agriculture to nutrition and tries to address the longer term, root causes of hunger and malnutrition, whereby agricultural production, stability of supplies, livelihoods, incomes and access to food, health and care are addressed simultaneously (FAO, 2016).
- *The Right to Food* (RtF)

  > has more to do with modes of production and issues of distribution than with levels of food production alone ... the right to food is closely related to the right of access to resources such as land, water, forests and seeds, that are essential to those who produce food for their own consumption.
  >
  > (UN General Assembly 2013: 4)

More than 30 countries now include the RtF in their national legislation, constitutions, and legal frameworks.

- *Food sovereignty* is broadly defined as 'the right of nations and peoples to control their own food systems, including their own markets, production modes, food cultures and environments' (Wittman, Desmarais and Wiebe, 2010: 2). The concept emerged from peasant movements in the Global South (i.e. La Via Campesina) but has grown to become the dominant challenge to food security discourse at multiple levels (McMichael, 2014; Patel, 2009).
- *Food justice* refers to 'ensuring that the benefits and risks of where, what, and how food is grown and produced, transported and distributed, and accessed and eaten are shared fairly' (Gottlieb and Joshi, 2010: 6). Food justice discourses borrow from the concepts of environmental justice (Schlosberg, 2013) and food sovereignty (Gottlieb and Joshi, 2010) to highlight the range of inequalities created by and affecting the food system (such as class, race, gender, North-South, rural-urban), and emphasize the need for rights-based approaches to food system sustainability (see also Cadieux and Slocum, 2015).

These multiple framings of food system sustainability and transformation not only represent different food movements (e.g. Clendenning *et al.*, 2016), but are also 'fluid and changing discourses that define the problem of hunger' (Jarosz, 2014: 168). They have been interpreted as being relationally opposed (especially food security and food sovereignty – see Lee, 2013; Holt Gimenez and Shattuck, 2009; Wittman *et al.*, 2010), or as an ideological continuum between (i) addressing hunger within existing structures, as an outcome of improved access (e.g. food security) and (ii) deeper critiques of food regime transformation (i.e. right to food, food justice and food sovereignty) (Jarosz, 2014; McMichael, 2014). These concepts certainly illustrate the tensions identified by Carolan (2013) between dominant understandings (of food security as a thing, an end in itself) and its 'original spirit' – 'a process that ought to make people and the planet better off' (Carolan, 2013: 7). In this chapter, rather than take the above interpretations as definitive, I agree with Jarosz (2014: 168) that these discourses are:

> tied to distinctive political and economic histories, ecologies, and identities at the national and local levels. They are differentially deployed depending upon geographic context and the political economy of development and underdevelopment. [They are] dynamic and changing in relation to the wider political and cultural economies of food system dynamics across scale.

The approach developed here also draws from work in environmental justice around the plurality of conceptions of justice (and injustice). This perspective argues that understanding how multiple viewpoints or discourses are constructed and related to each other is central to the analysis of 'just sustainabilities' (Agyeman, 2005; Schlosberg, 2013). Sustainability is a 'nodal discourse' around which other discourses, such as justice, cluster (Dryzek, 2013: 16). One consequence of multiple framings of 'the food problem' is that solutions for a more sustainable and just global food regime are diverse and strongly contested (see Holt Gimenez, Patel and Shattuck, 2009). For some, market-based mechanisms (such as REDD+, payments for ecosystem services, fair trade, 'climate-smart agriculture') are compatible with a move towards rights-based development in a context of climate change and mounting pressure to produce more food. For others (such as climate justice and food sovereignty movements), achieving justice is more closely associated with agro-ecology, re-localizing production, shifting consumption patterns and land reform. Combining these with eco-social policy reform (i.e. universal or targeted social protection) is also increasingly supported, but challenging to implement. These examples illustrate the kinds of class-based conflicts constitutive of global food regimes, whereby 'food and hunger efforts tend to split ideologically between those that seek to stabilize the corporate food regime, and those that want to change it' (Holt Gimenez and Shattuck, 2011: 114). In a context where justice definitions are many, the challenge lies in how to acknowledge, recognize and include discursive diversity in ways that also address how some discourses are privileged and others are disadvantaged (see Schlosberg, 2002).

Discursive typologies

Scholars have sought to classify emergent food sustainability discourses, many with a common thread of juxtaposition of 'radical' and 'reformist' positions. For example, in Dryzek's (2013) classification, environmental discourses can be either (a) reformist or radical, or (b) imaginative or prosaic. Dryzek uses a matrix to assess environmental discourses that have arisen in the context of industrialism's 'overarching commitment to growth in the quantity of goods and services produced and to the material wellbeing that growth brings' (Dryzek, 2013: 14). This is because, despite competing ideologies that historically have accompanied industrialism (e.g. liberalism, conservativism, socialism, Marxism, fascism), growth and the suppression of environmental concerns have always been common threads. As such, Dryzek (2013) argues that environmental discourses *must* depart from the terms of industrialism. These departures may be:

- Prosaic: Industrialism's political economy is taken as a given, albeit with problems to be overcome; requires action, but this can be reformist or radical.
- Imaginative: Seek to redefine relationship between environment and economy/society; action can be reformist or radical.

Sustainability discourses are described by Dryzek (2013) as imaginative-reformist: imaginative in their attempt to dissolve conflicts between environmental and economic values embedded in industrial society, and seeking reform via the redefinition of growth and development. They are not seen as radical per se, because weak sustainability discourses (the norm) do not reject industrialism.

Similarly, Clapp and Dauvergne (2011) argue that there are four competing environmental 'worldviews' that shape perspectives on global environmental change. These are: (1) market liberals; (2) institutionalists; (3) bio environmentalists; and (4) social greens. Each view has its own logic that fits with its assumptions, and each can be associated with specific institutional actors and power relations. The most radical of these are the bio environmentalists and social greens, who argue for fundamental changes in the human-nature relationship, such as revaluing collectivity (over individualism) and eco-centrism over econocentrism. Clapp and Dauvergne's worldviews typology helps to understand markedly different interpretations of the condition of the global environment, as well as the political tensions between different approaches to sustainable development. For example, applying their framework to an analysis of the ideological contestation around 'green economy,' UNRISD concluded that market liberal approaches dominate while the most radical viewpoints – bioenvironmentalists calling for limits to growth – are the most marginalized (Cook, Smith and Utting, 2012).

Applying a discursive lens to food system restructuring specifically, Holt Gimenez and Shattuck (2011) have also presented a typology with four main positions: neoliberal, reformist, progressive and radical. In this work, the authors compare the 'politics, production models and approaches to the food crisis' of the corporate food regime and global food movements. Neoliberal and reformist trends are associated with the corporate food regime, while progressive and radical trends are associated with food movements:

- Neoliberal politics support and strengthen the existing model of industrialization, corporatization and market deregulation, thus reflecting a 'food enterprise' discourse.
- Food security discourse represents a politics of reform, similar to neoliberal approaches that emphasize development and food aid, markets and governance but with minor adjustments.
- Food justice discourses reflect progressive politics based on empowerment, rights and re-regulation of sustainable production and consumption.
- Food sovereignty is seen as a radical ideology challenging corporate power, and re-envisaging democracy, rights, sustainable livelihoods and ecological wellbeing.

(Holt Gimenez and Shattuck, 2011)

Table 6.1 categorizes food discourses according to each of the conceptual typologies described above. In the remainder of this chapter, I seek to identify how these four categories are present in Zero Hunger documentation. In line with discourse analysis, I am particularly interested in how food security, food and nutrition security, right to food, food sovereignty and food justice concepts

Table 6.1 Typologies of food discourses

Discursive content	*Categorization of politics/discourse*		
Food security, productionism, eco-efficiency, green growth, green markets, agro-fuels, market deregulation, self-regulation, corporate agriculture, PPPs, international food aid; access.	Neoliberal	Market liberal	Prosaic –reformist
Food and nutrition security, mainstreaming fair/ethical trade, market-led land reform, bio/fortified/climate-resistant crops, capacity building, development finance, agricultural aid tied to 'green' production, resilience, access.	Reformist	Institutionalist	Imaginative – reformist
Food justice, right to food, food/land redistribution, social safety nets, localization, agro-ecology, green economics, environmental and social justice, gender equality, empowerment, social economy, collectives/ cooperatives, equality of access, entitlements.	Progressive	Social green	Imaginative – radical
Food sovereignty, food as a human right, right to development, sustainable livelihoods, rights to/ regulation of global commons, agro-ecology 'to cool the planet', human right to 'food sovereignty', capacities and entitlements, Nth-Sth solidarities, democratic decision making, social-solidarity economy, alter-globalisation, post-capitalism.	Radical	Bioenviron-mental	Imaginative – radical
Authors	Holt Giménez and Shattuck (2011)	Clapp and Dauvergne (2011)	Dryzek (2013)

are constructed, communicated and maintained, and by whom. I will assess the use of language, narratives and rhetorical devices (e.g. metaphors, imagery), as well as identify key agents and their motives (Dryzek, 2013), drawing on a range of publicly available documents and relevant academic literature.

Findings

Brazil

Under the Lula government, Brazil established *Fome Zero* in 2001 as a national food security programme to connect goals of hunger reduction, community mobilization for food security, and multi-stakeholder engagement led by the State (Marques *et al.*, 2013; Rocha, 2009). It was designed and supported institutionally by civil society (experts in nutrition, agronomy and sustainable development), social movements (around land reform, race and waste-picking) and multi-lateral development agencies (e.g. IMF, World Bank, and the FAO) (Ansell, 2015). Official government documents state that Fome Zero was:

> A strategy driven by the federal government to ensure the *human right* to adequate nourishment to people with limited access to food. This strategy is part of the promotion of *food and nutritional security*, seeking for social inclusion and for the achievement of the population's most vulnerable to hunger citizenship.
>
> (President of Brazil, 2012 cited in Marques *et al.*, 2013: 40 [my emphasis added])

Its mechanisms for implementation include programmes for family farming, maternal nutrition, child labour, government food procurement, and school feeding/attendance, as well as a cooking gas subsidy and conditional family allowances (Rocha, 2009). This was a consolidation of existing government initiatives. The key element was its conditional cash transfer (CCT) programmes, which are today mainstream social policy in Brazil and across Latin America (Hall, 2006; Rocha, 2009). Lula consolidated and expanded CCTs, first under Fome Zero and later under *Bolsa Familia*,[2] now the largest such scheme in the world. Brazil's preference for targeted social safety nets emerged from a rejection of values of universalization and citizenship rights (despite the fact that this had undermined social protection infrastructure during the market liberalism in the region since the 1980s), in place of which, development finance was thought to be best distributed via programmes targeting the poorest and most vulnerable (Hall, 2006). In addition to key ideological aspects described below, Fome Zero also aimed to foster corporate social responsibility and alliances between public and private sector actors, involving over 100 corporate actors (such as Unilever, Ford and major supermarket chains) in the scheme (Hall, 2006). Fome Zero still continues officially but has been replaced by Bolsa Familia in practice (Ansell, 2015).

In-depth ethnographic research by Ansell (2015) has investigated the construction of the Fome Zero discourse. He argues that strong political will to 'radicalize democracy' (Ansell, 2015: 1268) was supported by alliances between the state and well-organized local social movements, encouraging a shared vision between society, government and non-government actors to emerge. The metaphors of *mutirão* (traditional forms of collective labour) and *quilombos* (slave descendent identity) were mobilized to align Zero Hunger with existing local race and class-based social movements. Thus, Zero Hunger was framed as a state-supported network of diverse struggles, rather than a government-led social and economic policy. Proponents (state and civic) used the language of 'reclaiming rights' and 'state indebtedness' in a 'cultural engineering mission aimed at the political empowerment of marginal populations' (Ansell, 2015: 1279).

Results have been mixed. On one hand, Fome Zero has been estimated to have lifted 70 million people out of poverty, has promoted access to food, income-generating activities, partnership and social mobilization, and has strengthened family farming. Its success has been attributed to (a) strong legal basis for the right to food, (b) public participation and intersectorality in governance reform, and (c) the fact that it was funded by the state, not international donors (Oxfam, 2014). Through mobilizing historical discourses of local class and gender-based inequalities, local participants have shifted identities and created new alliances between state and civic organizations. On the other hand, it was limited by over-complexity and the inability to reconcile competing interests (Ansell, 2015), implementation inefficiencies, high costs and problems with targeting (Hall, 2006).

West Africa

Zero Hunger West Africa was launched in March 2014 by the UNFAO and ECOWAS. It has been described as a 'West African adaptation of a global goal,' and provides a framework linking sustainable agricultural investment with social protection programmes (FAO and ECOWAS 2014). Modelled on Latin American Zero Hunger efforts, the West African programme reflects the narrative that multiple and recurrent interconnected crises (food, climate, finance) must be met by addressing their structural causes. Zero Hunger West Africa aims to specifically address the right to adequate food, gender empowerment, family farming and sustainable food systems via state-led social protection safety nets, climate change adaptation, nutrition sensitive agriculture, credit/ market reform, and support for family farming (CEDEAO/ECOWAS, 2012). States, markets and technologies are key actors. Governance reform (including of financial investment) is also a main goal.

The official position paper indicates a blurring of discourses within Zero Hunger West Africa:

> The general framework of the ECOWAS agricultural policy clearly posts *food security* as the ultimate goal, with a view to ensuring *food sovereignty* of

> the region, focused on the development of the region's agricultural potential to reduce its *dependence on food imports*. This strong option must be seen in the context of global uncertainties about the world food outlook.
>
> (CEDEO/ECOWAS, 2012: 4 [my emphasis added])

Documents published by CEDEO/ECOWAS (2012) frame the need for a regional agricultural policy against (i) the context of structural and cyclical factors influencing global/regional food crises, and (ii) high levels of household level vulnerability to natural, economic or political shocks. The 2020 vision for the region (see ECOWAS 2016a) proposes to achieve overall food security by transforming agriculture (via 'sustainable intensification,' diversification and climate-smart agriculture), reducing dependence on food imports, engaging surplus rural labour, addressing price volatility, building up regional emergency food reserves, improving supply chains and reforming agricultural investment (see also ECOWAS 2016b; FAO 2016). Household resilience is framed around improving access to food and services, lowering consumer prices, and the provision of safety nets. The latter 'if they are judiciously chosen and negotiated with the beneficiaries,' can help the most vulnerable 'to get out of the poverty trap … and address the multidimensional nature of malnutrition' (CEDEO/ECOWAS, 2012: 6). Gender equality is also a key theme, to be addressed via the implementation of gender-responsive budgeting (ECOWAS, 2016a). Right to food rhetoric is also strong, despite few West African countries (Cote D'Ivoire, Niger, Mali and Senegal) having enshrined the right to food in their constitutions or framework laws (Oxfam, 2014).

A strong narrative of community engagement positions the regional body ECOWAS (rather than specific nation states) as responsive to local concerns and multi-stakeholder engagement. This involves building a framework for state and non-state actors to work together to implement sustainable agricultural investment and social protection programmes in order to eradicate hunger and build resilience (CEDEO/ECOWAS, 2012), although the exact mechanisms through which civil society can participate are unclear. For example, what role local communities and civil society actors (producer groups, grassroots producer organizations and women's associations in particular) should play in shaping or managing social safety nets was raised as a key challenge in early documentation (CEDEO/ECOWAS, 2012: 7). Promotional material is strong on emphasizing the collective nature of the project:

> What exactly is the Zero Hunger Initiative? It is not a programme but an invitation to action. It unites all who are working for food security and an end to hunger, encourages them to identify how they can contribute effectively from their own perspective, and stimulates a dramatic scale-up of collective efforts. There are many pathways to achieving Zero Hunger … Who is involved in this initiative? In short: Everyone!
>
> (FAO and ECOWAS, 2014: 2)

Oxfam (2014) has outlined reservations about the capacity of Zero Hunger West Africa to be truly participatory, however. First, due to the regional nature of the programme, coordinating governance across West African countries will be a major challenge. Second, progress on citizen participation varies across countries, due to an instrumental vision of participation and lack of political will, despite the existence of strong civil society organizations and rural peasant movements. No country-specific evaluation of Zero Hunger West Africa has yet been published. The FAO has reported that since 2014, the programme has increased awareness of the need to: translate the right to food into concrete actions; promote nutrition-sensitive and climate-smart agriculture; and promote gender sensitive interventions and social protection strategies. It has also developed a Zero Hunger index to assist with monitoring, advocacy and communication (FAO, 2016: 14).

Sustainable development goals

Zero Hunger appears as a distinctive term in the UN 2030 Agenda for Sustainable Development in two inter-related forms: (1) as *SDG#2* ('to end hunger, achieve food security and improved nutrition and promote sustainable agriculture'); and (2) in the *Zero Hunger Challenge*. As a goal, SDG#2 positions global hunger and malnutrition as 'a barrier to sustainable development [that] creates a trap from which people cannot easily escape' (UN 2016a: 2). In breaking from the tendency for sustainable development discourse (since its inception in 1992) to privilege economic or environmental goals (Cook *et al.*, 2012), SDG#2 connects hunger, food security, nutrition and sustainable agriculture with poverty, livelihoods, economies, health, education, gender equality, rights, and social development; all deeply social objectives. Indeed, all of the 17 SDGS[3] are intended to be integrated, with social goals either explicit (1–5, 10, 16, 17) or implicit (6–8, 11–13) in most (Lockie, 2016). This is elaborated in the 'five elements of the Zero Hunger Challenge:'

1 All food systems are sustainable: from production to consumption.
2 An end to rural poverty: double small-scale producer incomes and productivity.
3 Adapt all food systems to eliminate loss or waste of food.
4 Access adequate food and healthy diets, for all people, all year round.
5 An end to malnutrition in all its forms.

(UN, 2016b: 2)

The Zero Hunger Challenge uses a series of circular visualizations to operationalize these five foci, each illustrating how the central 'element' (sustainable systems, rural poverty, waste, diets, malnutrition) intersects not just with SDG#2, but with others as well. For example, 'sustainable food systems' is aligned with goals 6 (gender inequality), 7 (affordable and clean energy), 11 (sustainable cities and communities), 12 (responsible production and consumption), 13 (climate action), 14 (life below water), 15 (life on land) and 17 (partnerships for the goals). This framing of Zero Hunger builds a narrative of 'beyond business as usual,' by addressing complexity and associating solving hunger with the values of universality, inclusiveness, transparency, integrity, accountability, sustainability, human rights, gender sensitivity, and global cooperation (UN, 2016a: 2–3). Using a transformation narrative, the SDGs reimagine progress on global hunger.

The SDGs emerged from a wide consultative process between UN entities, governments, civil society groups and citizens, formalized into nine sectors of society through which participation was facilitated. These official 'Major Groups'[4] were invited to make submissions to the UN Conference on Sustainable Development (UNCSD) as part of the post-MDG and Rio+20 process; suggest suitable goals and priorities for the 2030 Agenda; and comment on progressive drafts of the SDG goals and text. The UN also initiated an online 'global survey' for citizens to add their 'voice' to the process via the *worldwewant2015.org* website. Civil society participation was actively sought to redress the restricted participation in the UN General Assembly in the past, and it resulted in numerous successful inclusions of Major Groups' negotiated language, concepts and agendas in the final SDG documentation (Gabizon, 2016).

UN documents from the SDG-setting and monitoring process provide a rich source of qualitative data on how food security, nutrition and justice-related sustainability themes were framed and contested during and after the construction of the Zero Hunger goal. Table 6.2 presents findings from a content analysis of key UN documents. The sample of documents was selected on the basis of (i) relevance to hunger and food negotiations, (ii) representation of the UN 'expert groups' most active in shaping SDG#2 (including the Major Groups, citizens, FAO, IFAD and the WTP), and it was deliberately kept small to enable in-depth narrative analysis. Comparing 'input' documentation, official Zero Hunger 'output' documents and 'post-process' reports that have since evaluated the first year of the SDGs provides insight into its trajectory.

Taken together, these texts reflect the narrative that achieving food and nutrition security is multidimensional, while simultaneously ensuring that economic growth, food production (via climate-smart sustainable agricultural intensification) and market access are prioritized. Empowerment and improved participation are acknowledged as prerequisites to achieving Zero Hunger. But beyond educating the rural labour force and investing more in agricultural development and extension, there is little elaboration on the socio-political shifts in power, resources or knowledge that might be needed. For example, social safety nets and social protection are associated with 'scaling up proven solutions' (see FAO, IFAD and WFP, 2014b), but with no further elaboration as to their role in alleviating the poverty or inequality that affect food access.

The right to adequate food and nutrition was present in Major Groups' inputs and the elaboration of the Zero Hunger Challenge, but surprisingly absent from the SDG#2 targets themselves and the 2016 evaluative reports analysed here. The more politicized language of food justice or food sovereignty is also mostly absent, with language on rights kept general and conservative. For example, the

Table 6.2 Content analysis of official SDG documents

Document for analysis	*No. times key word appears in the document*				
	Food	*Food (and/or) nutrition security*	*Justice (food or other)*	*Right to food*	*Food sovereignty*
Major groups' and citizens' amendments (UNCSD Major Groups 2014; 2015) [input]	13	1	3	2	2
FAO, IFAD and WFP amendments (2014a) [input]	60	9	0	0	0
Transforming our World: the 2030 Agenda (UN 2015a) [output]	8	6	7	0	0
SDG 2 Targets (UN 2015b) [output]	6	0	0	0	0
Zero Hunger Challenge (2015c) [output]	45	9	2	9	0
Global Sustainability Report (UN 2016b) [post-process]	36	12	12	0	0
FAO report on SDG goals for food and agriculture (FAO 2016) [post-process]	107	21	0	1	0
TOTALS	275	58	24	12	2

Zero Hunger Challenge is 'people centered and based on human rights and social justice' (UN, 2016b). Climate justice appears in the document once, but food justice does not. The FAO (2016) report mentions the right to food once, but specifies women's rights to land more often. The 2030 Agenda does not specify the right to food at all, but makes 29 references to human rights (general) and more specific rights (such as to safe drinking water and sanitation, women's rights, intellectual property rights, labour rights, childrens' rights and the right to development).

By contrast, inputs into the SDG process by civic actors reveal deeper levels of theoretical and political contestation around themes of food production, markets, nutrition, rights, justice and participation. For example, in their final response to the SDG working document, the farmers' Major Group states:

> We must also acknowledge food sovereignty as a key demand of the poor … We are concerned that smallholder farmers as the critical 'agents of change' have not been given the necessary prominence in the document … We note with concern that the right to adequate food is not explicitly mentioned.
>
> (UNCSD Major Groups, 2015: 12–13)

Major Groups' amendments (which included citizen responses) to goals and targets on SDG#2 requested that they include (a) 'ensure the progressive realization of the right to adequate food' as a specific 2030 sub-goal; (b) doubling the incomes of small-scale producers via securing land tenure and the promotion of cooperatives and producer organizations; and (c) limits on using human food crops for non-human purposes (UNCSD Major Groups, 2014: 6–7). These last two amendments closely align with food justice and food sovereignty discourses around land reform, biofuels, protein production and consumption, and solidarity economy. None of these themes was included in the final goals or targets. This is in line with Gabizon's (2016) findings about the exclusion of progressive language about gender and justice put forth by the Women's Major Group.[5]

Discussion: the evolution of discourse and implications for sustainability

Are the discourses informing *Zero Hunger* progressive or radical, or do they reflect more subtle reforms to the global food regime? Table 6.3 summarizes the discursive content and construction of Zero Hunger in the three case studies presented. Gender empowerment, smallholders and social protection were common themes across the case studies, as was the question of how best to produce food. More variation was found in the use of food security and food justice discourses, and in the use of narratives around rights, inequality, resilience and access. Food sovereignty was the least clearly discernable discourse.

First, I find elements of all four discursive categories (neoliberal, reformist, progressive and radical) being expressed in the narratives embedded in Zero

Table 6.3 Comparative analysis of Zero Hunger discourses

	Discursive content	*Rhetorical devices, narratives and metaphors*	*Agents and motives*	*Conclusion*
Fome Zero Brazil	Agroecology Decentralised decision-making Gender Right to food (+ law) Smallholders Social and food justice Social protection Redistribution (material and power)	Identity and history State indebtedness Redemocratization Localization Class analysis	State – build local alliances, democracy and citizenship, strengthen leftist ideology Social movement activists – claim rights, empower Local people – material goods	Radical, Social green Imaginative
Zero Hunger West Africa	Civic participation Climate-smart agriculture Food sovereignty Gender empowerment Smallholders Governance reform Household level food security Right to food (-law) Social protection	Regional cooperation Governance Pro-resilience Poverty alleviation	FAO & ECOWAS - regional governance reform Civil society – stakeholder engagement Funding bodies – economic development	Reformist/progressive Institutionalist Prosaic
SDGs	Climate-smart sustainable agricultural intensification Environmental justice Fairness and equity Food and nutrition security Gender Market reform Resilience and climate adaptation Rights (but not RtF) Social protection Universalism Cooperation	Justice as 'access' Growth Resilience Food systems	UN/global governance – generate global vision and consensus, specify measurable goals and targets Civil society – recognition of rights, meaningful participation, highlight interconnections	Neoliberal/reformist Institutionalist Imaginative

Hunger across time and scale. This supports Clapp and Dauvergne's (2011) reminder that their typology is analytical in the way of 'ideal types,' which cannot always be so neatly separated in the real world of environmental politics. Food politics is likewise complex.

In its initial conception, Zero Hunger Brazil was radical (according to Holt Gimenez and Shattuck, 2011), reflects a social green worldview (Clapp and Dauvergne, 2011) and was imaginative (Dryzek, 2013). Although food sovereignty was not stated explicitly in Fome Zero, the ideology was evident in the prioritization of solutions such as agroecology, cooperatives, land reform, localization and redistribution, and there was strong state support for the politicization of peasants' rights and citizen empowerment. Although many of these characteristics also reflect the concerns of food justice, Fome Zero went further by reimagining how local actors could play a role in redistributing the commons and regaining control over capacities and entitlements (especially land and identities). By constructing an historical narrative in which local farmers' rights needed to be claimed back from the state, a key motivation of the state was to shift the public ideology of governance from one of patronage towards one of solidarity and claims about the public good; a 'redemocratization discourse' (Ansell, 2015: 1269). Zero Hunger also introduced a class analysis that resonated with social movements' narratives of loss, particularly around rural peoples' displacement from land by agribusiness, commercial development and assimilation policies. Entitlements that marginalized groups had to land and other material goods, as well as strong rights embedded in laws, were at the centre of a re-imagination of the relationship between the state and local people's experiences of inequality. For these reasons, Zero Hunger in Brazil displays many of the characteristics of a radical-imaginative sustainability discourse. It may not reflect a bioenvironmental worldview (which in Table 6.1. is aligned with a radical, food sovereignty discourse), but none-the-less provides a critique of social-environmental relations, and thus, of sustainability.

Zero Hunger West Africa has been much more reformist in its discursive content around market-led solutions and reliance on international financial institutions. For example, although food justice discourse (which would indicate a progressive approach, according to Holt Gimenez and Shattuck, 2011) is mobilized at the local and household-level (i.e. economic empowerment and rights of women and family farmers), the promotion of climate-smart agriculture for achieving national and regional food security suggests support for corporate agribusiness and international development financiers. This is at odds with food justice and food sovereignty (see Patel, 2009, Wittman *et al.*, 2010), but completely in line with a food and nutrition security approach. The right to food is present but limited by the lack of legal frameworks or accountability. A narrative of reform emerges in which poverty alleviation and responding to complex food system challenges (hunger, climate change, financing) are mutually reinforcing policy priorities requiring regional vertical and horizontal cooperation to address this complexity. This is a key tenet of Clapp and Dauvergne's (2011) institutionalist worldview. The position paper also uses the language of capitals, capabilities, vulnerability and resilience to construct a 'pro-resilience' vision for food security, distinguishing Zero Hunger as a necessary reform to existing Regional and National Agricultural Investment Plans (RAIPs and NAIPs) (CEDEO/ECOWAS, 2012). While a growing body of literature argues that resilience thinking can promote transformations in addition to responding to shocks and reducing vulnerability (Olsson *et al.*, 2014), in the case of Zero Hunger West Africa, resilience tends to be equated unproblematically with mainstream (or prosaic, see Dryzek, 2013) meanings of sustainability.

It is difficult to distinguish the more imaginative use of rights and resilience discourse amidst the strong voice of the development and finance organizations that have created the programme. Further research on the role and influence of civil society, and especially the more 'radical' or 'progressive' social movements, is required.

In the SDGs, a neoliberal discourse (Holt Gimenez and Shattuck, 2011) is present, alongside a weaker reformist discourse on rights. Despite the rhetoric of addressing inequality, this is framed mostly as a function/outcome of addressing the vulnerability and resilience of marginalized groups, such as women and smallholder farmers. Emphasis is on facilitating these groups' *access* to sufficient and nutritious food, improving productive efficiencies, strengthening resilience to shocks, building skills and 'empowering.' Lockie (2016:115) suggests that the SDGs represent 'a radical departure from the ways in which sustainability has largely been operated' in the past (although, with some reservations). Weber (2014, 2015) disagrees, arguing that although the goals claim to place people, poverty and livelihoods at the centre of global development, both fail to challenge long-established assumptions about economic growth. This is a prosaic response, in terms of Dryzek's (2013) theory. The privileged position of the market and existing institutional players is difficult to challenge within the confines of high-level diplomatic language and negotiation processes; in Clapp and Dauvergne's (2011) typology, market-liberal and institutionalist worldviews are closely connected, which resonates with the case of the SDG process. For Weber (2014: 131; see also Weber, 2015), the SDGs have not diverged from the hegemonic (neoliberal) narratives of modernization and the 'development ladder' that underpinned earlier strategies (e.g. the MDGs), and their continued framing around goals rather than entitlements has further entrenched market-based measures and goals conceived in terms of market access.

Finally, in mapping this discursive trajectory, a key element has been the political dilution of the language of justice. For example, at global and regional levels, the right to food does not appear, or is mobilized in aspirational terms only. This is likely due to the already well-established (and hard won) consensus between the dominant (i.e. neoliberal) decision-makers around the legitimacy of the language of human rights, poverty alleviation, equality, resilience, inclusion, empowerment, participation, restructuring (economic and political), transparency and accountability. The core notion of justice is addressing inequity through 'access' to a fairer distribution of outcomes (negative or positive). These goals are compatible with market-oriented neoliberalism, and do not necessitate justice being more adequately integrated (as it is in food sovereignty or food justice movements). Progressive and radical notions of justice – such as entitlements, justicability (i.e. the rule of law) or a critical analysis of class-based inequalities – are still subject to conflict and non-consensus; an inevitable characteristic of any conception of justice, according to Schlosberg (2002). But they are needed in order to 'open up those pathways for social and political action,' despite missing 'a central metaphor to situate the ways in which the food system could be transformed' (Gottlieb and Joshi, 2010: 5–6).

Conclusion

What does justice, and its multiple discourses and experiences (of actors who participate in framing *Zero Hunger* policies and programmes, for example), matter for the transnational governance of sustainability? This chapter has illustrated that Zero Hunger, *as a framework for the development or reform of the global food regime*, has evolved from its beginnings as a progressive ideological exercise of the Brazilian state into something much more mainstream. This may present a limitation to the transformative promise of Zero Hunger, as hoped in the SDGs. But discourse analysis has also shown that some sustainability narratives – such as food as a human right, market and agricultural reform, collective social action and participation – are present across all of the case studies explored here.

In this chapter, I have applied insights from typologies of food regime/movement discourses and worldviews in an attempt to untangle the messy ways that sustainability and justice have been constructed and contested, gaining currency for some narratives over others. Specifically, I have described how some sustainability discourses – namely food and nutrition security – have been embedded in governance for 'sustainable and just' food policy more readily than the progressive or radical discourses of food sovereignty and food justice. In the case of Zero Hunger, this has resulted largely from politicizing or de-politicizing the issues of rights, entitlements, access, markets and participation. The dominant framing of hunger is one that acknowledges the complex interconnections between environmental (e.g. climate change, agricultural production, resource use, pests and diseases) and social (e.g. health, livelihoods, distribution, markets) determinants of vulnerability, access and resilience. Such framings do not necessitate the adoption of progressive or radical discourses around rights that are taken up more centrally by food sovereignty and food justice movements.

While the SDGs certainly reflect a continuing neoliberal development discourse, Zero Hunger still involves a good deal of imaginative redefinition. Its broad framing of rights and justice may be exactly what is needed to be more politically inclusive of the growing plurality of justice (ideas and actors) underpinning the dynamics of contestation between the corporate food regime and progressive/radical food movements. As Dryzek writes (2013: 15), discourses are not set in stone, and so have the potential to 'dissolve old dilemmas, treating environmental concerns not in opposition to economic ones, but potentially in harmony … The thinking is imaginative, but the degree of change sought can be small and reformist, or large and radical.' Investigations of how successful different food system actors are in mobilising their preferred food discourses for specific ends, whether to extend neoliberal market ideology or to seek radical reform of structural inequalities, provides rich ground for further research.

Notes

1 www.zerohungerchallenge.org; see also UN 2015b

2 Since 2003, spending on Bolsa Famı´lia has risen to over one-third of the social assistance budget for the poorest sectors. Its four sub-programmes (educational stipends to boost school attendance, maternal nutrition, food supplements and a domestic gas subsidy) benefit some 30 million of Brazil's poorest people, with a target of 44 million by 2006 (Hall, 2006).

3 The SDGs are a set of 17 goals with 169 associated targets (UN, 2015a:12). They are:

Goal 1 End poverty in all its forms everywhere
Goal 2 End hunger, achieve food security and improved nutrition and promote sustainable agriculture
Goal 3 Ensure healthy lives and promote well-being for all at all ages
Goal 4 Ensure inclusive and equitable quality education and promote lifelong learning opportunities for all
Goal 5 Achieve gender equality and empower all women and girls
Goal 6 Ensure availability and sustainable management of water and sanitation for all
Goal 7 Ensure access to affordable, reliable, sustainable and modern energy for all
Goal 8 Promote sustained, inclusive and sustainable economic growth, full and productive employment and decent work for all
Goal 9 Build resilient infrastructure, promote inclusive and sustainable industrialization and foster innovation
Goal 10 Reduce inequality within and among countries
Goal 11 Make cities and human settlements inclusive, safe, resilient and sustainable
Goal 12 Ensure responsible consumption and production patterns
Goal 13 Take urgent action to combat climate change and its impacts
Goal 14 Conserve and sustainably use the oceans, seas and marine resources for sustainable development
Goal 15 Protect, restore and promote sustainable use of terrestrial ecosystems, sustainably manage forests, combat desertification, and halt and reverse land degradation and halt biodiversity loss
Goal 16 Promote peaceful and inclusive societies for sustainable development, provide access to justice for all and build effective, accountable and inclusive institutions at all levels
Goal 17 Strengthen the means of implementation and revitalize the global partnership for sustainable development.

4 The UNCSD Major Groups are: women, children and youth, indigenous peoples, non-governmental organizations, local authorities, workers and trade unions, business and industry, the scientific and technological community, and farmers. In addition, governments invited other stakeholders, including local communities, volunteer groups and foundations, migrants and families, as well as older persons and persons with disabilities, to collaborate with the Major Groups.

5 Despite being present in negotiations and having numerous successes (i.e. gender equality is recognized as both a separate goal, #5, and across the other SDGs), the

Women's Major Group could not gain inclusion for the language of women's rights (instead opting for 'empowerment' in the goal itself), the gendered dimensions of climate change, or for explicit connection of gender inequality with redistributive policies and justice (Gabizon, 2016).

References

Agyeman, J. (2005). *Sustainable Communities and the Challenge of Environmental Justice*. New York: New York University Press.

Allen, P. (2008). Mining for justice in the food system: perceptions, practices, and possibilities. *Agriculture and Human Values* 25: 157–161.

Ansell, A. (2015). Lula's assault on rural patronage: Zero Hunger, ethnic mobilization and the deployment of pilgrimage. *Journal of Peasant Studies* 42(6): 1236–1283.

Cadieux, K. V. and Slocum, R. (2015). What does it mean to do food justice? *Journal of Political Ecology* 22: 1–26.

Carolan, M. (2013). *Reclaiming Food Security*. London and New York: Earthscan.

CEDEO/ECOWAS (2012). *Accelerating the ECOWAP/CAADP implementation, Position Paper: Towards local-level food security in West Africa: 'Zero Hunger in West Africa.'* Abuju, Nieria: ECOWAS Commision/Commission de la CEDEO.

Clapp, J. and Dauvergne, P. (2011). *Paths to a Green World: The Political Economy of the Global Environment*, 2nd edn. Cambridge Massachusetts and London, England: MIT Press.

Clendenning, J., Dressler, W. and Richards, C. (2016). Food Justice or Food Sovereignty? Understanding the rise of urban food movements in the USA, *Agriculture and Human Values* 33: 165–177.

Cook, S., Smith, K., and Utting, P. (2012). *Green Economy or Green Society? Contestation and politics for a fair transition*. UNRISD Occasional Paper No. 10, Geneva: UNRISD.

Dryzek, J. S. (2013). *The Politics of the Earth: Environmental Discourses*, 3rd edn. Oxford: Oxford University Press.

ECOWAS (2016a). Agriculture. www.ecowas.int/ecowas-sectors/agriculture/. Accessed 15 July 2016.

ECOWAS (2016b). FAO and ECOWAS to continue supporting women in agriculture. www.ecowas.int/fao-and-ecowas-to-continue-supporting-women-in-agriculture/. Accessed 17 March 2016.

FAO (1996). *World Food Summit: Rome Declaration of World Food Security*. www.fao.org/wfs/index_en.htm. Accessed 18 April 2018.

FAO (2016). *Food and Agriculture: Key to achieving the 2030 Agenda for Sustainable Development*. FAO: Rome. www.fao.org/3/a-i5499e.pdf. Accessed 17 March 2017.

FAO and ECOWAS (2014). Zero Hunger Initiative for West Africa. www.fao.org/fileadmin/templates/righttofood/documents/news/140224_ecowas/leaflet_ecowas_en.pdf. Accessed 3 March 2014.

FAO, IFAD and WFP (2014a). *Food security, nutrition and sustainable agriculture in the post-2015 agenda: priority targets and indicators identified by FAO, IFAD and WFP*. Rome: FAO, IFAD, WFP. www.fao.org/fileadmin/user_upload/post-2015/FSN_-Targets_and_indicators_-RBAs_joint_proposal_proofed_clean__2_.pdf. Accessed 17 March 2017.

FAO, IFAD and WFP. (2014b). *The State of Food Insecurity in the World 2014: Strengthening the enabling environment for food security and nutrition*. Rome: FAO. https://sustainabledevelopment.un.org/content/documents/1717The%20State%20of%20Food%20Insecurity%20in%20the%20World%202014.pdf. Accessed 10 January 2017.

Friedmann, H. (2005). From colonialism to green capitalism: Social movements and emergence of food regimes. In: F. Buttel and P. McMichael (eds). *New Directions in the Sociology of Development*, Vol. 11. UK: Elsevier.

Gabizon, S. (2016). Women's movements engagement in the SDGs: lessons learned from the Women's Major Group. *Gender and Development* 24(1): 99–110.

Goodman, D., Du Puis, M. and Goodman, M. (2014). *Alternative Food Networks: Knowledge, Practice and Politics*. London and New York: Routledge.

Gottlieb, R. and Joshi, A. (2010). *Food Justice*. Cambridge: The MIT Press.

Hall, A. (2006). From Fome Zero to Bolsa Familia: Social Policies and Poverty Alleviation under Lula. *Journal of Latin American Studies* 38(4): 689–709.

Holt Gimenez, E. and Shattuck, A. (2011). Food crises, food regimes and food movements: rumblings of reform or tides of transformation? *Journal of Peasant Studies* 38(1): 109–144.

Holt Gimenez, E., Patel, R. and Shattuck, A. (2009). *Food Rebellions! Crisis and the Hunger for Justice*. Oakland, California and Oxford Food First Press and Pambazuka Press.

Jarosz, L. (2014). Comparing food security and food sovereignty discourses. *Dialogues in Human Geography* 4(2): 168–181.

Lang, T. (2010). Crisis? What crisis? The normality of the current food crisis. *Journal of Agrarian Change* 10(1): 87–97.

Lee, R. P. (2013). The politics of international agri-food policy: discourses of trade-oriented food security and food sovereignty. *Environmental Politics* 22(2): 16–234.

Lockie, S. (2016). Editorial: Beyond resilience and systems theory: reclaiming justice in sustainability discourse. *Environmental Sociology* 2(2): 115–117.

Marques, R. D., Ganem, A. M. and Marques Junior, J. C. (2013). The Brazilian Zero Hunger social programme implications in achieving sustainable development. *Future of Food: Journal on Food Agriculture and Society* 1(2): 39–45.

McMichael, P. (2005). Global development and the corporate food regime. In: F. Buttel and P. McMichael (eds), *New Directions in the Sociology of Global Development: Research in Rural Sociology and Development*, vol. 11. UK: Elsevier.

McMichael, P. (2008). The peasant as 'canary'? Not too early warnings of global catastrophe. *Development* 51(4): 504–511.

McMichael, P. (2009a). A food regime analysis of neo-liberalism's food crisis *Agriculture and Human Values* 26(4): 281–295.

McMichael, P. (2009b). A food regime genealogy. *Journal of Peasant Studies* 36(1): 139–169.

McMichael, P. (2014). Historicizing food sovereignty. *The Journal of Peasant Studies* 41(6): 933–957.

Olsson, P., Galaz, V. and Boonstra, W. J. (2014). Sustainability transformations: a resilience perspective. *Ecology and Society* 19(4): 1.

Oxfam International (2014). *Food for Thought: Learning from Brazil, A source of inspiration for the successful implementation of 'Zero Hunger' in West Africa*, www.oxfam.org/sites/www.oxfam.org/files/file_attachments/rr-brazil-experience-preface-editorial-190214-en_1.pdf. Accessed 10 March 2017.

Patel, R. (2009). What does food sovereignty look like? *The Journal of Peasant Studies* 36(3): 663–706.

Pritchard, B., Rammohan, A., Sekher, M., Rarasuraman, S. and Choithani, C. (2014). *Feeding India: Livelihoods, entitlements and capabilities*. London and New York: Earthscan.

Rocha, C. (2009). Developments in National Policies for Food and Nutrition Security in Brazil. *Development Policy Review* 27(1): 51–66.

Rosin, C., Stock, P. and Campbell, H. (eds), (2012). *Food Systems Failure: The global food crisis and the future of agriculture*. London and New York: Earthscan.

Schlosberg, D. (2002). *Environmental Justice and the New Pluralism: The challenge of different for environmentalism*. Oxford: Oxford University Press.

Schlosberg, D. (2013). Theorising environmental justice: The expanding sphere of a discourse. *Environmental Politics* 22(1): 37–55.

Smith, K. (2016). Food Systems Failure: Can we avert future crises? In: M. Shucksmith and D. Brown (eds), *Routledge International Handbook of Rural Studies*, London and New York: Routledge: 250–262.

UNCSD Major Groups (2014). *Final compilation of amendments to goals and targets by Major Groups and other stakeholders including citizen's responses to MY World 6 priorities, to inform the Thirteenth and last Session of the Open Working Group on Sustainable Development Goals*, 14–18 July 2014, New York: United Nations. https://sustainabledevelopment.un.org/content/documents/4438mgscompilationowg13.pdf. Accessed 17 March 2017.

UNCSD Major Groups (2015). *Compilation Major Groups Statements and Focus Areas Responses*. New York: United Nations. https://sustainabledevelopment.un.org/content/documents/3362Compilation%20Major%20Groups%20Statements%20and%20Focus%20Areas%20Responses_March%2025_final.pdf. Accessed 17 March 2017.

United Nations (2012). *Review of the contributions of the MDG Agenda to foster development: Lessons for the post-2015 UN development agenda*, Discussion Note, March 2012. UN System Task Team on the Post-2015 UN Development Agenda. New York: United Nations.

United Nations (2015a). *Transforming our World: The 2030 Agenda for Sustainable Development*, A/RES/70/1. New York: United Nations. https://sustainabledevelopment.un.org/content/documents/21252030%20Agenda%20for%20Sustainable%20Development%20web.pdf. Accessed 17 March 2017.

United Nations (2015b). *Zero Hunger Challenge: Partnering for Nourishing our People, Nurturing our Planet, Growing Prosperity, Harvesting Peace, Transforming our Food Systems to Transform our World.* New York: United Nations. www.un.org/en/zerohunger/pdfs/ZHC%20-%20Pathways%20to%20Zero%20Hunger.pdf. Accessed 17 March 2017.

United Nations (2016a). *Zero Hunger: Why does it matter?* New York: United Nations. www.un.org/sustainabledevelopment/wp-content/uploads/2016/08/2_Why-it-Matters_ZeroHunger_2p.pdf. Accessed 14 August 2017.

United Nations (2016b). *Global Sustainable Development Report 2016*, Department of Economic and Social Affairs, New York: United Nations. https://sustainabledevelopment.un.org/content/documents/2328Global%20Sustainable%20development%20report%202016%20(final).pdf. Accessed 17 March 2017.

United Nations General Assembly (2013). *Right to Food, Note by the Secretary-General: Interim report of the Special Rapporteur on the right to food*, 68th session, Item 69b of the United Nations General Assembly, A/68/288, 7 August 2013. New York: United Nations.

Weber, H. (2014). When Goals Collide: Politics of the MDGs and the post-2015 sustainable development goals agenda. *SAIS Review of International Affairs*, 34(2): 129–139.

Weber, H. (2015). Reproducing inequalities through development: The MDGs and the politics of method. *Globalizations* 12(4): 660–676.

Wittman, H., Desmarais, A. A. and Wiebe, N. (eds), (2010). *Food Sovereignty: Reconnecting food, nature and community*. Cape Town, Dakar, Nairobi and Oxford: Pambazuka Press.

7 Greenwashing the animal-industrial complex

Sustainable intensification and the Livestock Revolution

Livia Boscardin[1]

Introduction

Animal agriculture emits more greenhouse gases than the global transport sector and is the main driver of biodiversity loss. Compared to the environmental repercussions of the livestock industry and its use of grain for fodder (namely 35 per cent of the global harvest), fracking or the production of biofuels look like Disneyland fantasies. Moreover, the sector's working conditions are extremely precarious and risky. And last but not least, it kills almost 70 billion animals every year for profit. Nevertheless, international institutions such as the Food and Agriculture Organization of the United Nations (FAO) assert that global livestock consumption will double by 2050. This increase is called the 'Livestock Revolution' – in reference to the Green Revolution – and is deemed an inexorable evolution in the course of globalization and development.

This chapter focuses on sustainability discourses in the livestock industry. More concretely, it examines how the Livestock Revolution should be attained by sustainable intensification (agro-ecology lite). In addition, the perspective of critical animal studies posits the violent commodification of (nonhuman) animals as a political problem. To gain a holistic picture of the situation, the chapter first portrays the animal-industrial complex and its impact on animals, planetary boundaries, and society. Second, the neo-productivist discourse of the Livestock Revolution occupies centre stage. What were the causes of the Revolution, and what instruments are used to further it? Third, the 'necessity' and 'efficiency' of raising and slaughtering billions of animals every year are scrutinized, as is this practice's effect on the alleviation of poverty. Eventually, the chapter shows how this 'sustainable,' 'modern,' high-tech form of animal exploitation deepens the ecological, societal, and food crises.

The animal-industrial complex[2]

What is the animal-industrial complex? This rather knotty term is an adaptation of 'military-industrial complex,' a term coined by former U.S. President Eisenhower, and it aims to sum up the multiple dimensions of the animal industry – from the feed producers, hatcheries, ranchers and pharmaceutical corporations to the slaughterhouses, governmental inspectors and packaging companies, to the transnational retailers (Noske, 1989). As diverse as these stages of production might be, they have one unifying principle and purpose: not the production of food, but the generation of profit. The very origin of the word 'capital' is actually the Latin word *capita* (the plural of *caput*, Latin for 'head'), designating 'heads of cattle.' Cattle were one of the earliest forms of exchange equivalents (Gunderson, 2013: 261). Nonetheless, this is probably the sole remaining feature of ancient Roman cattle-raising practices. The animal-industrial complex has undergone deep-seated transformations, particularly in the second half of the twentieth century. From 1961 to 2014, global meat production quadrupled from 71 million tons to 317 million tons (FAOSTAT, 2016a). Both meat and egg consumption roughly doubled. Weis' concept of 'meatification' encapsulates the radical shift from the marginal role animal products occupied in the human diet throughout history to these products' current status as the cornerstone of most diets. Yet, these average figures conceal the highly uneven geography of meat. In 2009, the United States, Australia, New Zealand, Argentina, Canada, and Western Europe together accounted for 12 per cent of the world population, but were responsible for 34 per cent of global meat production by volume, and for 68 per cent of world meat exports. On the other hand, the population of South and Southeast Asia and Africa represented almost half of the world population, yet accounted for less than 16 per cent of total meat production (Weis, 2013b: 67–68). In terms of consumption, the yearly mean consumption of meat in the Majority World[3] is 16 kilograms per capita (FAO, 2011: 5); whereas in the Minority World, it is more than five times higher (Thornton *et al.*, 2011: 1). However, the trend is negative. The Minority World's reaction to stagnating sales is to (violently) take the bull by its horns by investing in new markets in the Majority World. Already in 1995, the 'centre of gravity' of meat production has shifted from the Minority World to the Majority World (Steinfeld and Chilonda, 2006: 3). An astronomic rise in meat production and consumption is currently happening in Eastern Asia (Allievi *et al.*, 2015: 146).

This tremendous increase in animal production in the twentieth century was only feasible via industrialization (Heinrich Böll Foundation and Friends of the Earth Europe, 2014: 10–12). Globally, around 80 per cent of the world's pig meat, chicken meat and egg production stems from intensive systems, or animal factories (Herrero *et al.*, 2013: 20889 cited in MacLachlan, 2015: 32). The technical term for such factory farms is *Concentrated Animal Feeding Operation* (CAFO). Emel and Neo (2011) define CAFOs as 'the concentration of live animals (pigs, poultry or cattle), manure, and urine to small spaces where feed is brought in' (Emel and Neo, 2011: 67). In the United States, a medium-sized CAFO counts around 50,000 chickens or 2,000 pigs (EPA, 2017). CAFOs continue to push smaller farms out of business, and while the factory farms are expanding in size, their total number has declined. Ninety per cent of U.S. pig raisers lost their jobs from 1980 to 2001 (Heinrich Böll Foundation and Friends of the Earth Europe, 2014: 10). Corporations are also consolidating in the

Majority World (Costales *et al.*, 2006: 21). Already by 1996, ten vertically integrated companies controlled 80 per cent of poultry production in Thailand (Khan, 2004: 117). Furthermore, it is not just the animal factories that grow, the animal 'output' also gets bigger. In corporatist jargon, this is called a higher 'offtake per unit of stock' (Costales *et al.*, 2006: 24). Between 1960 and 2010, the average carcass weight of 'beef cattle' and chickens increased by 30 per cent (Thornton, 2010: 2854). A small handful of high-yielding breeding lines dominate the market. Regarding the animal species killed, the slaughter of ruminants has been largely superseded by the killing of chickens and pigs – these species are 'good feed converters' and can be forcibly confined in minimal space (Steinfeld *et al.*, 2006: 75–76). Finally, without governmental support (most notably, exorbitant subsidies), the industry would be in decline. Such financial incentives are often significantly higher for animal products than for plant-based foods (Machovina *et al.*, 2015: 426). In the European Union, subsidies totaled 190 USD per cow in 2014; in China, they amounted to 47 USD per individual pig in 2012 (Bailey *et al.*, 2014: 9).

Animals as primary inputs

For the animals, the animal-industrial complex is, first and foremost, expressed as unimaginable horror and violence, on both a quantitative and qualitative scale. In 2013, more than 69 billion land animals were killed for food production (FAOSTAT, 2016b). This means these sentient individuals were raised, exploited and eventually slaughtered in order to sell their eggs, milk, particular organs, or mere flesh. The rather inconceivable number of 69 billion beings encompasses 61.2 billion chickens, 2.9 billion ducks, 1.4 billion pigs, 1.2 billion rabbits, one billion sheep and goats, 0.7 billion fowl, 0.6 billion turkeys, 0.3 billion cattle, 70 million other rodents, 4.8 million horses, 2.6 million camels, and countless other individuals (FAOSTAT, 2016b). Animal suffering has amplified on par with the rise of intensive animal farming (Gunderson, 2013). The animals' short existence is marked by incarceration, the inability to move, crowdedness, noise, sensory deprivation, malnutrition, and so on. These factors result in psychological trauma, self-mutilation, and cannibalism (Nibert, 2011). Factory farms can only perpetuate their violent and excruciating enterprise with the massive use of antibiotics, feed additives, chemical pesticides, and disinfectants (Gunderson, 2013). This 'overpowering' (Weis, 2013b: 76) is a very cruel experience for the farmed animals. The unnatural, rapid growth of the farmed animals' bodies cannot be borne by their fragile bones, weak organs, and joints, inducing a wide range of chronic illnesses and injuries. Yet, it is imperative to note that the culprits of these horrors are not a handful of malicious slaughterhouse workers, some avaricious corporations, or even the lack of law enforcement. The capitalist system itself is geared towards profit maximization which will always happen on the back of the animals – and of the employees as well (Gunderson, 2013). In point of fact, the term 'livestock' concisely denotes how farmed animals are perceived: as literally living stock, raw material, or 'primary inputs.'

Ecological crisis

Besides killing almost 70 billion land animals every year, the animal-industrial complex is the main driver of climate change, ocean acidification, and biodiversity loss, and it contributes to the crossing of other planetary boundaries as well. In 2009, and with an update in 2015, environmental scientists identified and quantified a set of nine 'planetary boundaries' within which humanity can continue to live for generations to come (Rockström *et al.*, 2009; Steffen *et al.*, 2015). Crossing these boundaries would engender irreversible environmental alterations. These boundaries are climate change, change in biosphere integrity, stratospheric ozone depletion, ocean acidification, biogeochemical flows, land-system change, freshwater use, atmospheric aerosol loading, and the introduction of novel entities. The boundaries of climate change, biosphere integrity, land-system change, and biogeochemical flows have already been traversed due to anthropogenic influence (Steffen *et al.*, 2015), and the role of the animal industry is explicated in the following.

Starting with climate change, the animal-industrial complex is responsible for 14.5 to 51 per cent of global greenhouse gas emissions. Whereas the exact percentage varies from study to study, it is agreed that it exceeds the emissions of the global transport sector (Goodland and Anhang, 2009; Gerber *et al.*, 2013) and the total yearly emissions of the United States of America (Bailey *et al.*, 2014: 4). Moreover, the exploitation of ruminants is the principal source of anthropogenic methane emissions (Ripple *et al.*, 2013: 2). The enormous 'climate hoofprint' (instead of footprint) is ever more problematic as there is widespread evidence of much more ecological, healthy, and cheap plant-based alternatives. A shift to a vegan diet, for instance, reduces greenhouse gas emissions and land use demand by up to 50 per cent (Hallström *et al.*, 2015: 3). Checking the boundary land-system change, livestock systems cover a virtually inconceivable 45 per cent of global land surface (Thornton *et al.*, 2011: 1) and 80 per cent of agricultural land (Steinfeld *et al.*, 2006: 74), making the animal-industrial complex the largest land user on Earth (Herrero and Thornton, 2013: 20880). Seventy per cent of Amazon deforestation can be assigned to the killing-for-profit sector (Emel and Neo, 2011: 70). Considering these numbers, it is not surprising that the industry is the single largest cause of species extinction due to deforestation for pasture and crop fields and subsequent land degradation (boundary biosphere integrity) (Machovina *et al.*, 2015). The enormous spatial extent of the animal-industrial complex is particularly worrisome as its spread does not match its productivity: four-fifths of agricultural land yield in terms of global consumption, a quarter of total proteins, and a mere 13 per cent of total calories (Global Agenda for Sustainable Livestock, 2014: 2). Moreover, animal agriculture is the main driver of nutrients cycles in agriculture and the primary user worldwide of reactive nitrogen as fertilizer (boundary biogeochemical flows) (Sutton *et al.*, 2011). Livestock production is inherently inefficient compared to crop production (Bouwman *et al.*, 2013: 20886): the production of one megacalorie of rice, wheat, or potatoes necessitates one gram nitrogen; conversely, producing one megacalorie of dairy, chicken, pig meat or eggs requires three grams of nitrogen, and one megacalorie of cow flesh demands 19 grams of nitrogen (Eshel *et al.*, 2014: 11998). With respect to the boundary of freshwater use, the complex consumes a third of global freshwater resources yearly (Mekonnen and Hoekstra, 2012: 3232). Animal products have an extremely high water footprint. On average, one megacalorie in the form of industrial meat necessitates 4,000 litres of water (Schneider, 2014: 615). Ruminant meat is the most water-intensive foodstuff, requiring 15,415 litres per one kilogram of cow meat. One kilogram of rice necessitates 2,497 litres, and the same quantity of potatoes merely requires 287 litres of water (Mekonnen and Hoekstra, 2011, 2012). The industry is also a top polluter of rivers and streams by run-off from fertilizers, pesticides, and other contaminants, and subsequent nutrient overabundance, which advances the crossing of the boundary of ocean acidification (Eshel *et al.*, 2014: 11996).

Societal injustice

The ecological consequences of the animal-industrial complex, the gigantic land area it requires, and its colossal inputs like feed, water, and fertilizer, have a vast impact on people's livelihoods. Primarily, they affect the Majority World, which has given rise to the concept of 'environmental racism' (DeMello, 2012: 275). Environmental racism refers to the condition that groups who are discriminated against because of their race disproportionately suffer from environmental hazards, be it landfills, pollution through resource extraction or factories, or the general consequences of climate change and environmental degradation (Pellow, 2016). With almost one billion people chronically suffering from starvation, an additional billion malnourished (FAO, 2016a), and a third of the world's population without access to a sufficient amount of fresh water (Thornton *et al.*, 2009: 117), the industry's massive intake of edible crops and water is sheer murder. Out of the total estimated global cereal utilization in 2016–2017, which amounted to 2,564 million tons, 920 million tons (or 35 per cent) were deployed as feed, while just a slightly higher amount – 1,105 million tons (or 43 per cent) – were processed as human food (FAO, 2016b). More than half of the global maize harvest and 85 per cent of the global barley harvest directly ends up in feed troughs (Weis, 2010: 140). The main part of soybean production – as soybean meal – goes to farmed animals as well (Heuzé and Tran, 2016). Metaphorically speaking, farmed animals are 'reverse protein factories' (Lappé, 2010): about 90 per cent of calories are lost in the conversion of plants into animal matter (Godfray *et al.*, 2010: 816; Machovina and Feeley, 2014). To put matters into perspective, it helps to compare the uses of grains for biofuels and for feed. If biofuels are objectionable, then animal feed equally is, insofar as it is the conversion of grains, a cheap and healthy food, into flesh, videlicet, a very expensive commodity for a societal elite. In 2009, 16 per cent of global corn production was converted into biofuel (Locke *et al.*, 2013: 52), whereas 60 per cent ended up in feed troughs (FAO Commodities and Trade Division, 2009).

In the same year, less than 1 per cent of global wheat production was used for biofuels (Locke *et al.*, 2013: 52), while 17 per cent was fed to farmed animals (FAO Trade and Markets Division, 2009).

Nevertheless, already today, the world could theoretically be fed with current levels of cereal production. Hunger is thus not only a problem of production, but a much more complex problem of distribution: hunger is a political issue. Holt Giménez and Shattuck (2011) describe the food crisis as the corporate food regime's 'creeping normality.' Thence, the emphasis in the 'food enterprise' regime should perhaps be not on *food*, but on *enterprise*. Correspondingly, Gunderson has argued that the killing-for-profit sector does not aim to '*create food*, but *to make money*' (2013: 260–261, original emphasis). Schneider does not speak of a *food* regime, but of an *industrial meat* regime, as industrial meat – as well as milk or eggs – is not for food, but 'for politics, class and capital' (2014: 614, 628).

From its very origins, the animal-industrial complex has been entangled with processes of colonization and the violent displacement of indigenous people (Nibert, 2013). Colonialism is also continued by climate injustice – for indigenous people, this is a form of so-called colonial déjà vu (Whyte, 2017). Settler-colonialism has manifested itself in containing and ultimately erasing indigenous lives, through the physical confinement and the destruction of their political and cultural systems. Global warming and concomitant rapid environmental change can be viewed as just another manifestation of containment (Whyte 2017). The International Organization for Migration estimates that by 2050, there will be about 200 million 'climate refugees' – people who have to leave their homes because of climate change – which is the same figure as the number of international migrants today (IOM, 2014). Apart from these cases of environmental racism, the animal industry also generates further social repercussions, such as extremely dangerous, precarious and traumatizing work conditions, increases in domestic violence, and health problems in regions with CAFOs, all realities which unfortunately cannot be elaborated in this chapter.

The Livestock Revolution and sustainable intensification

After this portrayal of the current livestock industry and its social and environmental impact, its future development, the 'Livestock Revolution,' and the means to attain it, namely sustainable intensification, occupy centre stage. The term 'Livestock Revolution' originated in a joint discussion paper by the International Food Policy Research Institute, the International Livestock Research Institute, and the Food and Agriculture Organization of the United Nations from 1999 (Delgado *et al.*, 1999) and has been developed since then. In essence, the Livestock Revolution is a 70 per cent upsurge in the demand for animal products by 2050, which will be caused by rising disposable income, population growth, urbanization, and changing life styles. This demand is mainly propelled in the Majority World, and it will presumably be met by industrial systems. The evolution is said to be 'one of the largest structural shifts to ever affect food markets in developing countries' (Delgado *et al.*, 1999: 4). What makes the Revolution so exceptional is that it is said to be demand-driven, and not supply-driven; further, its growth is much larger and more than twice as lucrative as the Green Revolution (MacLachlan, 2015: 24). Hence, big money is at play.

But how can the supposed upsurge in demand be met while simultaneously balancing its vast environmental effects? The 'Global Agenda for Sustainable Livestock,' a multi-stakeholder initiative launched by the FAO in 2013, has a simple yet sophisticated formula: 'The livestock sector needs to produce more, using fewer resources, with benefits to all' (Global Agenda for Sustainable Livestock, 2017). These suggested measures fit the 'faster, harder, better' framing (Boyd *et al.*, 2001: 564) of sustainable intensification, and the classic eco-modernist agenda. Concretely, producing more with less is translated into efficiency increase, intensification, technological innovation in breeding and feeding, regulation and management, and public-private partnerships. Another indispensable factor is the transfer of technology and knowledge from the Minority World to the Majority World, or, in the words of Global Agenda, 'from the world's most to least efficient production systems' (Global Agenda for Sustainable Livestock, 2016). Industrial systems have the lowest emission intensities for both ruminant and monogastric production (Gerber *et al.*, 2013). Pastoral systems, conversely, are blamed for their inefficiency and the degradation of communal land. To resolve this 'tragedy of the commons,' international institutions call for a change in property rights and institutions (Delgado *et al.*, 2000: 10), and a reorientation of pastoralists towards the provision of ecosystem services (Steinfeld *et al.*, 2006: 280).

Critiques of meatification

Development narratives

As technical as these suggestions of sustainable intensification may sound, they would constitute no less than a radical transformation of society. The feasibility and correctness of these claims must be questioned, but also, and this is the principal issue, so should their necessity. Animal products are the primary motive for the perceived 'requirement' to sustainably intensify agriculture in the first place, as they are very resource-intensive, prodigal goods. The production of plant-based calories is four to twenty times more efficient than the production of calories from animal bodies (Bajželj *et al.*, 2015). However, the obligation to consume the latter is seldom questioned. On the contrary, international institutions present the dietary substitution of meat and milk for grain, first, as a natural and even desirable process, and second as a means of providing new livelihoods for 'the poor.'[4] These two arguments combine symbolic and material aspects of the Revolution which will now be scrutinized in more detail. First, animal-based nutrition is globally promoted as a 'superior diet' and marker of modernity, 'class ascension,' and 'Western' lifestyles more generally (Weis, 2013a: 70). The 'nutrition transition' towards diets richer in animal protein, the 'convergence of diets,' the race up the 'animal protein ladder' – these metaphors of the Livestock Revolution discourse are all development and modernization narratives (Sumberg and Thompson, 2013: 18). Yet, the ongoing structural change in the Majority World is not a 'convergence' but rather a guided process in which powerful players have vested interests in aligning global diets to the contemporary excessive consumption patterns of the Minority World. Statistically speaking, the Minority World's diet is the outlier, not the average; it is the exception, not the norm. Still, the discourse of the Livestock Revolution naturalizes and universalizes this situated and recent experience of a small fraction of humanity. Sundry cultures and continents are put on a uniform, unidirectional, meaty ladder towards modernization with the Minority World's society on top, which adds to the imperial and colonial character of the Livestock Revolution. The 'convergence' of diets, hence, is thus better defined as the meatification of diets, and this meatification is not only a reflection of global injustice – for the reasons laid out above – but also an exacerbation and reproduction of this injustice (Weis, 2013a: 9).

The second justification for the Livestock Revolution is the alleged vital importance of livestock as the number one 'pathway out of poverty' (FAO, 2008: 1). To join the Revolution seems to be a humanitarian imperative. However, this romanticizing and patronizing assessment completely ignores the economic and political reality on the ground. The Revolution is characterized by industrialization, large retailers, and vertical integration. 'The poor' have no chance to compete against those giant producers who benefit from economies of scale and low transaction costs, tax holidays, governmental subsidies, and so on (Delgado *et al.*, 1999: 42). Intensification remunerates those who own land, animals, and feed, and who have access to technology and the market. Other disadvantageous tendencies for smallholders include trade liberalization, volatile prices on the world market, rising food quality standards, and booming urban markets favoring peri-urban intensified production. The shift from pastoral ruminant to industrial monogastric production also entails higher expenses for feed (Millar and Photakoun, 2008). Rather than providing livelihoods for the 'rural poor,' the Revolution in fact erodes them. This is a serious turning point, as smallholders constitute the majority of the sector (FAO, 2008: 4), generating most of the food consumed in the Majority World (both animal and plant-based) (Steinfeld, 2004: 32). In other words, the Livestock Revolution bypasses the bulk of its current producers and consumers in the Majority World, which underlines the fact that, in the industrial meat regime, 'sustainable' production is geared towards global value chains and not local food sovereignty. Indeed, a look at the symbolic and structural underpinnings of the Revolution reveals that it is neither natural, nor beneficial for people. In capitalism, this does not pose any problems though. The most important thing is the accumulation of profits. As described above, the Livestock Revolution entails an entire mode of consumption and production that has conquered the Minority World in the twentieth century, and is now intruding on the rest of the globe. Referring to Weis, MacLachlan frames it as the 'agro-industrial model based on economies of scale and the integrated industrial-grain-oilseed-livestock complex' (2015: 36). The big profiteers are agribusiness and exporters of feed and livestock products. Both markets are dominated by the United States and the European Union. European subsidized exports of frozen chickens, for example, destroy local production in African countries (Fritz, 2014: 10–11).

Violence against nonhuman nature

Weis estimates that by 2050, global meat consumption will probably equal the deaths of almost 120 billion animals per annum (2013b: 71), and the majority of them will be exploited in industrial settings which will even be justified with the green agenda: extensive systems reportedly have a bigger 'ecological hoofprint' (Weis) than animals reared in factory farms (FAO, 2011: 31).[5]

One concrete example of how sustainable intensification affects animal bodies is feeding practices. The industry is struggling to find a balance between maximal growth rates and feed conversion on the one hand, and minimal nutrient input on the other hand. The nutrient input must be minimized because corporations have a financial incentive to comply with environmental regulations. One effect is that the animals' bones are too weak to bear their weight and fast growth, and they easily break during transportation and slaughter, flawing the carcass and lowering profits (Clark, 2012). In addition, feed concentrates designed to boost productivity and reduce emissions lead to gastric ulcers in pigs and liver abscesses in cows (Shields and Orme-Evans, 2015: 364–365). Clark observes that so-called 'environmental nutrition' is not adjusted to the needs of the animal but solely to the 'needs of capital' (2012: 109).

Capital also seems to be ambivalent with regard to limits to growth. The projected Livestock Revolution alone would engender a global warming of two degrees Celsius by 2050 (and not 2100, as the IPCC scenario predicts) (Bajželj *et al.*, 2014). In terms of biogeochemical flows, the share of the safe operating space for reactive nitrogen mobilization will increase from 117 per cent, as of 2000, to 294 per cent by 2050 (Pelletier and Tyedmers, 2010: 18372). If the Livestock Revolution is to take place, agriculture will consume 13,500 cubic kilometres of water, almost twice the total amount of water withdrawn today (Herrero *et al.*, 2009: 116). It equally remains unclear where the land necessary to grow feed should come from. Mitigation options on the supply side will not suffice to stay within planetary boundaries (Pelletier and Tyedmers, 2010). The narrow-minded productivism of the industrial meat regime is ruining the planet. A paradigm change is needed.

Conclusion

The animal-industrial complex annihilates billions of animal lives, destroys natural resources, and undermines social justice. Despite the staggering evidence for its contribution to the crossing of planetary boundaries, and its violent, colonial history and present, international institutions proclaim the need to double this bloody business. The discourse of the Livestock Revolution remains uncontested, almost like a self-fulfilling prophecy. This should not come as a surprise because the 'nutrition transition' is depicted as a law of nature (Schneider, 2014: 616). However, the Revolution is a form of deliberate and directed change in global nutrition serving capital interests. Its winners and losers emerge in a critical scrutiny of the means of fulfilling the Revolution. Whereas the Minority World is responsible for the bulk of the ecological dilemma and animal consumption – consuming roughly three times more than the Majority World (Steinfeld and Chilonda 2006) – the proposed changes should chiefly take place in the Majority World (Steinfeld *et al.*, 2006: 6, 81). In the prevalent high-tech sustainable intensification approach, the extensive subsistence agriculture of 'the poor' is blamed for 'poverty-led' degradation and inefficiency (Steinfeld *et al.*, 2006: 114, 280). But environmental regulations, 'sophisticated' technology, and institutional changes are not neutral, apolitical instruments. They entail a world of private property and state or corporate access and control, on both the intellectual and physical level. In the words of Banerjee: 'The Third World, still in need of development, now needs to be told how to develop sustainably' (Banerjee, 2003: 174). And sustainability is here framed as agro-ecology lite, with little regard for social and environmental justice. Interestingly, the Livestock Revolution is additionally legitimized on the grounds of its importance for 'food security and poverty alleviation.' Yet, rather than philanthropically feeding and 'modernizing the poor' with animal products (Twine, 2012: 14), transnational corporations and lobbies are trying to capture emerging markets to compensate for stagnating turnovers in the Minority World (Sanbonmatsu, 2011: 23). Furthermore, the Revolution's representation as something natural, as an altruistic mandate, and, thanks to sustainable intensification, even as environmentally friendly, justifies an ever-increasing exploitation of animals.

Jarosz argues that 'the food crisis is a predictable outcome of an oil-dependent feedgrain-livestock complex supplying a meat-centric diet for those who can buy it' (2009: 2065). As shown above, this complex generates only a quarter of total proteins and a mere 13 per cent of calories. This is not a particularly strong performance, given the sector's occupation of 80 per cent of arable land and its use of 35 per cent of the global grain harvest for feed. But again, in capitalism, productivity is not defined in terms of satisfying needs but in monetary terms. The wasteful cycling of nutrients through animal bodies in order to sell meat, milk, or eggs, or the 'destruction of large volumes of useable nutrition in grains and oilseeds,' as Weis puts it (2013b: 76), thus makes sense in the eco-modernist, industrial meat regime.

It remains for researchers committed to fight for a more just food system to expose the destructive, violent, and colonial character of the animal-industrial complex, to denounce how the Livestock Revolution is greenwashed via sustainable intensification, and to highlight that its goal is, ultimately, not to sustain people or the planet, but profits.

Notes

1 The chapter is dedicated to all those hearts burning for freedom and fighting oppression in these times full of fear and hate. For resources, check out *itsgoingdown.org*, and *everydayfeminism.com*. Regarding positionality, the author is a white, able-bodied, economically privileged cis-woman holding a Swiss passport. This chapter, in particular the section on the ecological crisis and the conclusion, builds on Boscardin (2017).

2 The following paragraph is on animal *agri*culture. Many of the characteristics outlined here also apply to aquaculture, which currently produces almost half of the globally consumed dead marine individuals (FAOSTAT, 2014), and whose 'yield' has multiplied tenfold since 1980 (Weis, 2013a: 19).

3 The terminology 'Majority/Minority World,' brought forward by Alam, aims to replace Eurocentric, colonial, or unsatisfactory dichotomies like 'developing/developed,' or 'Global South/North.' The topical term 'defines the community in terms of what it has, rather than what it lacks. In time, the majority world will reaffirm its place in a world where the earth will again belong to the people who walk on it' (Alam, 2008: 87).

4 In the discourse, 'the poor' are always depicted as one homogenous category, and being 'poor' is portrayed as a natural, deficient state lacking structural reasons. Remarkably, the necessary counterpart of monetary poverty, wealth (Rist, 2014), is not seen as scandalous at all.

5 To illustrate this: pasture-raised cattle live longer, need more feed, emit more waste and gasses, and occupy more space. Still, the image of 'happy meat' with an animal welfare label is invoked to further the Revolution. Those green products are extraordinarily prominent and dominate the imagery in the Minority World, notwithstanding their limited empiric occurrence. In addition, not only is the percentage of animals reared in a 'happy, sustainable way' minimal, but their 'happy' lives involve all the usual, abhorrent practices, such as gassing male chickens at birth, specialized breeds that are prone to diseases, forced and repeated artificial insemination, separation of siblings, and so on, and it ends with being murdered (Stanescu, 2013). In sum, not only do animal welfare labels and products like 'sustainable meat' not fulfil their ecological and welfarist pledge, thus instrumentalizing their green label for the generation of profits, they also reform and legitimize animal exploitation in general.

References

Alam, S. (2008). 'Majority World: Challenging the West's Rhetoric of Democracy,' *Amerasia Journal*, vol. 34, no. 1: 87–98.

Allievi, F., Vinnari, M. and Luukkanen, J. (2015). 'Meat consumption and production – analysis of efficiency, sufficiency and consistency of global trends,' *Journal of Cleaner Production*, vol. 92: 142–151.

Bailey, R., Froggatt, A. and Wellesley, L. (2014). *Livestock – Climate Change's Forgotten Sector: Global Public Opinion on Meat and Dairy Consumption* [Online], Chatham House. The Royal Institute of International Affairs (Research Paper: Energy, Environment and Resources). Available at www.chathamhouse.org/publication/livestock-climate-change-forgotten-sector-global-public-opinion-meat-and-dairy (Accessed 17 July 2015).

Bajželj, B., Benton, T. G., Clark, M., Garnett, T., Marteau, T. M., Richards, K. S., Smith, P. and Vasiljevic, M. (2015). *Synergies between healthy and sustainable diets: Brief for GSDR 2015* [Online]. Available at https://sustainabledevelopment.un.org/content/documents/635987-Bajzelj-Synergies%20between%20healthy%20and%20sustainable%20diets.pdf (Accessed 24 July 2015).

Bajželj, B., Richards, K. S., Allwood, J. M., Smith, P., Dennis, J. S., Curmi, E. and Gilligan, C. A. (2014). 'Importance of food-demand management for climate mitigation,' *Nature Climate Change*, vol. 4, no. 10: 924–929.

Banerjee, S. B. (2003). 'Who Sustains Whose Development? Sustainable Development and the Reinvention of Nature,' *Organization Studies*, vol. 24, no. 1: 143–180.

Boscardin, L. (2017). 'Capitalizing on nature, naturalizing capitalism. An analysis of the livestock revolution, planetary boundaries, and green tendencies in the animal-industrial complex,' in Nibert, D. (ed.) *Animal Oppression and Capitalism*, Santa Barbara, Praeger Press.

Bouwman, L., Goldewijk, K. K., Van Der Hoek, K. W., Beusen, A. H. W., Van Vuuren, D. P., Willems, J., Rufino, M. C. and Stehfest, E. (2013). 'Exploring global changes in nitrogen and phosphorus cycles in agriculture induced by livestock production over the 1900–2050 period,' *Proceedings of the National Academy of Sciences*, vol. 110, no. 52: 20882–20887.

Boyd, W., Prudham, W. S. and Schurman, R. A. (2001). 'Industrial Dynamics and the Problem of Nature,' *Society & Natural Resources*, vol. 14, no. 7: 555–570.

Clark, J. L. (2012). 'Ecological Biopower, Environmental Violence Against Animals, and the "Greening" of the Factory Farm,' *Journal for Critical Animal Studies*, vol. 10, no. 4: 109–129.

Costales, A., Gerber, P. and Steinfeld, H. (2006). 'Underneath the Livestock Revolution,' in McLeod, A. (ed.) *Livestock report 2006*, Rome, FAO: 15–28.

Delgado, C., Rosegrant, M., Steinfeld, H., Ehui, S. and Courbois, C. (1999). *Livestock to 2020: The Next Food Revolution: Food, Agriculture, and the Environment Discussion Paper* 28, Food and Agriculture Organization of the United Nations, International Food Policy Research Institute, International Livestock Research Institute [Online]. Available at www.fao.org/Ag/againfo/resources/documents/lvst2020/20201.pdf (Accessed 24 March 2011).

Delgado, C., Rosegrant, M., Steinfeld, H., Ehui, S. and Courbois, C. (2000). *The coming livestock revolution* (Background Paper No. 6. UN Department of Economic and Social Affairs; Commission on Sustainable Development, Eight Session, 24 April–5 May 2000, New York), New York, FAO.

DeMello, M. (2012). *Animals and Society: An Introduction to Human-Animal Studies*, New York, Columbia University Press.

Emel, J. and Neo, H. (2011). 'Killing for Profit: Global Livestock Industries and their Socio-Ecological Implications,' in Peet, R., Robbins, P. and Watts, M. (eds) *Global Political Ecology*, London, New York, Routledge: 67–83.

EPA (2017). *Regulatory Definitions of Large CAFOs, Medium CAFOs, and Small CAFOs* [Online]. Available at www.epa.gov/npdes/npdes-afos-policy-documents-0 (Accessed 14 January 2017).

Eshel, G., Shepon, A., Makov, T. and Milo, R. (2014). 'Land, irrigation water, greenhouse gas, and reactive nitrogen burdens of meat, eggs, and dairy production in the United States,' *Proceedings of the National Academy of Sciences*, vol. 111, no. 33: 11996–12001.

FAO (2008). *Livestock Policy and Poverty Reduction* [Online], Livestock Policy Brief no 4, Animal Production and Health, FAO Agriculture and Consumer Protection Department. Available at www.fao.org/ag/againfo/resources/en/pubs_sap.html (Accessed 20 August 2015).

FAO (2011). *Mapping supply and demand for animal-source foods to 2030*. Animal Production and Health Working Paper, no 2.

FAO (2016a). *About FAO: Who we are* [Online]. Available at www.fao.org/about/who-we-are/en/ (Accessed 24 August 2016).

FAO (2016b). *FAO Cereal Supply and Demand Brief: Record cereal production to lift global inventories to an all-time high* [Online]. Available at www.fao.org/worldfoodsituation/csdb/en/ (Accessed 5 January 2017).

FAO Commodities and Trade Division (2002). *Agricultural commodities: Profiles and relevant WTO negotiation issues: Prepared by the Commodities and Trade Division as a background document for the Consultation on Agricultural Commodity Price Problems, 25–26 March 2002, Rome* [Online], Rome, FAO. Available at www.fao.org/docrep/006/y4343e/y4343e00.htm (Accessed 30 July 2015).

FAO Trade and Markets Division (2009). *Food Outlook: Global Market Analysis: Wheat* [Online]. Available at www.fao.org/docrep/012/ak341e/ak341e03.htm (Accessed 13 January 2017).

FAOSTAT (2014). *World fisheries production, by capture and aquaculture, by country (2012): Yearbook of Fishery Statistics, Summary tables* [Online]. Available at ftp://ftp.fao.org/FI/STAT/summary/default.htm (Accessed 7 July 2014).

FAOSTAT (2016a). *Compare Data, Production, Livestock Primary, World, Production Quantity, Meat, Total* [Online]. Available at /www.fao.org/faostat/en/ (Accessed 18 October 2016).

FAOSTAT (2016b). *Production: Livestock Primary: Producing Animals/Slaughtered, World total, Meat total, 2013* [Online]. Available at www.fao.org/faostat/en/ (Accessed 28 November 2016).

Fritz, T. (2014). *The illusory promise of the livestock revolution: Meatification of diets, industrial animal farms and global food security*, Berlin, FDCL.

Gerber, P., Steinfeld, H., Henderson, B., Mottet, A., Opio, C., Dijkman, J., Falcucci, A. and Tempio, G. (2013). *Tackling Climate Change Through Livestock: A Global Assessment of Emissions and Mitigation Opportunities* [Online], Rome, FAO. Available at www.fao.org/ag/againfo/resources/en/publications/tackling_climate_change/index.htm (Accessed 15 November 2013).

Global Agenda for Sustainable Livestock (2014). *Towards Sustainable Livestock: Discussion Document* [Online]. Available at www.livestockdialogue.org/en/ (Accessed 14 February 2016).

Global Agenda for Sustainable Livestock (2016). *Thematic Area Greening livestock sector growth: Closing the efficiency gap in natural resource use* [Online]. Available at http://www.livestockdialogue.org/action-networks/action-networks/closing-efficiency-gap/en/ (Accessed 25 August 2016).

Global Agenda for Sustainable Livestock (2017). *Home: Slideshow* [Online]. Available at www.livestockdialogue.org/ (Accessed 21 January 2017).

Godfray, H. C. J., Beddington, J. R., Crute, I. R., Haddad, L., Lawrence, D., Muir, J. F., Pretty, J., Robinson, S., Thomas, S. M. and Toulmin, C. (2010). 'Food Security: The Challenge of Feeding 9 Billion People,' *Science*, vol. 327, no. 5967: 812–818.

Goodland, R. and Anhang, J. (2009). *Livestock and Climate Change: What if the key actors in climate change are… cows, pigs, and chickens?* [Online] (World Watch Magazine November/December 2009). Available at www.worldwatch.org/files/pdf/Livestock%20and%20Climate%20Change.pdf (Accessed 10 October 2011).

Gunderson, R. (2013). 'From Cattle to Capital: Exchange Value, Animal Commodification, and Barbarism,' *Critical Sociology*, vol. 39, no. 2: 259–275.

Hallström, E., Carlsson-Kanyama, A. and Börjesson, P. (2015). 'Environmental impact of dietary change: a systematic review,' *Journal of Cleaner Production*, vol. 91: 1–11.

Heinrich Böll Foundation and Friends of the Earth Europe (eds) (2014). *Meat Atlas. Facts and figures about the animals we eat*, 2nd edn, Berlin, Brussels, Heinrich Böll Foundation; Friends of the Earth Europe.

Herrero, M. and Thornton, P. K. (2013). 'Livestock and global change: Emerging issues for sustainable food systems,' *Proceedings of the National Academy of Sciences*, vol. 110, no. 52: 20878–20881.

Herrero, M., Thornton, P. K., Gerber, P. and Reid, R. S. (2009). 'Livestock, livelihoods and the environment: understanding the trade-offs,' *Current Opinion in Environmental Sustainability*, vol. 1, no. 2: 111–120.

Herrero, M., Havlik, P., Valin, H., Notenbaert, A., Rufino, M. C., Thornton, P. K., Blummel, M., Weiss, F., Grace, D. and Obersteiner, M. (2013). 'Biomass use, production, feed efficiencies, and greenhouse gas emissions from global livestock systems,' *Proceedings of the National Academy of Sciences*, vol. 110, no. 52: 20888–20893.

Heuzé, V. and Tran, G. (2016). *Soybean (general): Feedipedia. Animal Feed Resources Information System by INRA, CIRAD, AFZ and FAO* [Online], INRA, CIRAD, AFZ and FAO (Feedipedia. Animal Feed Resources Information System). Available at www.feedipedia.org/node/753 (Accessed 10 August 2016).

Holt Giménez, E. and Shattuck, A. (2011). 'Food crises, food regimes and food movements: rumblings of reform or tides of transformation?' *The Journal of Peasant Studies*, vol. 38, no. 1: 109–144.

IOM (2014). *Migration, Climate Change and the Environment: A Complex Nexus* [Online]. Available at www.iom.int/cms/en/sites/iom/home/what-we-do/migration-and-climate-change/a-complex-nexus.html (Accessed 9 July 2014).

Jarosz, L. (2009). 'Energy, Climate Change, Meat, and Markets: Mapping the Coordinates of the Current World Food Crisis,' *Geography Compass*, vol. 3, no. 6: 2065–2083.

Khan, A. A. (2004). 'Livestock Revolution in India: Its Impact and Policy Response,' *South Asia Research*, vol. 24, no. 2: 99–122.

Lappé, F. M. (2010). *Diet for a small planet*, 20th edn, New York, Ballantine Books.

Locke, A., Wiggins, S., Henley, G. and Keats, S. (2013). *Diverting grain from animal feed and biofuels: Can it protect the poor from high food prices?* [Online], London. Available at www.odi.org/publications/7385-grain-animal-feed-biofuels-high-food-price (Accessed 30 July 2015).

Machovina, B. and Feeley, K. J. (2014). 'Taking a Bite Out of Biodiversity,' *Science*, vol. 343, no. 6173: 838.

Machovina, B., Feeley, K. J. and Ripple, W. J. (2015). 'Biodiversity conservation: The key is reducing meat consumption,' *Science of the Total Environment*, vol. 536: 419–431.

MacLachlan, I. (2015). 'Evolution of a revolution: meat consumption and livestock production in the developing world,' in Emel, J. and Neo, H. (eds) *Political ecologies of meat*, London, New York, Routledge: 21–41.

Mekonnen, M. M. and Hoekstra, A. Y. (2011). 'The green, blue and grey water footprint of crops and derived crop products,' *Hydrology and Earth System Sciences*, vol. 15, no. 5: 1577–1600.

Mekonnen, M. M. and Hoekstra, A. Y. (2012). 'A Global Assessment of the Water Footprint of Farm Animal Products,' *Ecosystems*, vol. 15, no. 3: 401–415.

Millar, J. and Photakoun, V. (2008). 'Livestock development and poverty alleviation: revolution or evolution for upland livelihoods in Lao PDR?' *International Journal of Agricultural Sustainability*, vol. 6, no. 1: 89–102.

Nibert, D. A. (2011). 'Origins and Consequences of the Animal Industrial Complex,' in Best, S., Kahn, R., Nocella II, A. J. and McLaren, P. (eds) *The Global Industrial Complex: Systems of Domination*, Lanham, Maryland, Lexington Books: 197–209.

Nibert, D. A. (2013). *Animal oppression and human violence: Domesecration, capitalism, and global conflict*, New York, Columbia University Press.

Noske, B. (1989). *Humans and other animals: Beyond the boundaries of anthropology*, London, Winchester, MA, Pluto Press.

Pelletier, N. and Tyedmers, P. (2010). 'Forecasting potential global environmental costs of livestock production 2000–2050,' *Proceedings of the National Academy of Sciences*, vol. 107, no. 43: 18371–18374.

Pellow, D. N. (2016). 'Toward a Critical Environmental Justice Studies: Black Lives Matter as an Environmental Justice Challenge,' *Du Bois Rev.*, vol. 13, no. 2: 1–16. doi:10.1017/S1742058X1600014X.

Ripple, W. J., Smith, P., Haberl, H., Montzka, S. A., McAlpine, C. and Boucher, D. H. (2013). 'Ruminants, climate change and climate policy,' *Nature Climate Change*, vol. 4, no. 1: 2–5.

Rist, G. (2014). *The history of development: From Western origins to global faith*, London, Zed Books.

Rockström, J., Steffen, W., Noone, K., Persson, Å., Chapin, F. S., III, Lambin, E., Lenton, T. M., Scheffer, M., Folke, C., Schellnhuber, H. J., Nykvist, B., Wit, C. A. de, Hughes, T., van der Leeuw, S., Rodhe, H., Sörlin, S., Snyder, P. K., Costanza, R., Svedin, U., Falkenmark, M., Karlberg, L., Corell, R. W., Fabry, V. J., Hansen, J., Walker, B., Liverman, D., Richardson, K., Crutzen, P. J. and Foley, J. (2009). 'Research Planetary Boundaries: Exploring the Safe Operating Space for Humanity,' *Ecology & Society*, vol. 14, no. 2.

Sanbonmatsu, J. (2011). 'Introduction,' in Sanbonmatsu, J. (ed.) *Critical Theory and Animal Liberation*, Lanham, Md, Rowman & Littlefield Publishers: 1–34.

Schneider, M. (2014). 'Developing the meat grab,' *The Journal of Peasant Studies*, vol. 41, no. 4: 613–633.

Shields, S. and Orme-Evans, G. (2015). 'The Impacts of Climate Change Mitigation Strategies on Animal Welfare,' *Animals*, vol. 5, no. 2: 361–394.

Stanescu, V. (2013). 'Why "Loving" Animals is Not Enough: A Response to Kathy Rudy, Locavorism, and the Marketing of "Humane" Meat,' *The Journal of American Culture*, vol. 36, no. 2: 100–110.

Steffen, W., Richardson, K., Rockström, J., Cornell, S. E., Fetzer, I., Bennett, E. M., Biggs, R., Carpenter, S. R., Vries, W. de, de Wit, C. A., Folke, C., Gerten, D., Heinke, J., Mace, G. M., Persson, L. M., Ramanathan, V., Reyers, B. and Sorlin, S. (2015). 'Planetary boundaries: Guiding human development on a changing planet,' *Science*, vol. 347, no. 6223: 736, 1259855.

Steinfeld, H. (2004). 'The livestock revolution – a global veterinary mission,' *Veterinary Parasitology*, vol. 125, 1–2: 19–41.

Steinfeld, H. and Chilonda, P. (2006). 'Old players, new players,' in McLeod, A. (ed.) *Livestock report 2006*, Rome, FAO: 3–14.

Steinfeld, H., Gerber, P., Wassenaar, T., Castel, V., Rosales, M. and Haan, C. de (2006). *Livestock's Long Shadow: Environmental Issues and Options* [Online], Rome. Available at www.fao.org/docrep/010/a0701e/a0701e00.HTM (Accessed 14 April 2018).

Sumberg, J. and Thompson, J. (2013). *Revolution Reconsidered: Evolving Perspectives on Livestock Production and Consumption: STEPS Working Paper 52*, Brighton, STEPS Centre.

Sutton, M. A., Oenema, O., Erisman, J. W., Leip, A., van Grinsven, H. and Winiwarter, W. (2011). 'Too much of a good thing,' *Nature*, vol. 472, no. 7342: 159–161.

Thornton, P. K. (2010). 'Livestock production: recent trends, future prospects,' *Philosophical Transactions of the Royal Society B: Biological Sciences*, vol. 365, no. 1554: 2853–2867.

Thornton, P. K., Herrero, M. and Ericksen, P. (2011). *Livestock and climate change*, Nairobi, Kenya, ILRI.

Thornton, P. K., van de Steeg, J., Notenbaert, A. and Herrero, M. (2009). 'The impacts of climate change on livestock and livestock systems in developing countries: A review of what we know and what we need to know,' *Agricultural Systems*, vol. 101, no. 3: 113–127.

Twine, R. (2012). 'Revealing the 'Animal-Industrial Complex' – A Concept & Method for Critical Animal Studies?' *Journal for Critical Animal Studies*, vol. 10, no. 1: 12–39.

Weis, T. (2010). 'The Ecological Hoofprint and the Population Bomb of Reverse Protein Factories,' *Review*, vol. 33, no. 2: 131–152.

Weis, T. (2013a). *The Ecological Hoofprint: The Global Burden of Industrial Livestock*, London, Zed Books.

Weis, T. (2013b). 'The meat of the global food crisis,' *Journal of Peasant Studies*, vol. 40, no. 1: 65–85.

Whyte, K. P. (2017). 'Is it Colonial Déjà Vu? Indigenous Peoples and Climate Injustice,' in Adamson, J. and Davis, M. (eds) *Humanities for the environment: Integrating knowledge, forging new constellations of practice*, Oxford, Routledge.

8 Are food quality schemes an alternative to the conventional food system?

Reflections on the EU metaphors on agrifood quality regulation

Josep Espluga-Trenc, Marina Di Masso, Marta G. Rivera-Ferre and Arantxa Capdevila

Introduction

The current hegemonic agrifood regime grounded on the development of industrial agriculture, based on massive monoculture food production and global market oriented distribution, has produced multidimensional impacts. Evidence of these impacts has been described in several spheres, as human health issues in terms of farmers' poisoning, food contamination, or nutritional disequilibrium; as environmental degradation by chemical pollution, loss of biodiversity, and climate change impacts; as socioeconomic losses due to the destruction of agricultural employment; as territorial impacts caused by rural exodus and the overpopulation of cities; as political effects weakening peasant agriculture and increasing the power of large landowners and agrifood corporations; and as geopolitical consequences by reinforcing North-South colony-dependencies (Di Masso *et al.*, 2014; Espluga, 2002, 2004; Holt Giménez *et al.*, 2009; McIntyre *et al.*, 2009; McMichael, 2009; Patel, 2007; Rivera-Ferre, 2014). Such multidimensional impacts have led to the emergence of numerous social mobilizations around the world, which in turn have prompted urgent political responses.

The food regimes framework proposed by Friedmann and McMichael (1989) is useful in explaining how particular relations of food production and consumption are central to the functioning and reproduction of global capitalism. According to their proposal, the global food system has followed three stages: the first one (1870–1930s) was characterized by cheap food and raw materials from tropical and temperate settler colonies fuelling European industrialization; the second one (1950–1970s) was characterized by the global spread of industrial agriculture through the 'Green Revolution,' using a large amount of subsidized agrochemicals, machinery and high technology irrigation systems; currently, the third one (1980s to present) rests on the neoliberal capitalist expansion and liberalized global food trade, characterized by the high market power and profits of oligopolistic agrifood corporations.

Following Polanyi's hypothesis (1944) that unregulated markets would have serious difficulties being socially and environmentally sustainable, an increasing number of social movements and protests against agrifood policies have arisen. Such social pressure can lead to new regulations on the part of the political system. In fact, according to Polanyi's hypothesis, alternating periods of unregulated markets followed by state intervention protective of social concerns are a cyclical part of capitalism. This has taken place in each of the three food regimes. Thus, neoliberal food regimes can undergo substantive regulatory policy changes under social pressure.

One of the main regulatory changes related to agrifood policies reflects the 'quality turn.' As part of the food question, this refers to a change in discourse from quantity to quality (Constance *et al.*, 2014). In policy terms, this turn was made evident beginning in the 1980s and 1990s, when 'alternative' policy strategies emerged highlighting distinctive features based on the geographical origin of food (e.g. protected designations of origin, etc.) or on the production system (e.g. organic farming). These policies result from the concern of policy makers who are trying to address some of the impacts of the current food regime (health, environmental, territorial issues) by pushing forward alternative agricultural models in opposition to the hegemonic and global market-focused model. One such alternative model is frequently conceptualized as 'non-anonymous' and/or 'territorial friendly' (Di Masso, 2013), taking the form of 'protected designations of origin' or 'protected geographical indications' policies or programmes. From a food regimes perspective, the quality turn indicates the challenge by the 'Food from Somewhere' to the 'Food from Nowhere' regime (Campbell, 2009; McMichael, 2005) and a key emerging question is whether the *'Food from Somewhere' agrifood regime can be conceptualised as a significant countermovement to the 'Food from Nowhere' regime* (Constance *et al.*, 2014).

In this chapter, we aim to show how territorial food quality has been defined by policy makers, in order to assess whether the policies resulting from such definition are capable of modifying the current neoliberal food regime, and to what extent. We will focus on the EU Regulation 1151/2012, approved by the European Parliament and the Council of the European Union, which all EU member countries must adopt. This regulation aims to help producers of agricultural products and foodstuffs to communicate the product characteristics and farming attributes of those products and foodstuffs to buyers and consumers, thereby ensuring: (a) fair competition for farmers and producers of agricultural products and foodstuffs having value-adding characteristics and attributes; (b) the availability to consumers of reliable information pertaining to such products; (c) respect for intellectual property rights; and (d) the integrity of the internal market. The measures set out in this Regulation are intended to support agricultural and processing activities and the farming systems associated with high quality products, thereby contributing to the achievement of rural development policy objectives. Our analysis shows that EU Regulation 1151/2012 on quality schemes for agricultural products and foodstuffs is mainly focussed on economic factors, and its conceptual framework is devoted to promoting special protection,

as a kind of social policy, for producers (farmers and others) at risk of being excluded from the current hegemonic agrifood regime, but it does not address the causes of such exclusion.

Defining food quality

The quality turn reflects a general discontent with the impersonal 'industrial food world' and a resolute turn to the 'interpersonal food world' where quality agreements embed trust and tradition within a moral economy of place and provenance (Goodman, 2003; Morgan *et al.*, 2006). Quality is defined broadly to include quality of food, quality of work, quality of health, and quality of life for humans and animals (Constance *et al.*, 2014). Several criteria of positive differentiation have traditionally been considered in the assignment of a food quality (Ilbery *et al.*, 2005):

- its nature (composition, nutritional value, organoleptic properties).
- its elaboration and distribution process (traditional, sustainable, fair, corporate, independent, direct).
- its geographical origin (which confers a differentiated identity linked to the qualities of the territory).

The difference is constructed by different actors in the food chain combining these attributes of 'product, process and origin' as main ingredients (Ilbery *et al.*, 2005). In fact, quality is a social definition in which different actors interact with different motivations and interests. The way of signifying it (the values it embodies, the concepts associated with it) is also a subjective exercise. Therefore, the definition of the concept is not a matter of purely terminological interest, and depending on how it is defined, this definition will have a different impact on the agrifood system (Di Masso, 2013). In this sense, it is necessary to evaluate how food quality can be established in ways that substantively challenge the conventional system (Murdoch *et al.*, 2000) and the hegemonic food regime.

The explosion of official certification and labelling schemes is a response to multiple factors (Ilbery *et al.*, 2005), from environment protection to the aim of adding value to agricultural production. As a result, a diversity of labels coexists, ranging from the defence of local production and the social and economic structure that supports it to the concerns about producing food through more equitable alternative means.

Within these certification practices, the EU origin labels have been identified as expressions of 'local,' 'quality' or 'endogenous' food systems (Marsden *et al.*, 2000; Ilbery and Kneafsey, 2000 cf. Holloway and Kneafsey, 2004). It has been pointed out that the attempt to connect product and place with the aim of adding value to agricultural products is one of the dominant features of alternative food networks, particularly in Europe (Maye *et al.*, 2007). The concept of locality is what distinguishes these alternative labels of product and origin, whose particularity and added value is its geographical specialization, the explicit allusion to the link with a given territory. As far as consumption is concerned, the presence of geographical identification labels implies specific messages on the mode and place of production, as opposed to simple labelling of the inherent qualities of the product (ingredients, nutritional information, etc.).

The oldest protection legislation at the European level dates back to 1970 and was exclusively limited to wine products. In 1992, the EU extended this protection to the remaining agrifood products through the creation of the quality schemes known as Protected Designations of Origin (PDOs), Protected Geographical Indications (PGI) and Traditional Specialties Guaranteed (TSGs) (Galdós, 2004). Taking the French 'appellation d'ôrigine controlee' (AOC) as a model, the European PDO and PGI systems enable regions to claim collective ownership over local food production modes. Contrary to trademarks, source labels 'belong' to regions and are administered by state governments to prevent fraud and oversee certification schemes (Holloway and Kneafsey, 2004). The European Commission has been increasing the formalization of labels that indicate the region of origin or system of production of different food products. Although it offers the advantage of harmonization, on the other hand this regulation offers little information to the consumer regarding relevant activities related to the production system, such as the type of farm or producer (Gilg and Battershill, 1998), and it has also resulted in unexpected trends, such as homogenization of production systems and exclusion of agrobiodiversity.

The PDO designates the name of a product whose production, processing and elaboration must take place in a given geographical area with a recognized and proven specialization. The PGI indicates the link with the territory in at least one of the stages of production, transformation or elaboration. The two levels of geographical reference are different: the link with the territory is stronger in the first case. As regards TSGs, this quality protection system does not refer to origin (i.e. it is not linked to the territory), but rather aims to emphasize a traditional product composition or a traditional mode of production (Barco Royo, 2007). Currently in the EU there are around 1,430 names of cheeses, meats, fruits, vegetables and other products registered as PDO, PGI and TSG (EC, 2018).

Metaphors' methodological approach

In order to know how territorial food quality has been defined by policy makers and assess if and to what extent these measures attempt to modify the current food regime, we have performed a metaphors analysis. Our underlying assumptions are that every social process is controversial as there are different actors fighting in the public arena to position their viewpoints (Cammaerts, 2012), and that metaphors play a key role in defining controversial issues (Capdevila, 2004; Moragas Fernández, 2016). Because of their ability to 'simplify and make understandable political events' (Mio, 1997: 121), metaphors are one of the mechanisms used by political actors in their struggle to impose a certain vision of contested issues.

Metaphor deals with two different domains of knowledge (Lakoff and Johnson, 1980) that relate to each other: one is the point of departure (source domain) for understanding the meaning of the other (target domain). Hence, metaphors function as political schema used as a mean 'to shape public discourse' and 'to present political "reality" in a specific way' (Cammaerts, 2012: 244), while they influence evaluation of the metaphorized topics (Charteris-Black, 2011) and thus are also persuasive. In this sense, metaphor plays a double role: on the one side, it has a cognitive dimension because it helps us know and understand a concrete issue; on the other side, it has a persuasive dimension, as it leads to action (or attitudinal pre-disposition). From this perspective, the act of choosing a 'source domain' to represent a 'target domain' is in itself a persuasive act. Besides, metaphors provide the main storylines or perspectives along which conceptual mappings are developed and extended (Musolff, 2004). As Charteris-Black (2011: 32) points out: 'In political rhetoric the primary purpose of metaphors is to frame how we view or understand political issues by eliminating alternative points of view.' This is by making a choice of which ideas will structure the speaker's discourse in order to convince the audience.

Metaphors are also an effective problem-solving tool, since they 'provide specific guidance on how to deal with novel situations' (Bougher, 2012: 146). According to Grey (2000), metaphor is not only a tool of conceptual economy, but also a tool of discovery, providing a way of imposing or discovering structure within novel or unfamiliar situations. In keeping with that, Mio (1997), states that metaphors are like devices that, at first sight, 'simply draw a comparison between one thing and another, but in a more subtle way, they usually imply a whole narrative story and prescription for action' (Mio, 1997: 120).

From a methodological level, Charteris-Black (2010: 31) states that 'metaphor arises from our expectations about meaning that are based on our knowledge of how words have previously been used.' Breaking the original meaning of words and, consequently, the ordinary understanding of the text, results in what semiotics defines as breaking isotopy (Greimas, 1971) or the 'direction that a text exhibits when submitted to rules of interpretative coherence' (Eco, 1980: 153). Arduini (1993) labels the breaking of the isotopy as a deviation from the normative use, which establishes a new one. This deviation provides some practical applications in the detection of metaphors. Arduini (1993) warns us about relegating them to deviations to the norm, which leads us to think that, though we detect metaphor in a superficial level of discourse, we cannot understand it completely without focusing on the analogy that takes place at the deep level of discourse, which lets us arrive at its meaning and the reasoning for its use. For instance, in the 'love is a journey' metaphor (Moragas Fernández, 2016: 66),

> if 'love' is the target domain and is understood in terms of the conceptual domain of 'journey' (source domain), then, the travellers will be the lovers, the vehicle its love relationship and the obstacles along the way are the difficulties encountered in the relationship.

Beyond these basic correspondences, source domains also map other ideas, which are named 'entailments' or 'inferences.' In Lakoff's words (1993: 210): 'each mapping defines an open-ended class of potential correspondences across inference patterns. When activated, a mapping may apply to a novel source domain knowledge structure and characterize a corresponding target domain knowledge structure.'

We have based our analysis on the current main European regulation on quality: the 'EU Regulation 1151/2012 of the European Parliament and of the Council of 21 November 2012 on quality schemes for agricultural products and foodstuffs.' This law is composed of 59 articles distributed in three chapters, plus two annexes, preceded by an extensive explanatory statement where precedents and motives (up to 65 motives) for approving the law are fully explained. We have focused our analysis on this explanatory statement because it is where policy makers have made explicit their ideas and arguments about the need for and purposes of this law.

Once the official text where the quality food schemes are explained and justified was selected, the analysis procedure (based on Capdevila, 2004; Mussolf, 2004) focused on identifying and interpreting the metaphors used in the text to refer to 'food quality' as our 'target domain.' This process consisted of looking for the semantic tension or isotopy, trying to identify in the text words referring to different semantic fields related with the one we are talking about. These words are called 'tokens' (signals) and they identify the metaphors. Once the metaphors in the text had been identified, the interpretative phase consisted of connecting them with their conceptual sources, that is, with their respective 'source domains.' In order to simplify the analysis we followed Musolff's proposal (2004) of 12 general source domains that would cover the vast majority of metaphors used in everyday life: Way-movement-speed; Geometry-geography; Technology-building; Group-club-class; School-discipline; Economy-business; Love-marriage-family; Life-health-Strength; War-fight-battle; Game-sport; Show-play; and Nature-time.

Our analysis covered all the 65 paragraphs of the above mentioned explanatory statement where precedents and motives for approving the law are fully explained. An exhaustive analysis going through the whole set of paragraphs has been done, and as a result we have identified the entire set of lexemes ('tokens') (the words to which its original meaning have been changed) connected with our target domain (food quality). A total of 36 tokens have been identified in the text, of which 33 can be inductively classified in three of the main source domains proposed by Musolff (2004) (Table 1). A fourth (residual) source has also been identified but with very few signals throughout the text (only three tokens).

Results

Food quality policies become an attempt to minimize or avoid some of the negative impacts of the neoliberal agrifood regime, in particular those related to

Table 8.1 Metaphors detected and their respective lexemes (tokens)

Source domain	*Conceptual elements*	*Lexemes (tokens)*
Economy-business	Competitive advantage	competitive advantage (§1)
	Reward	rewarded (§3)
		reward (§4)
		benefit (§4)
	Added value	adds value (§20)
		add value (§44)
		add value (§44)
		added value (§46)
Technology-building	Enhancing	enhances (§20)
	Support	support (§4)
	Securing / ensuring	securing (§18)
		ensures (§20)
		ensures (§20)
		guarantee (§47)
		guaranteed (§47)
		ensure (§48)
	Control	controls (§46)
		controls (§46)
		controls (§46)
		control (§50)
		control (§50)
War-fight-battle	Protection / Defence	protecting (§5)
		protects (§20)
		protecting (§23)
		protection (§24)
		protection (§24)
		protection (§24)
		protected (§24)
		protection (§29)
		protected (§29)
		protection (§55)
		protected (§60)
		protection (§60)
Life-health-Strength	Life	alive (§1)
		living (§1)
	Strength	strengths (§1)

Note:
Numbers in brackets identify the exact paragraph in which each lexeme has been found.

environmental and social impacts. In theory, by following the EU food quality rules, the characteristics of the agricultural products and the way they have been produced should be more sustainable and respectful of the environment and local traditions, and supposedly more attractive to a consumer willing to pay more for such characteristics, while protecting small farmers from being excluded from the market. The regulation analysed here, EU Regulation (1151/2012), aims to help producers of agricultural products and foodstuffs to better communicate these characteristics to buyers and consumers. But not all the actors playing in the global food market have to agree with these corrective policies, or at least not all them will benefit equally. So, we can assume that the definition of agrifood quality is embedded in a political struggle between those actors. As stated before, because of their ability to simplify and make political events understandable (Mio, 1997), metaphors are one of the mechanisms used by actors in their struggle for imposing a certain vision of contested issues. In this context, one would expect to find in the EU Regulation metaphors related to environmental and social sustainability conceptual domains.

The analysis carried out shows that the EU policy makers who produced the text used basically three types of metaphors related with the following wide source domains (according to Musolff's terminology): Economy-business (8); Technology-building (13); and War-fight-battle (12). A very few additional metaphors (3) related to the Life-health-strength source domain were found.

Economy-business metaphors

The 'Economy-business' source domain includes all those words (signals or token) associated with the production, use, and management of resources, distribution, trade, and consumption of goods and services by different economic agents, monetary transactions, economic credit or debit values, and the whole set of concepts related to economy and business.

The analysis shows that EU policy makers are interpreting food quality in terms of economy and business conceptual frames. In the regulation text food quality is presented as a 'competitive advantage' to producers (in relation to other producers):

> The quality and diversity of the Union's agricultural, fisheries and aquaculture production is one of its important strengths, giving a *competitive advantage* to the Union's producers and making a major contribution to its living cultural and gastronomic heritage.
>
> (§1)

Food quality also means (economic) 'reward' to producers (for their effort in doing something different):

> Producers can only continue to produce a diverse range of quality products if they are *rewarded* fairly for their effort.
>
> (§3)

Food quality is characterized also as a benefit to the 'rural economy':

> Operating quality schemes for producers which *reward* them for their efforts to produce a diverse range of quality products can *benefit* the rural economy.
>
> (§4)

Food quality represents a way of adding economic value to the products in the market (internal and external):

> Provision should be made for the development of designations of origin and geographical indications at Union level and for promoting the creation of mechanisms for their protection in third countries in the framework of the World Trade Organisation (WTO) or multilateral and bilateral agreements, thereby contributing to the recognition of the quality of products and of their model of production as a factor that *adds value*.
>
> (§20)

Food quality is interpreted as something optional or subject to voluntariness, and therefore is a regulatory proposal that fits perfectly with the free market framework:

> A second tier of quality systems, based on quality terms which *add value*, which can be communicated on the internal market and which are to be applied voluntarily, should be introduced.
>
> (§44)

> The optional quality term 'mountain product' has met the conditions up to now and will *add value* to the product on the market.
>
> (§44)

Food quality is considered as something depending on consumers' trust (as it would be in a kind of stock exchange market). Official controls (by third parties) should be provided to help in achieving consumer trust as an expected outcome:

> The *added value* of the geographical indications and traditional specialities guaranteed is based on consumer trust. It is only credible if accompanied by effective verification and controls.
>
> (§46)

Technology-building metaphors

The 'Technology-building' source domain refers to concepts and words (signals or tokens) related to a structure or system standing more or less permanently in one place, such as a house or a building, including elements and a basis that support the entire system, builders or makers of the system, the technologies and process of constructing and operating the system, situations of stability or collapse of the system, materials and building elements, and all kind of concepts related to this semantic area.

A second wide conceptual sphere to which EU policy makers refer when talking about food quality is the 'Technology-building' source domain. This would include such metaphors as security, support, guarantee or control, all them referring to the maintenance of the stability of strong structures (markets) where food quality will be found or produced.

From this perspective, food quality means a way of 'enhancing' the credibility of the products (in the market), and 'ensuring' fair competition among producers (in the market):

> A Union framework that protects designations of origin and geographical indications by providing for their inclusion on a register facilitates the development of those instruments, since the resulting, more uniform, approach *ensures* fair competition between the producers of products bearing such indications and *enhances* the credibility of the products in the consumers' eyes.
>
> (§20)

Food quality is understood as a kind of 'support' for the producers (to help them remain in the market):

> In this way quality schemes are able to contribute to and complement rural development policy as well as market and income *support* policies of the common agricultural policy (CAP).
>
> (§4)

Food quality is also conceived as a way of 'securing' fair (economic) returns for producers, 'ensuring' fair competition between them and 'ensuring' impartiality of the competent authorities, and 'guaranteeing' specific characteristics of the products. All these words recall security concepts:

> The specific objectives of protecting designations of origin and geographical indications are *securing* a fair return for farmers and producers for the qualities and characteristics of a given product, or of its mode of production …
>
> (§18)

> A Union framework that protects designations of origin and geographical indications by providing for their inclusion on a register facilitates the development of those instruments, since the resulting, more uniform, approach *ensures* fair competition between the producers of products bearing such indications and enhances the credibility of the products in the consumers' eyes.
>
> (§20)

> To *guarantee* to the consumer the specific characteristics of geographical indications and traditional specialities *guaranteed*, operators should be subject to a system that verifies compliance with the product specification.
>
> (§47)

> In order to *ensure* that they are impartial and effective, the competent authorities should meet a number of operational criteria. Provisions on delegating some competences of performing specific control tasks to control bodies should be envisaged.
>
> (§48)

Food quality also means having official 'control' over production, processing and distribution, and 'control' plans to be launched by governments:

> The added value of the geographical indications and traditional specialities guaranteed is based on consumer trust. It is only credible if accompanied by effective verification and *controls*. Those quality schemes should be subject to a monitoring system of official *controls*, in line with the principles set out in Regulation (EC) No 882/2004 of the European Parliament and of the Council of 29 April 2004 on official *controls* performed to ensure the verification of compliance with feed and food law, animal health and animal welfare rules (16), and should include a system of checks at all stages of production, processing and distribution.
>
> (§46)

> Information on *control* activities for geographical indications and traditional specialities guaranteed should be included in the multiannual national *control* plans and annual report prepared by the Member States in accordance with Regulation (EC) No 882/2004.
>
> (§50)

War-fight-battle metaphors

The 'war-fight-battle' source domain refers to concepts and words related to the military sphere in wide sense, characterized by the combat between opposing forces used to achieve specific objectives, with all kind of strategies, techniques, activities and tools used both to attack and defend.

Another wide source domain found in the analysis is that related with War-fight-battle metaphors, with several references to 'protection' used to mean 'defence.' In this sense, food quality is conceived as a way of 'protecting' producers against unfair practices (of other producers):

> Agricultural product quality policy should therefore provide producers with the right tools to better identify and promote those of their products that have specific characteristics while *protecting* those producers against unfair practices.
>
> (§5)

> A Union framework that *protects* designations of origin and geographical indications by providing for their inclusion on a register facilitates the development of those instruments, since the resulting, more uniform,

> approach ensures fair competition between the producers of products bearing such indications and enhances the credibility of the products in the consumers' eyes.
>
> (§20)

Food quality is also related to the 'protection' of the natural resources or landscape of the production area or improving farm animals' welfare:

> An agricultural product or foodstuff bearing such a geographical description should meet certain conditions set out in a specification, such as specific requirements aimed at *protecting* the natural resources or landscape of the production area or improving the welfare of farm animals.
>
> (§23)

Food quality is widely related to the 'protection' of products in the European markets (fighting, in some way, against products of other countries made with different criteria):

> To qualify for *protection* in the territories of Member States, designations of origin and geographical indications should be registered only at Union level. With effect from the date of application for such registration at Union level, Member States should be able to grant transitional *protection* at national level without affecting intra-Union or international trade. The *protection* afforded by this Regulation upon registration, should be equally available to designations of origin and geographical indications of third countries that meet the corresponding criteria and that are *protected* in their country of origin.
>
> (§24)

> *Protection* should be granted to names included in the register with the aim of ensuring that they are used fairly and in order to prevent practices liable to mislead consumers. In addition, the means of ensuring that geographical indications and designations of origin are *protected* should be clarified, particularly as regards the role of producer groups and competent authorities of Member States.
>
> (§29)

> The criteria by which subsequent trademarks should be refused or, if registered, invalidated on the ground that they conflict with a prior designation of origin or geographical indication should correspond to the scope of *protection* of designation of origin or a geographical indication laid down.
>
> (§55)

Food quality is also related to 'protection' against misleading labelling for the consumers. Policies, institutions and resources should be devoted to guarantee these protection activities (playing a defence role):

> The symbols, indications and abbreviations identifying participation in a quality scheme, and the rights therein pertaining to the Union, should be *protected* in the Union as well as in third countries with the aim of ensuring that they are used on genuine products and that consumers are not misled as to the qualities of products. Furthermore, in order for the *protection* to be effective, the Commission should have recourse to reasonable budget resources ...
>
> (§60)

Life-health-strength metaphors

Finally, three more tokens have been found in the analysis referring to the 'Life-health-strength' source domain. Although it is barely mentioned, from this perspective food quality means keeping farmer traditions alive (while taking into account the new production methods, referring to the second green revolution agricultural technologies):

> This is due to the skills and determination of Union farmers and producers who have *kept traditions alive* while taking into account the developments of new production methods and material.
>
> (§1)

Food quality is also referred to as a European strength, which would offer a competitive advantage to the EU and a major contribution to the producers' living cultural heritage:

> The quality and diversity of the Union's agricultural, fisheries and aquaculture production is one of its important *strengths*, giving a competitive advantage to the Union's producers and making a major contribution to its living cultural and gastronomic heritage.
>
> (§1)

Discussion and conclusion

The analysis carried out gives us indications about the expected role a type of policy measure on food quality, as expressed by EU policy makers, plays in EU agrifood policy. The analysis focused on the metaphors appearing in the text of the European law on the subject of territorial food quality (EU Regulation 1151/2012).

As stated above, metaphors are seen from our theoretical approach as one of the mechanisms used by political actors in their struggle to impose a certain vision of political issues (i.e. food quality). Metaphor deals with two different domains of knowledge that relate to each other: from one of which we depart (source domain) in order to understand the meaning of the other (target domain). In our case, the meaning of food quality (target domain) is understood through other concepts belonging to three main source domains: Economy-business; Technology-building; and War-fight-battle; and the few that do not fall into these categories seem to be related to the Life-health-strength source domain. As it has been an exhaustive analysis, there are no more metaphors related to the target domain in the analysed text.

In terms of the 'Economy-business' source domain, agrifood quality means a competitive advantage for producers (in relation to other producers), an (economic) reward for producers (for their effort in doing something different), a benefit also extensive to the 'rural economy' in general, a way of adding value to the products (in the internal and external markets), something to reach through producers' voluntariness (in a context of free market), and something depending on (anonymous) consumers' trust (that is, depending on other wills beyond the producers themselves).

From the 'Technology-building' source domain, agrifood quality means a way of enhancing products' credibility (in the market), a way of ensuring fair competition between the producers (in the market), a way to support producers (helping them to remain in the market), a way of securing a fair (economic) return for producers, and the necessity of an official (state or third party) control over production, processing and distribution.

Agrifood quality is related also to the 'War-fight-battle' source domain, in the form of a strong defensive attitude aimed at protecting producers against unfair practices (from other producers), protecting the natural resources or landscape of the production area, or improving animal welfare, and also protecting the products in the context of European markets, and protecting consumers against misleading labelling practices.

Finally, only three lexemes related to the 'Life-health-strength' source domain have been found in the first paragraph of the analysed text, which indicates that agrifood quality would also mean keeping farmer traditions alive (but, interestingly, taking into account the new production methods coming from the technological agroindustrial system and promoting homogenization). It also represents a European strength. These are the only references to social factors related to food quality, and they appear only in the initial paragraph of the analysed EU normative but without any further development throughout the text, evidencing the little weight that is given to this dimension.

In sum, territorial strategies to promote food quality through the geographical origin of food (e.g. PDO or PGI), in the way they are promoted by EU policy makers, are mainly focused on economic and market spheres. All the three main emerging source domains deal with economic relations and factors, but other metaphors related to environmental or social concepts are almost absent (only three of 36 lexemes could be related to them). Most of the conceptual framework embedded in the EU regulation is devoted to promoting special protection for specific producers: those needed for protection against the turbulences of the capitalist food markets. From this viewpoint, the analysed EU Regulation could be understood as a kind of 'social policy' addressed to farmers in risk of being excluded from the current hegemonic agrifood regime, but without addressing the causes of exclusion, which would rest on the normal functioning of global food markets. These policies do not propose other relationships between producers and consumers than the ones framed in the standard capitalist market. On the contrary, given the heavy reliance of food quality on labelling and certification; they deepen even more farmers' and consumers' dependency on third party certification, which is a weak form for addressing the impacts of the agrifood regime (Bain *et al.*, 2013; Busch, 2011; Sodano *et al.*, 2008). This would support Polanyi's hypothesis (1944) that unregulated markets would have serious difficulties being socially and environmentally sustainable, engendering social pressures and leading to new regulations on the part of the political system, alternating periods of unregulated markets and other periods of protective state intervention. These cycles would be part of the normal functioning of capitalism (Polanyi, 1944). Thus, neoliberal food regimes can undergo substantive regulatory policy changes under social pressure. In this sense, the analysed EU policy would be part of governments' protective interventions to cope with the social pressure on small and medium farmers and consumers in Europe. But it seems to have little transformative impact and rather contributes to the maintenance of the neoliberal agrifood regime.

In this sense, one would expect to find metaphors related to environmental and social sustainability in the EU Regulation, but the analysis shows how it is focused mainly on economic frameworks: environmental and social sustainability is intended to be achieved primarily through market mechanisms, which is consistent with the axis of the typical mode of accumulation of capital of the neoliberal agrifood regime. For this reason we argue that the EU food quality policies are good in correcting some market impacts, but are far from solving the sustainability problems generated by the global agrifood system.

Food policies can be framed in the debate about ecological modernization and the new ecological paradigm (see Foster, 2012), in the sense that the former can be seen as a way of maintaining the capitalist mode of production and consumption with some cosmetic corrections. From the ecological modernization theory perspective, the unlimited growth of capitalist industrial society is conceived as possible and the ecological crisis can be surmounted through the incorporation of nature within the capitalist economic process via technological changes and market mechanisms (Mol, 1995; Spaargaren *et al.*, 2000). Ecological modernization theorists argue that there is evidence showing that capitalist societies of the late twentieth and early twenty-first centuries are progressively introducing effective measures for environmental protection (Mol *et al.*, 2009). These measures are at least of two types: on the one hand, political measures (local agendas 21, industrial ecology, product lifecycle study, green labelling, recycling policies, organic farming, and other environmental and sustainability policies at various levels) and, on the other hand, market measures (green capitalism, environmental responsibility, greenhouse gas emission quotas and markets, etc.). In addition, thanks to technological innovation, such a degree of eco-efficiency of products, processes and services could be achieved, which could reconcile economic growth without damaging the environment or depleting natural or energy resources. Ecological modernization theory is a normative rather than descriptive proposal that fits perfectly with the political proposal of 'sustainable development' and, therefore, with the model of progress-growth-development that has dominated the capitalist expansion of the last century.

The food quality schemes proposed in the analysed EU Regulation fit very well with the ecological modernization approach, as it intends to correct certain dysfunctions of the market economy with the intention of protecting certain actors and practices threatened by the global economy, but without altering or questioning the operation of this economic model, based on the universalization of export agriculture, and not altering the power relations that occur there and that cause those dysfunctions. In sum, this policy does not properly address 'the undemocratic and unsustainable impact of the contemporary food trade and investment regime' (as said by McMichael, 2016: 469). According to our analysis, a more holistic perspective (in terms of agrifood system or regime) would be needed to properly address sustainability issues from a policy perspective. Alternative agrifood quality schemes should take into account not only economic dimensions from a reductionist neoliberal market framework (as the EU Regulation seems to do), but also environmental and social dimensions from an alternative framework more oriented to the restructuring of the production and consumption of food rules on a world scale and overcoming the tension between transnational and local economies.

Acknowledgements

This text has benefited from its previous discussion in the XIV World Congress of Rural Sociology, Sustainable and Just Rural Transitions: Connections and Complexities, held in Toronto (Canada) in August 2016. The methodology about analysis of metaphors is based on a research project financed by the Spanish Ministry of Economy and Competitiveness titled 'The Role of Metaphor in the Social Perception and definition of Conflicting Issues: Institutions, media and the citizens' (ref. CSO2013–41661-P).

References

Arduini, S. (1993). 'La figura retórica como universal antropológico de la expresión,' *Castilla: Estudios de Literarura*, vol. 18: 7–18.

Bain, C., Ransom, E., and Higgins, V. (2013). 'Private Agri-food Standards: Contestation, Hybridity and the Politics of Standards,' *International Journal of Sociology of Agriculture and Food*, vol. 20, no. 1: 1–10.

Barco Royo, E. (2007). 'Denominaciones de origen. La incidencia de la localización y deslocalización,' *Distribución y Consumo*, vol. 96: 27–39.

Bougher, L. (2012). 'The case for metaphor in political reasoning and cognition,' *Political Psychology*, vol. 33, no. 1: 145–163.

Busch, L. (2011). 'Food standards: the cacophony of Governance,' *Journal of Experimental Botany*, vol. 62, no. 10: 3247–3250.

Cammaerts, B. (2012). 'The strategic use of metaphors by political and media elites: The 2007–11 Belgian constitutional crisis,' *International Journal of Media and Cultural Politics*, vol. 8 nos 2&3: 229–249.

Campbell, H. (2009). 'Breaking new ground in food regime theory: corporate environmentalism, ecological feedbacks and the "food from somewhere" regime?' *Agricultural Human Values*, vol. 26, no. 4: 309–319 DOI 10.1007/s10460–009–9215–8.

Capdevila, A. (2004). *El discurso persuasivo. La estructura retórica de los spots electorales en television*, Tecnos, Madrid.

Charteris-Back. J. (2011). *Politics and Rhetoric: The Persuasive Power of Metaphor*, Palgrave Macmillan: New York.

Constance, H. D., Friedland, W. H., Renard, M. C. and Rivera-Ferre, M. G. (2014). 'The Discourse on Alternative Agrifood Movements,' in D. H. Constance, M. C. Renard, and M. G. Rivera-Ferre (eds) *Alternative Agrifood Movements: Patterns of Convergence and Divergence* (Research in Rural Sociology and Development, Volume 21) Emerald, Bingley (UK): 3–46. DOI 10.1108/S1057–192220140000021024.

Di Masso, M. (2013). 'Redes alimentarias alternativas y soberanía alimentaria. Posibilidades para la transformación del sistema agroalimentario dominante,' PhD thesis, Universitat Autònoma de Barcelona, Bellaterra (Spain). www.tdx.cat/handle/10803/107853 (Accessed 28 February 2017).

Di Masso, M., Rivera-Ferre, M. G. and Espluga, J. (2014). 'The Transformative Agrifood Movement in Catalonia: Operational Divergences in the Construction of Food Sovereignty,' in D. H. Constance, M. C. Renard, and M. G. Rivera-Ferre (eds) *Alternative Agrifood Movements: Patterns of Convergence and Divergence*, Emerald, Bingley (UK): 159–181.

Eco, U. (1980). 'Two problems in textual interpretation.' *Poetics Today*, vol. 2, no. 1: 145–161.

Espluga, J. (2002). 'Percepción del riesgo y uso de pesticidas en la agricultura.' *Ecología Política, Cuadernos de Debate Internacional*, vol. 22: 17–30.

Espluga, J. (2004). 'La exposición laboral a pesticidas en la agricultura. Un caso de aparente subestimación del riesgo.' *Sociología del Trabajo*, vol. 51: 33–60.

EU Regulation 1151/2012 of the European Parliament and of the Council of 21 November 2012 on quality schemes for agricultural products and foodstuffs, http://eur-lex.europa.eu/legal-content/en/ALL/?uri=CELEX:32012R1151 (Accessed 28 February 2017).

European Commission (EC) (2018), 'Database of Origin and Registration.' European Commission – Agricultural and Rural Development. http://ec.europa.eu/agriculture/quality/door/list.html. (Accessed 16 April 2018).

Foster, J. B. (2012). 'The Planetary Rift and the New Human Exemptionalism: A Political-Economic Critique of Ecological Modernization Theory.' *Organization Environment*, no. 25: 211–237.

Friedmann, H. and McMichael, P. (1989). 'Agriculture and the state system,' *Sociologia Ruralis*, vol. 29, no. 2: 93–117.

Galdós, R. (2004). 'La intervención pública en la promoción de la calidad agroalimentaria: normativa comunitaria, española y vasca,' *Investigaciones Geográficas*, vol. 34: 47–62

Gilg, A. W. and Battershill, M. (1998). 'Quality farm food in Europe: a possible alternative to the industrialised food market and to current agri-environmental policies: lessons from France,' *Food Policy*, vol. 23, no. 1: 25–40.

Goodman, D. (2003). 'The quality "turn" and alternative food practices: reflections and agenda,' *Journal of Rural Studies*, vol. 19: 1–7.

Greimas, A. J. (1971). 'Narrative Grammar: Units and Levels,' *Comparative Literature*, vol. 86, no. 6: 793–806.

Grey, W. (2000). 'Metaphor and meaning,' *Minerva – An Internet Journal of Philosophy*. www.minerva.mic.ul.ie//vol. 4/metaphor.html (Accessed 28 February 2017).

Holloway, L. and Kneafsey, M. (2004). 'Producing-Consuming Food: Closeness, Connectedness and Rurality in Four "Alternative" Food Networks,' in L. Hollowayand M. Kneafsey (eds) *Geographies of Rural Cultures and Societies*, Ashgate, Aldershot.

Holt Giménez, E. (2009). 'EEUU: Crisis alimentarias, movimientos alimentario y cambio de régimen,' *Ecología Política. Cuadernos de Debate Internacional*, vol. 38: 73–79.

Holt Giménez, E., Patel, R. and Shattuck, A. (2009). *Food rebellions: crisis and the hunger for justice*, Food First, Oakland.

Ilbery, B. and Kneafsey, M. (2000). 'Registering regional speciality food and drink products in the United Kingdom: the case of PDOs and PGIs,' *Area*, vol. 32: 317–325.

Ilbery, B., Morris, C., Buller, H., Maye, D. and Kneafsey, M. (2005). 'Product, Process and Place: An Examination of Food Marketing and Labelling Schemes in Europe and North America,' *European Urban and Regional Studies*, vol. 12, no. 2: 116–132 doi:10.1177/0969776405048499.

Lakoff, G. (1993). 'The Contemporary Theory of Metaphor,' in Ortony, A. (ed.) *Metaphor and Thought*, Cambridge University Press, Cambridge.

Lakoff, G. and Johnson, M. (1980). *Metaphors we live by*, University of Chicago Press, Chicago.

Marsden, T., Banks, J. and Briwtow, G. (2000). 'Food supply chain approaches: exploring their role in rural development,' *Sociologia Ruralis*, vol. 40: 424–438.

Maye, D., Kneafsey, M. and Holloway, L. (2007). 'Introducing Alternative Food Geographies,' in D. Maye, L. Holloway, and M. Kneafsey (eds) *Alternative Food Geographies. Representation and Practice*, Emerald, Bingley.

McIntyre, B., Herren, H., Wakhungu, J. and Watson, R. (2009). *Agriculture at the crossroads. Global Report. International Assessment of Agricultural Knowledge, Science and Technology for Development, 2009*. Available at: www.fao.org/fileadmin/templates/est/Investment/Agriculture_at_a_Crossroads_Global_Report_IAASTD.pdf (Accessed 27 November 2017).

McMichael, P. (2005). 'Global development and the corporate food regime,' in *New directions in the sociology of global development. Research in rural sociology and development*, vol. 11: 269–303. Elsevier, Amsterdam.

McMichael, P. (2009). 'A food regime genealogy,' *The Journal of Peasant Studies*, vol. 36, no. 1: 139–169.

McMichael, P. (2016). 'Commentary: Food regime for thought,' *The Journal of Peasant Studies*, vol. 43, no. 3: 648–670.

Mio, J. (1997). 'Metaphor and Politics,' *Metaphor and Symbol*, vol. 12, no. 2: 113–133.

Mol, A. P. J. (1995). *The refinement of production: Ecological modernization theory and the chemical industry*, International Books, Utrecht.

Mol, A. P. J., Sonnenfeld, D. A. and Spaargaren, G. (eds) (2009). *The ecological modernisation reader*, Routledge, London.

Moragas Fernández, C. M. (2016). 'Claiming independence in 140 characters. Uses of Metaphor in the construction of Scottish and Catalan political discourses on Twitter,' PhD thesis, Universitat Rovira i Virgili (Spain) www.tesisenred.net/handle/10803/387319 (Accessed 28 February 2017).

Morgan, K., Marsden, T. and Murdoch, J. (2006). *Worlds of food: Place, power and provenance in the food chain*, Oxford University Press, Oxford.

Murdoch, J., Marsden, T. and Banks, J. (2000). 'Quality, Nature, and Embeddedness: Some Theoretical Considerations in the Context of the Food Sector,' *Economic Geography*, vol. 76, no. 2: 107–125, doi/10.1111/j.1944–8287.2000.tb00136.x/abstract.

Musolff, A. (2004). *Metaphor and Political Discourse. Analogical reasoning in debates about Europe*, Palgrave Macmillan, London.

Patel, R. (2007). *Stuffed and starved*, Portobello Books, London.

Polanyi, K. (1944). *The great transformation*, Beacon Press, Boston.

Rivera-Ferre, M. G. (2014). 'Impacts of Climate Change on Food Availability: Distribution and Exchange of Food,' in B. Freedman (ed.) *Global Environmental Change, Volume 1 of the series Handbook of Global Environmental Pollution*, Springer, Dordrecht: 701–707.

Sodano, V., Hingley, M. K. and Lindgreen, A. (2008). 'The usefulness of social capital in assessing the welfare effects of private and third-party certification food safety policy standards: trust and networks,' *British Food Journal*, vol. 110, no. 4/5: 493–513.

Spaargaren, G., Mol, A. P. J. and Buttel, F. H. (eds) (2000), *Environment and global modernity*, Sage, London.

9 Discourses on sustainability in the French farming sector

The redefinition of a consensual and knowledge-intensive 'agroecology'

Jessica Thomas

Introduction

Even though the governance of the food system (McMichael, 2009) is now discussed in international arenas, national governments and programmes remain decisive spaces within which the sustainability of agriculture has become a challenge on the political agenda. Many governments align with the *'reformist trend'* (Holt Giménez and Shattuck, 2011) in order to respond to the environmental, sanitary and social concerns of citizens or professionals of the agrifood sector. Within some activist and professional spheres, calls are increasingly being made for public authorities to be accountable for fostering multi-sectoral transitions. Public policies are expected to respond to socio-economic, environmental and social damage after years of monocultures, intensification and standardization of practices. The consequential economic pressures, massive use of inputs, natural resources depletion, extensive pollution, and some farmers' social exclusion (Pearse, 1980), are denounced. Public authorities therefore need to attest to their commitment to tackle the economic, environmental and social challenges threatening the sustainability of the food system.

This chapter contextualizes the institutionalization of 'agroecology' in France. It presents an analysis of diverse discourses in the context of the implementation of a national programme supporting agroecology since 2013. The first part of the chapter explores *strategies underlying government's rhetoric and interventions*. The second part focuses on discourses of various actors of the French farming sector when they refer to *knowledge valued as necessary to support sustainable practices*. The French government recently promoted the 'agroecological transition' as a systemic evolution for agricultural development. It established agroecology as a new 'paradigm' (Wezel *et al.*, 2009), in order to mobilize the actors of the farming sector. This led to a redefinition of agroecology, and therefore to many reactions, appropriations, adaptations, rejections and debates on agroecology as reframed by the government. The analysis presented here shows that the launch of the French national programme supporting agroecology is an emblematic case of the constrained political strategies undertaken by reformist governments in countries where agriculture is highly intensified. In this chapter, this strategic process is named as the *'institutionalization'* of agroecology, because officials themselves and researchers use this term. In France, 'agroecology' appears to be incorporated into the reformist framework of ecological modernization of the agricultural sector (Mol and Sonnenfeld, 2000; Buttel, 2000).

Some social movements and academics have pointed out the gap between official discourses and concrete political change in the reformist trend (Levidow *et al.*, 2014). The analysis presented in this chapter draws on these civic and academic concerns, and deepens the analysis of how official discourses produce particular framings of issues and specific actions. The objective is to discuss explicit and implicit political choices which partially circumscribe possibilities for change. It appears necessary to disclose unrevealed values and logics underpinning what discourses present as unquestionable objectives, proofs and methods to improve the sustainability of agriculture. A challenge is to question the arguments that the official discourse presents as self-evident, because they make the contingent nature of choices less visible and the issue or solution framing harder to contest.

This research focuses on discourses' 'normativity' and 'performativity,' considering discourses as tools used to justify governments' reformist interventions. When official discourses define common values and objectives, establishing them as incontestably associated with societal progress, they contribute to excluding some options and opportunities for change. This is why the proposition in this chapter is to supplement criticisms of a political cooptation of agroecology by the French government, first with an analysis of institutional discourses and practices, second with an analysis of *valuation* principles and activities. To this end, the chapter describes discourses and practices contributing to hybridization of values shaping singular visions of agroecology.

A particularity emerging from the analysis of this study case is the need to consider the normativity and performativity of a specific topic of discourses on sustainability: the role of *knowledge* in sustainable transitions. In the interviews, some French officials and actors of the agricultural knowledge system (AKS) have argued that ecological transitions require the production of knowledge, innovations, and changing knowledge production processes. The chapter describes how the studied actors shape the *values* and the *practices* attached to *knowledge for sustainable farming practices*. It shows that discourses on knowledge for sustainability are likely to be vigorously normative. This chapter therefore argues that, in the analysis of contested sustainability discourses, the *valuations* (Stark, 2000) of knowledge for sustainable practices should be carefully studied.

The methodology adopted here is the combination of a study of policy documents constituting the national programme, with the analysis of approximately 20 interviews held with different groups of actors:

- Public officials working on policy instruments to promote the agroecological transition.
- Scientists involved in the reorientation of research policies towards agroecology.
- Professionals of the Agricultural Knowledge System (AKS) including groups of farmers receiving subsidies to conduct agroecological experiments,

advisers from agricultural cooperatives and Chambers of Agriculture, members of technical research institutes.

- Educational personnel in agricultural secondary schools, engaged in programmes to teach sustainable practices.

The first part of the chapter presents an analysis of the French government's discourse, and some of its interventions, supporting sustainability of the agrifood system, and what it ultimately produces in terms of issue and solution-framing. The aim is to discuss the normativity and performativity underpinning these institutional practices as they contribute to the framing of public issues and solutions for agricultural development. In this way I intend to disclose political choices that the official discourse sometimes transforms into self-evidences, thus making them difficult to challenge. The second part draws attention to specific references to valued knowledge, and knowledge production processes, to support the agroecological transition. By unfolding debates on knowledge between various stakeholders of the French agricultural sector, this part intends to reveal the actors' choices and hybridizations underpinning knowledge valuation activities.

Characterization of the normative and performative effects of the French government's discourse on agroecology: a strategy to relegitimize established actors' interventions?

The analysis in this first part of the chapter suggests that, in order to understand the French government's strategic choice to promote agroecology, it is relevant to explore the strategies that reformist governments develop to integrate criticism (Boltanski and Chiapello, 1999) of public policies. The government's harnessing of the agroecology concept is understood as an attempt to improve the acceptability of the state's move towards the ecological modernization of the agrifood system. The following sections describe two tactics that the French government employs to build consensus in favour of the agroecological transition, thus contributing to a reframing of agroecology.

A discourse legitimizing agro-environmental policies by opening opportunities for economic valorization of voluntary commitments to green practices

In the wide diversity of discourses on sustainability that have been flourishing since the 1970s, a significant trend is the proliferation of governments' discourses on their reformist ambitions to foster the ecological transition of various sectors. It is from this perspective that the French government's choice to support agroecology can be understood as a consensual political programme targeting as many actors of the agrifood sector as possible. This objective of 'massification of sustainable practices' is a first explanation for the choices made in redefining the vision of agroecology defended by the French government.

Indeed, the first characteristic of the government's discourse is to redefine agroecology through the consensual framework of 'sustainable development.' It recycles the reconciliation of economic, environmental and social issues, to define agroecology as an approach ensuring '*the triple performance*' of agriculture. In the most recent framework law on agriculture, enacted in 2014, and which establishes a definition of agroecology, the reconciliation of these three dimensions is a founding principle:

> Public policies promote the sustainability of agroecological systems of production, including organic farming, that combine economic, social … environmental and sanitary performances. These systems promote farms' autonomy and the improvement of their competitiveness, by maintaining or increasing economic profitability, by improving the added value of produce and reducing the consumption of energy, water, fertilizers, plant protection products … antibiotics. They depend on biological interactions, ecosystem services, the potential of natural resources … They contribute to the adaptation and mitigation of climate change.[1]

This is partially due to the necessity to enrol in the transition a wide range of key players in the agricultural sector. During the interviews held for this study, French decision-makers developed a narrative around the necessity for a '*change of paradigm*' in agriculture. This ambitious goal implied targeting as many farmers as possible, along with other economic actors of the agro-food industry. As a consequence, officials envision a widespread acceptability of public intervention as crucial. The interviews showed that agroecology was presented as a solution when officials anticipated – and wanted to prevent – opposition to agro-environmental policies by leading farmers' representatives in the existing context. In 2013, the largest farmers' trade union (FNSEA[2]) routinely won the elections for professional representatives, with 55 per cent of the votes. Since the 1960s the leaders of this trade union have had persistent advantages in power relations that influence public policies for the French farming sector. They are largely represented in public and private organizations that play critical roles in French Agricultural Knowledge Development (Chambers of Agriculture, technical institutes, agricultural cooperatives, etc.). Moreover, French agricultural policies are historically based on policy-making processes associating the influential representatives of professionals. Therefore strong corporatist interests have had an impact on agricultural public policies for decades (Keeler, 1987). These dominant actors were at first reluctant to promote agroecology, perceived as an additional governmental injunction to respect environmental norms. The substantial work of officials on the reframing of agroecology to build a consensus on the agroecological transition with these dominant actors was to redefine it as an economic policy. Presenting agroecology as an opportunity for an economic valorization of the respect for agro-environmental norms facilitated its recognition in conservative professional spheres. In reaction, social movements defending more radical visions of agroecology warned about the absence of a systemic understanding of agroecology in the governmental approach. Social dimensions of agroecology, renewed forms of farmers' empowerment, or new forms of consumers' commitment were marginalized themes in the debates occuring during the institutionalization of agroecology (Lamine, 2015). As for the concept of sustainability (Neumayer, 2003), activists expressed their concerns about the political choices resulting in a 'weak' agroecology (Duru *et al.*, 2014).

The analysis of the officials' interviews and communication documents attests to the authorities' efforts to give a positive image to agro-environmental policies. To defend an ecological modernization of the agricultural sector, government officials considered that they had to contain farmers' concerns that environmental practices would be an additional source of economic burdens. Decision-makers explained that, by referring to the concept of agroecology, they intended to change farmers' mentalities and particularly their conception of agro-environmental policies as coercive measures. They sought to demonstrate that the policies supporting the agroecological transition were primarily economic, rather than environmental. This discursive strategy had some performative effects in state administration. While the environmental department of the Ministry of Agriculture had initially been in charge of agroecological policies, the national programme implementation was taken over in 2014 by the department in charge of the 'economic and environmental performance of farms.'[3]

Even though the characteristics and consequences of agricultural 'modernization' have constantly been criticized since the 1970s, the modernist rationale persists and is being rehabilitated through the framing of agroecology. In official discourses, agroecology is linked to 'ecological intensification' as well as 'intellectual intensification.' According to this narrative, it is progress in knowledge and control of natural interactions that affords new opportunities for farmers who give up fossil fuel dependent agriculture. The redefinition of agroecology thus shares common logics with the concept of 'sustainable intensification' (Levidow *et al.*, 2014). By devising a narrative on 'ecological intensification,' 'optimization of ecosystem services.' and 'maximization of natural interactions,' the French government set a new framework for ecological modernization. This discourse renewed the notion of agricultural 'progress' and set out the future of agriculture. Decision-makers wanted to convince reluctant farmers that an economic added value resulted from the optimization of natural cycles in production processes. Moreover, it emerged that the reformist discourse of the French government also incorporated characteristics of the '*neoliberal trend*' (Holt Giménez and Shattuck, 2011). It disseminated a narrative shared within neoliberal and reformist governmental spheres: the relevance of market-based approaches to prompt economic actors to green their practices.

From this perspective, the French government's programme consistes mainly of incentives for farmers' voluntary commitment. There are no constraining regulations directly linked to the French agroecological project, which is hardly binding. Government interventions are presented as flexible, officially to respect 'the agroecological spirit.' However, the study case reveals that these political choices are a strategy of legitimization of loose policies likely to satisfy conventional actors, rather than respect for the inherent characteristics of 'deep agroecology.' Decision-makers themselves described the national plan as a 'meta-political programme.' They justified these choices by the nature of agroecology itself, which requires a 'non-normative approach,' a term explicitly used during some interviews. This claim has orientated the research questions of the analysis presented in this chapter. My approach is designed to point out the paradoxical nature of the assertion of 'non-normativity' of agroecology, and ultimately to refute it. Because a development goal for the ecological modernization of a sector cannot be a neutral objective, I describe how the government necessarily undertakes political work on values to build a shared project.

A rapid analysis of the instruments of public action identified in discourses and effectively implemented in France reveals performative effects of the discourse on the economic benefit of good environmental practices. First, the tools of the national project consist mainly of a set of existing public subsidies, with slightly adapted selection criteria, to prioritize agroecological approaches. Only a limited number of existing subsidies have new agroecological requirements. The farmers declaring agroecological practices are prioritized in the selection process for access to two substantial public subsidies: aid for young farmers starting out, and investment support for compliance with modernization norms on farms. But few subsidies specifically for agroecological commitment have been offered. Second, the government subsidizes some collective research and development projects that experiment with innovative agroecological practices. Third, a specific tool – the only new instrument supporting agroecological policy – was recently created: the Economic and Environmental Interest Grouping (EEIG). This is a label through which public authorities recognize a group of farmers' collective commitment to agroecological practices. The objective is to help them to get additional public and private funding to experiment with new practices collectively. There has not been any structural reorganization of the agricultural administration, or serious reform of agricultural policies, resulting directly from the intention to develop agroecology. The interventions are ad hoc and part of an already established trend of ecologization of agricultural policies.

These rhetorical and implementation choices attest to governmental justifications for incremental reformist strategies. For the state, the agroecological discourse serves a strategy of compromise between society's expectations of the ecological modernization of the sector, and economic actors' resistance. It raises questions on the limited leeway of the national government in a context of strong professional protest, based on the belief that environmental norms represent a threat to farms' viability. The processes of 'institutionalization' and 'normalization' in ecological modernization appear to be deeply dependent on particular contexts, on path dependency logics, and on the 'acceptability' of public interventions by influential actors.

A discourse creating consensus by valorizing the 'coexistence' of agricultural models

Another manifestation of an official discourse that is intended to build consensus among actors of the agrifood sector is the narrative describing agroecology as the '*coexistence*' of diverse agricultural models, systems and practices. Decision-makers frequently refer to this concept of 'coexistence' to describe the future of French agriculture and the flexibility of the agroecological approach. Once again, I interpret this discursive framing as a redefinition of agroecology aimed at securing the acceptability of reformist public interventions. This section reveals how the references to 'coexistence' of agricultural models tone down the radical character of agroecology. It is interpreted as an attempt to 'pacify' the power relationships among the actors.

The official agroecology is presented as a list of coexisting technical practices used to effectively implement agroecology. In practice, the French administration supports agroecology by segmenting it into selected techniques and legitimate practices. A technical vision of agroecology prevails in policy instruments and the incompatibility between various models is not questioned. This vision marginalizes the claims about the political implications of agroecology aimed at transforming the asymmetric power dynamics of the food system. Some dimensions characterizing more radical activists' definitions of agroecology are excluded. Examples are the absence of actions to challenge the monopolies of input suppliers, or the lack of agroecological policies supporting transformations in modes of consumption (Lamine, 2015).

I argue that when the government institutes 'coexistence' as an objective, it circumvents the identification of potential 'losers' and 'winners' in the players of the agrifood sector. 'Coexistence' allows a persisting presence of conventional actors – alongside former marginal ones – even though their continuing intensive practices are regularly contested. The consensual concept of coexistence is a way to re-legitimate their participation in agricultural development. In the institutionalization process, public authorities do not recognize the persistence of unequal opportunities for developing alternative practices, and they ignore the risk of strengthening dominant actors that develop green-washing strategies. Although they do not imply the same degrees of transformation, risk, or economic loss, diverse practices and technologies are put on an equal footing with 'agroecological innovations:' conservation agriculture; direct drilling; biocontrol; methanization; measures to protect pollinators; organic farming; diversification of crop rotation, species and varieties; introduction and combinations of leguminous plants in crops; substitution or decrease of inputs; agro-forestry; genetic improvement; local seed sharing; improved feed autonomy; alternatives to antibiotics; etc. are all considered as agroecological practices despite their different greening effects, and their various impacts on natural and social environments.

I posit that the discourse on 'coexistence' of various agricultural models is a manifestation of the general tendency of reformist governments to promote multi-actor governance (Kooiman, 2003) as a consensual solution to societal challenges. This approach theorizes possible horizontal relations among a variety of actors participating in policy-making and sectorial management. It promotes 'deliberative' or 'participatory' democracy within multi-stakeholder arenas in charge of sectorial public action. In the case of agroecology, the valorization of 'coexistence' officially implies that a variety of actors in the agrifood sector would be associated with decision-making for the orientation of the food system. Following the analysis of Mouffe (1999), I draw attention to the lack of the 'dimension of the political' in these logics of governance. Public participation and multi-actor governance are also strategies to improve the acceptability of public interventions. Two main risks can be identified. First, the narrative on 'coexistence' could attenuate criticism of the persisting monopoly of socio-economic actors that have the capacities to effectively influence the policy-making process. Second, this narrative denies the transformative potential of conflicting alternatives. This is a concern in light of the central role of conflict in compelling governments to undertake ecological modernization. Awareness of the necessity to reform states' interventions in agricultural development stemmed from the conflicting claims of social movements denouncing intensive agriculture, and from the perception of conflicting interests between citizens and some professionals of the agrifood sector. This transformative potential of conflicts calls for a vigilant analysis of discourses on the 'coexistence' of practices for sustainable development, because of the strategy of dilution of political struggles that it may entail.

In conclusion, it appears that the French government actively redefines agroecology through discourse and practices. Agroecology becomes normative and has performative effects because the government attaches selected values, practices, and operational tools to this concept. Two main strategic moves in this redefinition of agroecology have been identified in this first part. The first is the redefinition of agroecology as an opportunity to economically valorize sustainable agricultural practices. The political appropriation of agroecology serves the justifications of state-led ecological modernization, in a context of vigorous expressions of professionals' reluctance to embrace the government's normalization activities to attenuate the environmental impacts of agriculture. A second strategic reframing of agroecology is to establish it as the common horizon for a large panel of coexisting agricultural models and practices. It participates in attenuating the 'politics of agroecology,' in order to build some consensus for undertaking limited reforms of the sector. The transformations of power dynamics that agroecology would require are excluded from discussions, to convince influential actors to take part in the transition. This relatively loose conceptualization of agroecology can both reflect and explain the conflicting debates on the nature of agroecological practices. The second part of this chapter studies this controversy by focusing on a key dimension identified after the analysis of the interviews and regulation: the controversy on agroecology is based partially on divergent valuations of knowledge for sustainable practices.

Knowledge valuation for sustainability by the actors of the French farming sector: legitimization strategies through assemblages of definitional principles of 'agroecological knowledge'

The analysis presented in this second part focuses on how various actors of the agrifood sector 'valuate' relevant knowledge for sustainable practices. As one of the research objectives is to analyse debates on knowledge for sustainability, I extend the discourse analysis to include the arguments of professionals, scientists, academics and other socio-economic or educational actors that are associated with the implementation of the national programme. I refer to 'valuation' as defined in pragmatic sociology (Boltanski, Thévenot, 2006). 'Valuation' theory does not simply reduce the analysis of values to the examination of actors' representations, but also investigates their concrete practices that are considered as revealing and constructing values (Stark, 2000). The focus of this research on knowledge valuations results from the finding that the actors of the agrifood sector mentioned above frequently refer to their production, mobilization and diffusion of knowledge to justify their intervention in agricultural development. During the interviews, many actors involved in the implementation of the national programme characterized agroecology as 'intellectual intensification' for farmers and advisors. Knowledge production and innovations processes were identified as key levers to induce systemic transformations in many commonly shared theories of change (Voß and Freeman, 2016). These theories influence decision-makers and executors engaged in transition management. Actors willing to consolidate or strengthen their role in power relationships among stakeholders of the agrifood system are therefore willing to demonstrate the relevance of their intervention in the production and diffusion of knowledge. The proposition here is to discuss the strong normative content of discourses when the debates are about knowledge production and diffusion for sustainable transitions.

A hypothesis emerging from this case study is that the institutionalization of agroecology is a time of revival of ongoing debates on the role of knowledge in agricultural development. For decades, actors of the agricultural knowledge system have been discussing the relevance of various kinds of knowledge to meet agricultural development objectives. Oppositions and hybridizations between different types of knowledge have been identified, asserted, contested and sometime reified (Kloppenburg, 1991). Some examples of these debates are: the status that should be granted to farmers' experiences compared to scientific knowledge; the justifications for knowledge co-construction between farmers and experts; the need for contextualization of knowledge; the different roles of applied versus fundamental knowledge, of factorial models versus multi-factorial or systemic experiments, and of participative research versus fundamental science, etc.

In the French context, the calls for and official recognitions of the necessity for a 'paradigm' change resulted in a questioning of epistemic evolutions potentially facilitating the agroecological transition. The responsibility of scientific authorities in the development of intensive farming practices has been brought to light by social movements, and progressively acknowledged within political, scientific and professional arenas. The whole ecosystem of agricultural research and development structures is criticized for having participated in the diffusion of practices, technologies and inputs that have had destructive consequences on the environment. As a consequence, the legitimacy of alternative modes of knowledge production is being revived. The French government communicates on its willingness to support the production and diffusion of farmers' knowledge and experience, in order to green agricultural practices. The official discourse thus celebrates bottom-up and collective experiments, participative research and development projects, and various initiatives to produce 'co-constructed' agricultural knowledge between professional, technical, scientific and socio-economic actors.

In the debates around valuable knowledge to improve the sustainability of farming practices, actors of the agrifood sector can express convergences on several characteristics that knowledge should have in order to improve sustainability, even though their vision of what agroecology actually is differs. But these debates also reveal potentially irreconcilable conceptions of valuable knowledge and relevant processes of knowledge production. My objective is not systematically to classify criteria of knowledge valuation within antagonistic categories according to actors' interests. Even though some actors disagree on the most relevant orientations of agricultural practices, they can converge on some criteria of value. The interpretation proposed here is intended to identify the points of convergence and divergence among actors. The idea is to show that the value of knowledge depends on a complex set of agreements and disagreements on a variety of criteria (e.g. the role of technologies or of traditional knowledge, the place of politics, the participation of actors, the value of scientific proof, etc.). The complexity of the arrangements constituting actors' valuation logics and practices is exemplified by the conflicting conceptions on the role of technologies, or the relevance attached to generic versus local knowledge, even though they all recognize the value of professionals' contextualized experiences.

Identification of five ideal-type narratives that actors refer to and assemble when they value relevant knowledge for a sustainable agriculture

First set of criteria: valuing agroecological knowledge embedded in farmers' non-intensive practices, based on a singular and ethical relation to nature

NGOs and pioneer farmers introduced the concept of agroecology in France during the 1990s. They referred to a version of agroecology based on criticisms of intensive agriculture, and were inspired by social movements advocating for agroecology in South America. They supported a demanding definition of agroecology, considering that farmers' knowledge – elaborated in a local context and through personal experience – should be valued and developed to support sustainable practices. The actors that refer to this first set of principles denounce the standardization of knowledge produced by scientific authorities, which they consider as largely responsible for environmental degradation. They seek to promote the elaboration of alternative knowledge, and they identify observation of agro-systems and local experiments as key learning processes. Testimonies of personal experience are also circulated and considered as sources of knowledge. Traditional practices are presented as the most resilient solutions to adapt to changing environments. Knowledge is considered to be more relevant when it is adapted to a particular ecological region and when it is based on a respectful relationship with nature. Knowledge also has a particular social dimension as it is shared and transmitted among a group of actors with a common vision and objectives. The actors that value this type of knowledge want to demonstrate the contribution of small farmers to agricultural development, in a system where these farmers are marginalized. They can confer a political status on traditional, professional, experiential, and local knowledge.

This first set of principles can be used to determine which type of knowledge is considered as supportive of agroecology and which is not, by activists, alternative farmers, alternative scientists, and some actors of the educational system who agree with a value-oriented version of agroecology. The actors following this first set of principles of knowledge evaluation typically push for a profound transformation of agricultural systems. As a result, they tend to denounce the recent institutionalization of agroecology as a weakening of its initial ambitions, and they challenge the misappropriation of this concept by more conventional scientists and by some reformist governments.

Second set of criteria: valuing agroecological knowledge embedded in professional practices for knowledge capitalization and transmission

This second approach, which determines the types of knowledge that can support agroecological practices, is in many ways similar to the first approach. One of the central criteria to attach value to knowledge is also whether it emerged from professional practices or not. Farmers' experience is perceived as a rich source of information. However, the actors that refer to this second approach argue that knowledge has to be more objective, and should not be too influenced by ethical or political considerations. Another key difference with the first approach is the acceptance of formalization – and sometimes of institutionalization – of knowledge in order to accelerate its transmission. Knowledge will be considered as particularly relevant when it is connected to farmers' needs and directly applicable in a particular environment, in accordance with natural cycles. But in this second approach, knowledge transmission to accelerate change in practices implies attempts to 'reveal' and sometimes formalize tacit knowledge embedded in actions (Nonaka and Takeuchi, 1995). The underlying idea is to collect and capitalize on professionals' knowledge, in order to diffuse it through training, education, and advisory mechanisms.

Activists, 'alternative farmers,' and some scientists – especially social science researchers – agree with this conception of agroecology and work on tools to value knowledge emerging from action. They consider that social science tools can be mobilized in order to elicit knowledge (through interviews, field research, etc.). They also develop participative research programmes that try to go further than simply consultation with social actors. In this approach, research programmes are valuable when they integrate social actors' questions and investigate the need for direct application, to produce '*trans-disciplinary knowledge*.' In addition, some teachers in technical agricultural schools also refer to this vision. Finally, actors with very different priorities – for instance farmers and developers considering that the main objective should remain productivity – can mobilize some of these principles, as they also strongly value farmers' experience. They will use these principles to argue that farmers' practices which slightly reduce input use should be considered 'agroecological.' Utilitarian conceptions of knowledge produced on farms for productivity can nourish this knowledge valuation approach, when they are combined with other sets of principles described in the following categories.

Third set of criteria: valuing knowledge for commitment to the institutionalized agroecological transition

This approach presents knowledge as the main instrument for reaching the 'triple performance' of agricultural practices with regard to economic, environmental and social aspects. Knowledge is a tool to foster 'ecological intensification.' In order to do so, the combination of agronomy and ecology is supposed to generate new knowledge to adapt practices to changing environments. This vision supports the transition promoted by the French government. It is anticipated that it will help to keep yields high in a context of increasingly constraining agro-environmental regulations. Expectations regarding the degree of 'ecologization' are not as high as in the first two approaches, for new technologies can be considered as relevant solutions as soon as they enable a reduction of inputs. In this approach, knowledge valuation depends notably on its social acceptability. The French government puts out this vision to integrate a large number of actors even though their intensive practices are only slightly greened.

In this approach, knowledge can be both collected from professionals and produced by experts and scientists. Political programmes and communication encourage 'pioneer farmers' to share their experience and to produce 'references' – mainly technical and economic. These references are designed to prove the superiority of particular practices or models. In contrast to the two previous approaches, innovation is valued through techno-scientific references, and new technologies and 'precision farming' are put forward as agroecological solutions. Decision-makers and administrations are not the only actors that refer to this set of principles to value knowledge. Before the political injunction to develop agroecology, the main public research institute in agronomy (INRA, the National Institute for Agricultural Sciences) chose agroecology as one of its strategic research objectives. Many scientists, as well as some technicians and some advisors of development organizations, have also adopted this approach. Here, in contrast to the two previous approaches, knowledge emerging from farmers should be validated through scientific methods. Echoing this concept, interviews with some public servants, scientists and professionals of traditional development organizations (Chambers of Agriculture, farming cooperatives, technical institutes) revealed a powerful trend: the adaptation of classic organizations to the 'agroecology paradigm' in order to maintain and re-legitimate their authority in the French agricultural knowledge system. Some of these actors develop and commercialize complex indicators and data analysis software for farmers to help them to comply with agro-environmental regulations. A look at their commercial practices shows the discursive work and the marketing practices that label their products or commercial services as 'agroecological.' This supports the finding of a knowledge privatization trend in agriculture (Labarthe, 2006; Byerlee *et al.*, 2002).

Fourth set of criteria: valuing innovation for 'ecologically-intensive farming'

In this approach, the main criteria for assessing knowledge are the economic benefits, the yield outcomes, and compliance with the agro-environmental regulations of the new practice or technology adopted. The actors who have adopted this concept value new technologies for sustainable intensification. They use biotechnologies and other technological tools that analyse complex environmental, economic and technical data to manage agricultural systems. They are opposed to the actors attached to the first two approaches, who strongly criticize the lack of farmers' autonomy in this 'knowledge externalization' trend.

To reaffirm the legitimacy of their role in agricultural development after being strongly criticized for the effects of their intensive practices, partisans of conventional farming have partially consented to greening their techniques. They argue that they are key actors of the '*ecological and intellectual intensification*' that should replace chemical intensification. Behind the expression of '*intellectual intensification*,' they are reviving the myth of salutary technological solutions to increase food production. The framework of 'cognitive capitalism' (Gorz, 2003) informs this vision by promising an economic valorization of ideas, creativity or knowledge. Partisans of this approach denounce restrictive regulations, even when they aim to protect natural resources, arguing that knowledge creation requires a liberal environment, where voluntary commitment is the most efficient way to promote innovations. They consider that agroecology could be economically valorized through eco-labels or eco-certificates for products, and commercialization of innovations.

In this vision, knowledge is evaluated through standards inspired by the strategies of corporations' R&D departments. It can be 'detected' on farms, or created by experts. The innovations detected on farms need to be legitimized by expert evaluations. Scientific methods therefore produce the 'proof' that a technique or a practice is innovative and relevant to meeting economic and environmental goals. This is however not the only way to validate innovations. Engineers and technicians of farming cooperatives, Chambers of Agriculture, and technical institutes use their expertise to validate and valorize professional knowledge. The classical hierarchy between experts' knowledge and knowledge acquired through practice persists, as well as top-down relations between farmers and advisors.

Fifth set of criteria: valuing agroecology as a science

In this approach the superiority of scientific methods remains a key criterion to rank knowledge. Scientists also face strong public criticisms denouncing the responsibility of agronomy in the reign of intensive farming practices. Criticism focuses on the de-contextualized and standardized solutions put forward by agronomists. Many scientists now claim that they are trying to reorient their research towards environmental protection, to consider new research subjects, and to transform their methods. They value interdisciplinary studies blending agronomy and ecology. Most of these scientists recognize that agroecology is an opportunity to 'green' their research but, depending on their discipline, values, and work environment, they modify their research methods to varying degrees. Some of them value strong transformation, like the 'alternative scientists' who follow the first two approaches. They increasingly value systemic experiments, to enrich scientific methods. Others, who are partisans of this 'scientific agroecology,' consider that the evolution towards 'multi-factorial' analysis and modelling, to describe the complexity of interactions within agro-systems, will be enough to overcome criticism regarding the reductionism of factorial analysis. A growing number of these actors value knowledge co-construction with socio-professional actors. However, they consider that scientific validation remains essential to guarantee an objective evaluation.

Several conclusions can be drawn from the identification of these five narratives. The research presented here shows that the institutionalization of agroecology coincides with debates on relevant knowledge for agricultural practices. Therefore actors develop discourses to demonstrate that they have a role to play in knowledge production, mobilization and diffusion. They try to position themselves as relevant participants of a 'knowledge chain,' from elaboration to remobilization by professionals. References to knowledge and the theme of '*intellectual intensification*,' replacing intensive practices and use of chemicals, is a leitmotiv for various actors, even though they differ on epistemic definitions of 'agroecological knowledge.' The analysis of the controversies on agroecology reveals that 'knowledge' and 'sustainability' are issues heavily charged with values in actors' discourses and practices. The references to particular types and processes of production of 'agroecological knowledge' are strategic arguments of justification for their intervention in agricultural development. These justifications, through singular knowledge valuations, contribute to strategic legitimization of power relationships among the actors of the agrifood sector. Following on from these observations, research is needed to analyse carefully the debates on relevant knowledge for sustainable practices in contested discourses on sustainability. The study of the activities of knowledge valuations undertaken by concerned actors in sustainable transitions could reveal some explicit and implicit strategies of legitimization of power relations between the actors. It could also improve the understanding of the power asymmetries, as it shows that the actors do not have the same intentions, resources, tools, spaces, and 'onus of proof' to justify their vision of relevant knowledge for sustainable practices.

Conclusion

The research presented here shows two key on-going processes in the discourses on sustainability of the actors involved in the institutionalization of agroecology in France.

First, agroecology is being redefined through the government's discourses and interventions for agricultural development. More than a constraining environmental policy, the support for agroecology is presented as an opportunity to ensure the economic valorization of professionals' voluntary commitments in various agricultural models and practices with unequal degrees of sustainability. To use the terminology of Holt Giménez & Shattuck (2011), the French government's discourse and intervention for the institutionalization of agroecology is interpreted as a manifestation of '*the reformist trend*' in the governance of the agrifood sector. Embedded in this reformist ambition of the French government are also some characteristics of '*the neoliberal trend*,' as competitiveness and productivity expectations continue to determine the legitimization of agricultural practices. The performativity of the French government's neoliberal and reformist choices is evidenced in the operational tools chosen: professionals' voluntary commitments are supported through incentives and opportunities for market valorization. The choice of agroecology as a development goal for the French agricultural sector is an interesting example for analysing the capacity of neoliberal and reformist policies to take on the challenges facing the food system. Indeed, the French government has chosen a concept that was at first put forward by actors of '*the radical trend*.' Agroecology was initially a dissenting alternative proposed by South American and international social movements defending small farmers against the domination of hegemonic agro-business players (Altieri, 2005). The harnessing of this concept by the French government shows the capacity of neoliberal and reformist policies to integrate the criticisms of radical actors. In this case it has toned down the transformative potential of agroecology. The analysis of Holt Giménez and Shattuck also helps to interpret the official discourse on agroecology as an attempt to pre-empt radical claims. In the cycle of successive liberal and reformist orientations of the capitalist agrifood system that the authors describe, the institutionalization of agroecology in France would be an articulation of *neoliberal* and *reformist* objectives, with a particular strategy: declaring the integration of *radical* claims on agroecology in order to undermine protest against intensive farming. These compromises constructed around the concept of agroecology are classic political negotiations in reformist activities. The agroecology promoted is the result of these political activities building consensus among various actors positioned in established power relationships. It requires the reconciliation of diverging, unequally represented interests to contribute to the definition of agricultural development goals.

Second, this research has presented a variety of knowledge valuations analysed as strategic tools to legitimize interventions in the agrifood sector. Knowledge valuation appears as a crucial issue in contested discourses on sustainability. In this chapter, a key trend identified in the French context is the adaptation of major socio-economic actors to a redefined agroecology. Facing extensive criticism for their role in the prevalence of intensive farming, some agro-business firms, cooperatives, Chambers of Agriculture, and public or private research institutes adopt the agroecological narrative to legitimize their persisting influence in the agricultural knowledge system. As a result, various discourses, framings, and actions for agroecology conflict with one another. Martinez-Alier (2002; 2008) shows the link between inequalities in contemporary environmental conflicts and the unequal power of the actors' languages of valuation. According to him, one of the causes of the lack of environmental justice is the impossibility for some actors of sharing publicly and fairly their languages of valuation. They do not necessarily have the resources and opportunities to compare their valuation principles with those of dominant languages on an equal footing.

A research proposal follows from these findings: *the analysis of knowledge valuation in discourses on sustainability could improve the understanding of explicit and implicit strategies of power legitimization among the actors of ecological transitions.* Further analysis of discourses on sustainability could therefore extend the investigations on language of valuation, especially considering references to knowledge, because of their strong normative content.

Notes

1 French 'Agricultural Framework Law' of 13 October 2014. *Loi n° 2014–1170 du 13 octobre 2014 d'avenir pour l'agriculture, l'alimentation et la forêt*

2 FNSEA: *Fédération Nationale des Syndicats des Exploitants Agricoles*. The National Federation of Farmers' Unions is the first farmers' union in France.

3 Since 2013, the 'Agroecological Project for France' has been implemented by a specific directorate of the French Ministry of Agriculture: the '*General Directorate for the Economic and Environmental Performance of Enterprises*.' This directorate was recently renamed to reflect its similar positive rhetoric around the 'performance' of farming activities.

References

Altieri, M. A. (2005). The myth of coexistence: why transgenic crops are not compatible with agroecologically based systems of production. *Bulletin of Science, Technology & Society* 25(4): 361–371.

Boltanski, L., Chiapello, E. (1999). *Le nouvel esprit du capitalisme*. Paris: Gallimard.

Boltanski, L. and Thévenot, L. (2006). *On Justification: Economies of Worth*. Princeton, NJ: Princeton University Press.

Buttel, F. H. (2000). Ecological modernization as social theory. *Geoforum* 31(1): 57–65.

Byerlee, D., Alex, G. and Echevarria, R. G. (2002). 'The Evolution of Public Research Systems in Developing Countries: Facing New Challenges.' In Byerlee, D. and Echevarria, R. G. Agricultural Research Policy in an Era of Privatization Wallingford: CABI: 19–34.

Duru, M., Fares, M., Theron, O. (2014). Un cadre conceptuel pour penser maintenant (et organiser demain) la transition agroécologique de l'agriculture dans les territoires. *Cah Agric* 23: 84-95.

Gorz, A. (2003). *L'immatériel. Connaissance, valeur et capital*. Paris: Ed. Galilée.

Holt Giménez, E. and Shattuck, A. (2011). Food crises, food regimes and food movements: rumblings or reform or tides of transformation? *The Journal of Peasant Studies* 38(1): 109–144.

Keeler, J. T. (1987). *The Politics of Neocorporatism in France: Farmers, the State, and Agricultural Policy-making in the Fifth Republic*. New York: Oxford University Press.

Kloppenburg, J. (1991). Social theory and the de/reconstruction of agricultural science: local knowledge for an alternative agriculture. *Rural Sociology*, 56(4): 519–548.

Kooiman, J. (2003). *Governing as Governance*. Thousand Oaks (Calif.), Sage.

Labarthe, P. (2006). *La privatisation du conseil agricole en question: evolutions institutionnelles et performances des services de conseil dans trois pays européens*. Doctoral dissertation, Université de Marne-la-Valée.

Lamine, C. (2015). Sustainability and resilience in agrifood systems: reconnecting agriculture, food and the environment. *Sociologia ruralis* 55(1): 41–61.

Lascoumes, P. and Le Galès, P. (2004). *Gouverner par les instruments*. Paris: Presses de Sciences po.

Levidow, L., Pimbert, M. and Vanloqueren, G. (2014). Agroecological research: Conforming—or transforming the dominant agro-food regime? *Agroecology and Sustainable Food Systems* 38(10): 1127–1155.

Martinez-Alier, J. (2002). The environmentalism of the poor: a study of ecological conflicts and valuation. Edward Elgar, Cheltenham.

Martinez-Alier, J. (2008). Languages of valuation. *Economic and Political Weekly* 43(48): 28–32.

McMichael, P. (2009). A food regime genealogy. *The Journal of Peasant Studies* 36(1): 139–169.

Mol, A. P. and Sonnenfeld, D. A. (2000). *Ecological Modernisation around the World: Perspectives and Critical Debates. Environmental Management and Health* 11(5): 475–476.

Neumayer, E. (2003). *Weak Versus Strong Sustainability: Exploring the Limits of Two Opposing Paradigms*. Northampton, MA: Edward Elgar Publishing.

Pearse, A. (1980). *Seeds of Plenty, Seeds of Want: Social and Economic Consequences of the Green Revolution*. London: Oxford University Press.

Stark, D. (2000). For a sociology of worth. *Working Paper Series, Center on Organizational Innovation*. New York: Columbia University.

Voß, J. P. and Freeman, R. (2016). Introduction: Knowing Governance. *Knowing Governance*. Palgrave Macmillan UK: 1–33.

Wezel, A., Bellon, S., Doré, T., Francis, C., Vallod, D. and David, C. (2009). Agroecology as a science, a movement and a practice. A review. *Agronomy for Sustainable Development* 29(4): 503–515.

10 Duelling discourses of sustainability

Neo-conventional and organic farming on the Canadian Prairies

Michael Gertler, JoAnn Jaffe and Mary A. Beckie

Contextualizing contested cropping systems

Farming on the Canadian Prairies is a recent experiment. After eons of soil formation under native grasses and millennia of inhabitation by nomadic hunters, the resident Indigenous peoples were dispossessed to make way for settler ranches and farms. This colonization project nearly foundered thanks to predatory railroads, merchants, and banks, and inappropriate farming technologies. After the starvation and displacement of Indigenous populations associated with colonial expansion and the collapse of the bison herds in the 1870s (Daschuk, 2013; Waiser, 2016), the next regional eco-economic disaster came with the droughts, dust, and Depression of the 1920s and 1930s (McManus, 2011). The crisis was alleviated by precipitation, new tillage tools, public investment in infrastructure, wartime prices, and innovations such as co-ops, community pastures, and the Canadian Wheat Board (CWB), but ongoing restructuring generated winners and losers in terms of households, enterprises, and settlements (Jaffe and Quark, 2006). Declines in rural population mirrored declines in farm numbers. Rather than addressing broader rural development issues, however, researchers focused on technical refinements that would increase yields and labour productivity as well as profits for agribusiness firms and some farmers (see Müller, 2008).

In the face of sustainability issues, farmers and scientists also struggled to develop farming systems that were soil-conserving and cost-effective. Inventors worked on tillage equipment that left straw on the surface, on seeders that performed well in crop residues, and on cropping systems that allowed sufficient storage of moisture and nutrient provision without summer fallow (Lindwall, 2009). After World War II, herbicides such as 2,4-D controlled many weeds in cereal crops. Other herbicides were introduced as new modes of action were discovered and greater weed control was desired in crops to which commercial fertilizers were applied. An important watershed event was the development of glyphosate (RoundUp), a systemic herbicide that became commercially available in 1976. Spraying stubble fields with glyphosate prior to planting became a key practice supporting direct seeding into cereal stubble (Holm and Johnson, 2009). This technological package was soon married to the products of agricultural biotechnology, particularly to herbicide-resistant varieties of canola that were sold to farmers starting in 1995. With a push from agrichemical and seed companies, cropping systems uniting reduced tillage, new herbicides, and herbicide-resistant cultivars of canola, lentils, soybean, and corn have been added to the crop mix and have contributed to the transformation of farming and debates about sustainable agriculture.

We propose 'neo-conventional' as a label for this regionally dominant approach to crop production and environmental management as it can be distinguished from earlier 'conventional' cropping systems that depended more on tillage, summer fallow, and seeds that were the product of conventional plant breeding. With respect to logics and priorities, there are continuities with earlier approaches championed by input suppliers and public-sector researchers. However, newer technological developments tend to be further removed from the purview of producers. Frequently, they are patented, 'black-box' technologies that deskill farmers by reducing their ability to understand and control production systems (Diaz and Stirling, 2003). Finally, 'neo-conventional' seems apt because this approach to agriculture and sustainability has hallmarks of other neo-liberal projects: central roles for corporations, supporting roles for universities and governments, and new organizations preoccupied with defending the legitimacy of this 'neo-conventional' regime.[1]

In the 1970s, while cropping systems on the Prairies were being retooled in the neo-conventional (NC) mould, some producers were also investigating organic production. Organic farming relies on extended crop rotations (including legumes for nitrogen fixation), green manure crops and intercropping, strategic tillage, selection of competitive cultivars (e.g. taller rather than semi-dwarf varieties), adjusted seeding rates and dates, and on other methods of weed control such as mowing and chaff collection (Frick, 2013; SCIC, nd). To qualify for certification, organic farmers cannot use synthetic pesticides and fertilizers, genetically modified (GM) seeds, antibiotics as feed additives, or hormone growth promoters.

Despite price premiums reflecting growing demand, organic production on the Canadian Prairies has not expanded as much as in jurisdictions where it has gained more support from publics and governments. The organic sector competes for adherents with more strongly promoted framings of agricultural sustainability (see Selfa *et al.*, 2015), including what we call NC agriculture. Our goal is to explore the contest between organic and NC crop production on the Canadian Prairies, and to understand what has supported or inhibited adoption of these practices. To target the inquiry, we focus on the experiences of farmers in the Province of Saskatchewan. Our investigation is based on select interviews with farmers, traders, researchers, and organization staff.[2] It also incorporates observations from workshops, tradeshows, conventions, and conferences. The project gains depth as well from our collective experience in studies of alternative farming systems in several countries.

As a site of agricultural contestation, the Canadian Prairies are worth studying, in part because this is a region where large-scale production, long-distance trade, sparse settlement, and ecological limits challenge visions of re-localization and smaller-scale intensification frequently advanced by advocates of agricultural and food system restructuring (see Lamine, 2015). This important locus of innovation also stimulates fresh thinking about agribusiness, neo-liberal governments, multi-stakeholder partnerships, corporatization of universities, and forms of producer and consumer mobilization. Furthermore, it epitomizes the ways that the politics of knowledge take centre stage in contemporary development debates. The study of contending production systems in their regional context reveals complex interactions of natural and social systems. While such interactions create path dependencies, complexity makes knock-on effects hard to predict. Farming systems favoured in the present conjuncture may not prevail under the influence of climate change, environmental reregulation, or new patterns of consumption.

To make sense of the developments in evidence, we reengage with 'agro-food complexes' as a way to conceptualize the agglomerations of actors and interests that align with particular production systems. We proceed to consider significant aspects of the NC and organic agro-food complexes, highlighting in the NC case the interaction of reduced tillage technologies, new herbicides and herbicide-resistant seeds, and land markets. Focusing on organic markets, the micro-dynamics of organic farm reproduction, and gender dynamics in large-scale crop farming, we consider how organic farming on the Prairies has emerged as an alternative, and what has kept organics from expanding more quickly in the context of burgeoning demand. To conceptualize the ways that agricultural risks are generated and addressed, we resurface 'ecologies of risk' as a shorthand for such multifaceted, social-ecological processes. In the concluding section, we revisit complexities associated with these competing agro-food complexes and the significance of their agronomic and epistemological differences.

Agro-food complexes

As agribusiness mergers proceed and governments reorient their spending, neo-liberal policies, partnerships, and worldviews become more entrenched. The promotion of NC agriculture is embedded in networks of formal and informal collaboration. At the same time, and partly in response to issues surrounding NC approaches, a cluster of actors has come together around alternative systems of production. These assemblages or constellations (Deleuze and Guattari, 1980) can be thought of as agro-food complexes. While similar terminology has been advanced by other researchers (see Constance and Heffernan, 1991; Winson, 1993; Murray, 1998), we propose to amend and extend the concept. For us, an agro-food complex connotes the actors, institutions, resources, and discourses, as well as the rules, norms, and values that are invoked and that coalesce around specific agricultural 'projects' or efforts to (re)orient the development of agro-food systems.[3] While not strictly local or regional, such agro-food complexes are fairly specific in terms of history and biography, territory and ecology, production system and politics – in terms of their materiality in the deeper sense of this concept.

Agro-food complexes include key actors from corporations, academe, and governments, and from organizations that defy easy categorization given that they draw on both state and corporate funding. These complexes include farmers who are early adopters and champions of new approaches. Individuals and organizations within such agro-food complexes act separately and jointly; competition is combined with co-operative and coordinated initiatives. This conception of agro-food complexes incorporates understandings from the 'worlds of production' literature, which sees such synchronization relying on shared 'definition of the situation' and common reading of the structural context (Morgan *et al.*, 2008; Storper and Salais, 1997).

The identification of specific agro-food complexes does not lead us to regard these as always, and in all ways, separate and distinct. Agro-food complexes evolve in response to other contending aggregations and larger forces such as globalization. Individuals, enterprises, and organizations can have connections to more than one complex. At different moments, a person may be more oriented towards the goals and evaluative frameworks of one or another complex (Thevenot, 2001). Sometimes this leads to hybrid forms that combine or mimic elements of erstwhile competing approaches and coalitions. Yet, while the concerns and practices of different agro-food complexes may overlap, their diagnoses of situations and their ideas about appropriate responses generally do not.

As suggested above, the farming philosophies of individuals do not necessarily align with only one approach (Rud, 2013). Some farmers farm organically due to health concerns or environmental convictions. Others may be more financially motivated and will switch to NC practices in the absence of earnings advantages with organic crops. Some producers are transitioning land to organic production and have tangible connections to both systems. Many NC farmers have concerns about sustainability and show interest in practices associated with organic farming such as extended crop rotations, multi-species cover crops, and intercropping. Formerly known as the Manitoba North Dakota Zero Tillage Farmers Association, The Northern Prairies Ag Innovation Alliance (www.mandakzerotill.org/) promotes such exploration and has received support from agrochemical companies that understand the need to take steps to address environmental issues related to pesticide and synthetic fertilizer use, and to keep farming profitable in the face of rising input prices.

The organic and NC agro-food complexes discussed here have some similarities in terms of actors and processes. Nevertheless, they are also distinctive in terms of development logics and potentials. The NC complex is dominant and its character has been heavily influenced by corporate sponsors – primarily agribusiness companies selling inputs – and by allies of these firms in universities, commodity groups, and purpose-built organizations. The organic complex has fewer corporate sponsors and these are located mainly downstream on the processing and marketing side. Though not strongly evident locally, the organic sector has significant consumer support and is linked to environmental movements challenging corporate-industrial development agendas. It has received less backing from governments and lacks the network of researchers and institutes that promote NC agriculture. While this is a challenge, it also leaves more space for the perspectives and priorities of farmers. While primary producers are important actors in both these constellations of power, the power relations and politics of knowledge are different; knowledge in the NC complex is much more the purview of consultants, communications specialists, and scientific experts supported by agribusiness.

In the context of intensifying (agri)cultural struggles, the terrain of engagement includes a contest for hearts and minds. With publics more and more concerned about health and environmental sustainability, it is important to seek legitimacy by occupying – and redefining – the moral high ground. Hence, the effort by NC advocates to associate with projects on sustainable farming and food security. Hence, also, the efforts to rally supporters and to build identities around NC approaches and knowledge networks. An example is a recent collaboration between the Canola Development Commission, the Government of Saskatchewan, the Government of Canada, and allied members of the NC complex to produce the film, *License to Farm* (licensetofarm.com/). They also co-operate to publish *The Real Dirt on Farming* (www.realdirtonfarming.ca/) and to support an advocacy organization known as 'Agriculture More Than Ever' (www.agriculturemorethanever.ca). At the University of Saskatchewan, they have promoted the development of supportive institutional infrastructure including the Global Institute for Food Security and the Centre for the Study of Science and Innovation Policy.

The introduction of reduced-tillage and direct seeding systems depended on the collaboration of multiple participants: Monsanto, which had patented the glyphosate herbicide, Roundup, in 1974, and which went on to develop herbicide-resistant crops; machinery companies eager to sell a new generation of equipment; university researchers who needed funding and corporate partners; governments seeking growth and ways to address income and resource management problems; soil conservation groups funded by governments and corporations; and farmers interested in new possibilities for profitable and sustainable production. Proponents of agricultural biotechnology allied (and overlapped) with the proponents of reduced tillage and direct seeding systems. These associated actors have more recently endorsed and promoted 'precision farming,' 'sustainable intensification,' and 'climate-smart agriculture.'

Organic farming has a long history but its more recent development is in part a response to perceived limitations of conventional and NC approaches (Lockeretz, 2007). On the Canadian Prairies, organic farming was originally championed by a small number of farmers, environmentalists, researchers, and marketers. Today, the organic agro-food complex continues to reflect these origins; its politics oscillate between seeking acceptance as an agronomically productive and economically viable sector, and functioning as a movement that provides critique and resistance from the margins. Like its NC counterparts, the organic sector mounts conferences and tradeshows where purveyors of inputs, processors and marketers, and specialized support organizations are sponsors and exhibitors; however, the scale of activity is much more modest.[4] Moreover, as mentioned above, with less corporate buy-in, farmers have had a more central role in the organic agro-food complex in terms of advocacy, agronomic innovation, and market development.

Farming in Saskatchewan

Canada's Prairies occupy the northern portion of the Great Plains of North America. They extend from the foothills of the Rocky Mountains in Alberta, across Saskatchewan, to the Red River Valley in Manitoba. Bordering on North Dakota and Montana, Saskatchewan is the province with the most farmland. Even with its semi-arid climate, harsh winters, and a frost-free growing season that averages only 100 days, Saskatchewan's fertile soils and land base make it an important grain exporting region. With 61.6 million acres of farmland, it accounts for 39 per cent of the agricultural land and 46.8 per cent of the total area in field crops in Canada (Statistics Canada, 2017a). Until recently, wheat was the main field crop but canola, almost all of it genetically modified and herbicide-resistant, accounted for 11.1 million acres in 2016, an increase of 78 per cent over 2006. The 6.7 million acres of spring wheat grown in 2016 represented a decrease of 64 per cent from 2006 (Statistics Canada, 2017a). Lentils were the third most important crop in terms of acreage (Statistics Canada, 2017a). A variety of other cereals, oilseeds, and pulses are grown, including barley, durum wheat, mustard, flax seed, canary seed, chickpeas, and field peas.

Saskatchewan farms tend to be large and specialized. In 2016, there were 34,523 farms, a 22 per cent decrease from 2006. Over the same decade, average farm size increased 19 per cent to 1,784 acres. While farms of 400–1,599 acres have declined in number, farms with 3,500 or more acres have become increasingly common.[5] Many farms operate on a mix of owned and rented land, with rented farmland accounting for 38.5 per cent of the total in 2016 (Statistics Canada, 2017a). Although independently owned and operated family farms dominate numerically, the number of incorporated farms has increased. In Saskatchewan, incorporated farms of all kinds accounted for 27.2 per cent of census farms in 2016 (Statistics Canada, 2017b). Total farm receipts were $13.8 billion (CAD) in 2016, and 22 per cent of farms had gross receipts of $500,000 or more. In 2016, Saskatchewan farmers spent $1.6B on fertilizers, $1.1B on pesticides, and $0.6B on seed (Statistics Canada, 2017a). Even with low interest rates, they paid close to $0.5B in interest charges (Statistics Canada, 2017a). Farming in Saskatchewan is an annual mega-project, with large amounts of money flowing to suppliers, traders, lenders, and some farmers.

With volatile commodity prices and rising input costs, trade and transportation challenges, interest and exchange rate fluctuations, droughts and flooding, frosts and hail, insects and fungi, Saskatchewan farmers operate in a high-risk, boom-and-bust economy. In the past, farmers responded to asymmetries in bargaining power by supporting collective arrangements such as co-operatives and marketing boards. However, the demutualization of the Saskatchewan Wheat Pool in 2007 (now Viterra, a subsidiary of Glencore) along with the loss of 'single-desk' selling capacity at the CWB in 2012 has led to increased marketing margins and transportation charges, and has accentuated differences related to market power and geographic location.[6]

Neo-conventional crop production

In the 1970s and 1980s, concern about soil erosion and salinity prompted research on cropping systems that leave residues on the surface. Replacing crop-fallow rotations with continuous cropping was made possible by equipment that minimized soil disturbance and allowed direct seeding into stubble. Collectively known as conservation tillage, reduced tillage practices range from minimum-tillage to no-till or zero-till (ZT).[7] Direct seeding systems open the soil, place seeds, and band fertilizer and herbicide in a single pass. Air seeders or hoe drills may have a working width of 100 feet and are typically paired with large-capacity seed and fertilizer carts. Seed and fertilizer placement is precise, evaporation is reduced because only a narrow strip of soil is disturbed, and earlier seeding is facilitated – all of which can translate into higher yields. The development of glyphosate, a broad-spectrum herbicide, was instrumental in the widespread adoption of no-till systems. The herbicide can be used to 'burn-off' weeds in the spring or fall, to kill crops and promote drying before harvest, and for weed control in standing crops such as herbicide-resistant canola. It is typically applied with high-clearance sprayers that have a boom span of 80–120 feet.

Canada's Prairie region has been a leader in the development and adoption of no-till/zero-till (ZT) systems. With firms such as Bourgault, Morris, Friggstad, Flexi-Coil, Conserva Pak, Seed Hawk, and SeedMaster, Saskatchewan has been an important centre for the design and production of air seeders and air drills. In 2016, 74.4 per cent of the cropped land in Saskatchewan (just under 28 million acres) was reportedly planted using ZT technologies (Statistics Canada, 2017a).[8] Large farms, a generally dry climate, and erosion-prone soils have helped to make ZT attractive (Awada, Gray and Nagy, 2016). Other factors that have influenced its adoption include promotion by agribusiness, governments, universities, and organizations such as commodity groups and the Saskatchewan Soil Conservation Association, along with significant private and public investment in the development of equipment, herbicides, and herbicide-resistant crops (Beckie, 2000). Low interest rates have supported the adoption of new equipment and spending on inputs (Awada *et al.*, 2016). Because it reduces labour requirements, allows for faster land preparation and seeding, and facilitates continuous cropping, the NC package including ZT has also been associated with expansion of farms and greater competition for cropland.

Although advantages are sometimes overstated, NC farming systems featuring ZT are frequently profitable in spite of higher input costs. They appeal to many producers because they mitigate some agronomic risks associated with short growing seasons, droughts, and weeds. With positive impacts on soil erosion and tractor fuel consumption, they also allow producers to claim improved environmental sustainability (see Kassam *et al.*, 2014). The organic sector thus faces competition from a technological package that has enjoyed significant economic, agronomic, environmental, political, and discursive success. This has been a challenging context in which to develop and promote alternative approaches.

The situation is dynamic and the picture continues to evolve. Despite decades denouncing tillage, NC practitioners have begun to add more cultivation in an effort to deal with emerging agronomic issues. Although combines are equipped with straw choppers, buildup of residues has led to interest in more aggressive tillage. This about-face also reflects problems with herbicide-resistant weeds, residue-borne pathogens, and soil compaction, as well as the need to dry out soils in wet years and to smooth ruts left by sprayers (Agriculture and Agri-Food Canada, 2014; Barker, 2015). Tillage that was declared unnecessary and unwise a few years ago is being reconsidered as part of updated NC cropping systems. Meanwhile, organic farmers and researchers have been developing no-till and reduced-tillage approaches applicable to organic cropping systems (Légère *et al.*, 2013).

Genetically modified canola and land markets

Herbicide-resistant varieties of GM canola were introduced in the mid-1990s at about the same time that ZT technologies were being refined to make direct seeding more effective. In Saskatchewan, canola acreage increased from 2.8 to 6.1 million acres between 1990 and 1995. There was little growth over the next decade but an upward trend was again evident starting in 2007 (Statistics Canada, 2017a). By 2016, canola was grown on 11.1 million acres in Saskatchewan (Statistics Canada, 2017a). Farmers were able to plant GM canola on land considered marginal or not well-suited for previous varieties of canola; some farmers reported 'pushing their rotations' to grow more canola in order to make more money (Weiss, 2008: 91). Along with the prevalence of ZT, the emergence of GM canola as a dominant crop has affected the development of organic crop production in several ways. Because of the widespread use of GM cultivars, contamination of non-GM with GM seed, and pollen drift, it has become difficult to produce canola that tests negative for GMOs. GM contamination leads to rejection of organic crops and, for this reason among others, very few organic farmers plant canola (Bouchie, 2002).[9] GM canola is not welcome in Europe, which has closed this market to most farmers in Saskatchewan – even those purporting to grow non-GM varieties.

Notwithstanding the loss of European markets, profitable canola crops are one reason that farmers in Saskatchewan justify paying more for land; higher land prices, in turn, promote specialization in higher value crops and use of more inputs. In the 1990s, after a decade of decline, a rise in farmland prices coincided with a jump in canola acreage (FCC, 2017b; Henry, 2017; Statistics Canada, 2017a). Corresponding to a renewed upward trend in canola production, average farmland prices more than doubled in real terms from 2007 to 2016 (FCC, 2017a; Henry, 2017). Commodity prices, the pursuit of economies of size, the need for more land to support a farming household, low interest rates, relaxation of restrictions on farmland purchases by non-residents and corporations, and land values that had lagged those in adjacent provinces all contributed to this climb, but increased planting of GM canola as part of a NC-ZT production package also played a role.

As land prices rise, the space for experimentation and the 'room to maneuver' (Ploeg, 2009) shrinks. Organic farmers compete for land with neighbours that use NC technologies and who can avail themselves of new financing arrangements – including bank- and merchant-sponsored zero-interest schemes for purchasing inputs, and contracts providing payment for future canola crops.[10] They compete as well with non-farmer buyers (Desmarais *et al.*, 2015). Holdings as large as 180,000 acres have been reported (Holtslander, 2015). For investors, the goal is to hold land until it accrues in value and can be profitably sold. Rental income is a dividend (Fairbairn, 2014). This is one reason that rental rates have not increased as fast as purchase prices. Renting land is frequently an important farm expansion strategy, but this may be less true for organic farmers. To achieve certification and to establish the necessary rotations, they need more security of tenure than is afforded by even a three- or five-year rental contract (see Carolan, 2005).

Dominating landscapes and mindsets

Notwithstanding its oft-repeated claims, the NC agro-food complex is not primarily motivated by sustainability or by 'feeding the world.' It is propelled most centrally by the profit concerns of input suppliers who take steps to develop and deregulate markets, to reduce competition, and to block legislation that limits their actions. In Western Canada alone, the markets for seeds, agrichemicals, machinery, and agronomic services are worth billions. These revenues, together with the tax treatments triggered by research and marketing expenditures, underwrite investments in lobbying, product development, and 'partnering' with universities, governments, commodity groups, and advocacy organizations (Barker, 2014; Howard, 2016).

Neo-conventional production systems have become dominant by pathways both obvious and less evident. The interactions of production systems, working landscapes, and worldviews can be subtle. Promoters of NC systems use many kinds of communication and sponsorship to convince producers that NC agriculture represents their best option. NC approaches also happen to be compatible with a culture that defines good farming in terms of high yields and tidy fields (Burton 2004; 2012; Carolan, 2006; Burton *et al.*, 2008). Farmer identities and reputations are bound up with styles of farming that feature weed-free and uniform crops, and impressive machinery. It is hard to uphold an alternative aesthetic that accommodates mixed stands, weeds, and refuge areas for wildlife and beneficial insects. Organic farms are frequently seen as a threat to phytosanitation even though it is NC producers who promote herbicide-resistance in weeds and are more often implicated in livestock disease and food safety crises.[11]

In addition to 'neighbourly' pressures to conform, farmers face the power of agribusiness firms, scientists, commodity groups, and governments that dominate discourses of good farming and sustainability (Youngberg and DeMuth, 2013). The developmental trajectory of the system, including their own farms, leads to an infrastructure of production and marketing that makes other options seem less viable. Given the economic and organizational context in which they operate – including the increased extractive power of grain merchants and transport companies in the wake of the dismantling of the CWB – farmer capacity to make real choices is constrained. Despite the exploitation that they face in a neoliberal landscape, many NC farmers hesitate to disconnect from a familiar system that provides various kinds of support and has powerful backers (Arnason, 2015a).

Organic crop production

In the 1960s and 1970s, a small number of Saskatchewan farmers began using extended crop rotations, intercropping, green manures, and other cultural practices as alternatives to agrichemicals. The needs of such farmers were at first mostly ignored by scientific institutions, so it was the farmers themselves, with the aid of a few researchers, who led the development of relevant knowledge. Agricultural extension agents had little training in organic methods and, in order to maintain the third-party status legally required for certifying bodies, organic certification organizations were not permitted to provide training in organic farming. Support for agricultural alternatives coalesced with the establishment of Earthcare, an organization that hosted conferences and produced the first book on organic agriculture for the Canadian Prairies. In 1992, the Saskatchewan Research Council published a study of agronomic and economic aspects of organic farming systems (Rutherford *et al.*, 1992). Since 2001, some research on organic cropping systems has been carried out at public institutions as part of the Organic Science Clusters coordinated by the Organic Agriculture Centre of Canada and funded by Agriculture and Agri-Food Canada and industry partners. Compared to NC systems, however, research on organic agriculture has been underfunded.[12] Recent organic research has included cover crops and intercropping but the main focus has been on issues more directly affecting yields (e.g. seeding rates and weed control). More holistic, multi-factorial studies that take into account the context-specific character of organic farming are rare. Research is typically carried out in plot trials at research stations even though many practitioners see a need for studies that are methodologically distinctive and include multidisciplinary, participatory, on-farm investigations.

Over the past several decades, Canada has seen an increase in the number of organic farms but, overall, Canada's organic farming sector is small compared to many European countries where subsidies cover some costs of transition and there is a longer history of funding for organic research. According to the 2016 Census, there were 4,339 organic farms in Canada, representing 2.2 per cent of all farms (Statistics Canada 2017a). In Saskatchewan, the organic sector has also exhibited a general upward trend, albeit with some setbacks. The 2016 Census of Agriculture counted 879 organic farms representing 2.5 per cent of farms in the province (Statistics Canada, 2017a). This can be compared to 1,064 organic farms representing 2.9 per cent of farms as of the 2011 census. With 0.5 million acres of organic field crops, or 60 per cent of organic field crop acreage in Canada (COTA, 2017; Guerra, 2017), Saskatchewan is an important producer of organic commodities such as wheat, oats, flax, kamut, field peas, barley, lentils, mustard, faba beans, and hemp.

Organic markets and marketing

Drawing on industry sources, the Canada Organic Trade Association (COTA) reported that the number of certified primary producers in Saskatchewan peaked at 1,123 in 2009, fell to 764 in 2013, and partially recovered to 824 by 2015 (Guerra, 2017). The decline was attributed to reduced prices for organic products during the global recession and strong prices for non-organic commodities as a result of grain-based ethanol production, increased demand from emerging markets, and speculation (Arnason, 2015a; Statistics Canada, 2017a). Price declines for organics reflected in part the marketing strategies of large retailers that stocked organic products but also house brands with labels like 'natural' that were attractive to consumers looking to cut spending. After the recession, prices for organic crops rebounded with premiums of as much as 300 per cent; this spurred some increased interest but farmers still seemed to be slow to adopt the organic option (Arnason, 2015b).

Though frequently significant, organic price premiums may not be adequate to compensate for extra management efforts, lower yields, certification costs, and lean years associated with the conversion period when no premium can be earned. Grading systems that emphasize weight and protein content, while heavily downgrading for weed seeds even though processors have effective methods for grain cleaning, are not friendly to organic producers. A related problem is the paucity of local processing and distribution facilities for some organic crop and livestock products (Bromm and McRae, 2009). Given their export orientation and the small size of the local market, organic field crop farmers in Saskatchewan generally lack connection to a critical mass of local consumers who would provide feedback and political support. Moreover, given the dominance of chain stores, it is hard to gain access to local eaters; when purchasing organic produce and grocery products, supermarket buyers tend to look to the consistent supply and wider range of products available from large, mostly US-based companies that dominate in organics (Howard, 2016).

Around the world, demand for organic products has outpaced growth in the rest of the food industry (Howard, 2016). The global sector was estimated to generate sales of $82 billion (USD) in 2015 (Willer and Lernoud, 2017) and $4.7 billion (CAD) in Canada (Guerra and Martin, 2017). Demand for organic grains, pulses, and oilseeds has been strong (Guerra and Martin, 2017). In 2015, to expand Prairie organic grain production and strengthen marketing channels, the Canadian Government and industry partners provided $3 million to establish the Prairie Organic Grain Initiative (POGI) (prairieorganicgrain.org). With online resources (www.pivotandgrow.com), field days, and workshops, POGI encourages Prairie farmers to try organic agriculture.

Even with this initiative, however, the organic sector in Saskatchewan remains largely self-organized and dependent on volunteers; there are only small numbers of paid positions in organic sector organizations, certifying groups, and research institutions. Organic farmers have had to rely heavily on themselves and on peer networks, but also on processors and traders. These intermediaries provide valuable services. However, in the absence of accessible market intelligence and countervailing producer power, they have significant opportunity to increase marketing margins. In the words of one trader interviewed for this study: 'The whole organic market is very cowboy; it's the Wild West.' While some co-operative processing and marketing enterprises have been successfully established, few have survived (Heit and Gertler, 2009).[13] They compete with larger firms that spread costs over greater volumes and draw on wider connections, and whose organic operations may be one part of an established and diversified trading enterprise.

The 'double conversion' and organic farm reproduction

Organic farm numbers reflect decisions to enter or exit farming, and decisions to convert existing operations. It is easy to stop being an organic farmer but harder to transition to organic production or to regain certification; the former takes minutes while the latter normally takes three years. Established farmers who want to produce organically are taking on a new farming system and a new identity. They must rethink their production system and realign with the sensibilities and logics of organic farming. This dual process constitutes a kind of 'double conversion' in which new agronomic techniques are adopted and new worldviews and personal constructs are elaborated (see Lamine and Bellon, 2009).

Given the lead times and uncertainties involved, and given the challenge of gaining the requisite knowledge, many farmers are hesitant to switch to organic (Padel, 2001; Khaledi *et al.*, 2007; Khaledi *et al.*, 2010; Milligan, 2015). As discussed above, restructuring pressures also have implications for organic farm numbers. Farm households farming non-organically frequently feel a need to expand in order to spread costs, make use of productivity-enhancing technologies, and maintain a satisfactory income. Because of the need to service debt, it may be difficult to contemplate any change that puts cash flow at risk. On the other hand, some farmers switch to organic production as a way to generate adequate income on a moderate-sized farm – and may even downsize as part of this strategy.

The social reproduction of organic farms can present challenges. For the farm to continue as organic, the orientation to organic farming must be successfully passed on along with the farm itself. While some younger people are attracted to organic farming, it is not easy to achieve intergenerational harmony. There may be additional difficulties when the production system consigns you to minority status in the farming community. If there is no organically-inclined successor, retiring organic farmers may not be replaced by someone committed to managing the land organically. New entrants may bring with them an interest in alternative farming approaches but, with rising land and machinery prices, barriers to entry are significant. It also matters that some lenders are sceptical of business plans based on organic production. While some new and established organic farmers are seeking land that is certified or certifiable as organic, retiring organic farmers are not always able to find a buyer who plans to farm organically. Most land that goes on the market ends up being purchased by an established farmer with a larger land base or by an investor with outside sources of capital. Neither is likely to be an organic enthusiast.

Sharing knowledge is important for people working in the organic sector, in part because relevant information is rarely found in standard production guides. Nor is it widely available from input suppliers and consultants who take on advisory roles in NC agriculture (Bell *et al.*, 2015). Given their small numbers, however, organic farmers may face challenges when it comes to finding like-minded practitioners who can mentor and collaborate. Not surprisingly, organic farms often arise in clusters where neighbours assist and support each other. Many organic farmers also profit from participation in regional knowledge networks, such as SaskOrganics and Young Agrarians, where they find people ready to discuss organic farming.

Women entering farming are often attracted by opportunities to pursue forms of agriculture that add value and values through strategies such as diversification into horticultural and livestock specialties, processing, direct marketing, and educational or social farming projects (Johnson and Schnakenberg, 2017). In many parts of Saskatchewan, however, distances, combined with low population densities, make it harder to develop a farm business based on relational marketing or community engagement. Moreover, mixed farming is a not an easy option given the absence of necessary infrastructure at the farm and regional levels. In many parts of North America, organic farming has allowed women to combine farming with environmental and social concerns (Sachs *et al.*, 2016). However, the extent of women's involvement in organic farming varies with context and subsector (Hall and Mogyorody, 2007). Field crop production on the Prairies remains a male-dominated arena characterized by big machines and, in the case of non-organic production, reliance on industrial inputs. Women face cultural, financial, and knowledge barriers with respect to participation in such enterprises (Keller, 2014). Notwithstanding the lead roles that some women have in organic organizations, research, and farming, this too has likely slowed the growth of the organic sector.

Ecologies of risk and duelling discourses of sustainability

Interacting economic and environmental risks emanate from and are experienced in specific eco-social contexts. Ecosystems, technologies, institutional arrangements, and cultures all contribute to the constitution of specific ecologies of risk (EOR), which reflect and generate inequalities in terms of exposures and impacts. While local in key respects, EOR are open systems that interact with broader fields of politics, power, and development. Agricultural systems produce, reproduce, and manage EOR. Despite the salience of sustainability in contemporary discourses, dominant production and consumption arrangements and associated EOR focus attention on short-range fixes and a narrow set of evaluative criteria in which economic viability trumps other concerns (Jaffe and Gertler, 2008; Beckie, 2000).

The risks associated with particular production systems manifest unevenly at the individual, enterprise, and regional levels. Capitalist market-based economic systems generally reward 'risk-takers' who ignore or shift risks. Enterprises are more willing to engage in risky practices if institutional arrangements make it possible to avoid the consequences. Allowing producers to externalize ecological costs without triggering a regulatory response is one way that societies underwrite NC agriculture (see Hedley *et al.*, 2017). The Saskatchewan government also provides subsidies in the form of sales tax relief on fuels, machinery, seeds, pesticides, and fertilizers (Warick, 2017). Since pesticides and fertilizers are little used in organic systems, this disproportionately supports NC agriculture. Likewise, federal income stabilization programmes designed to offset the costs and risks of investing in agrichemicals asymmetrically support high-input agriculture.

Organic and neo-conventional farmers operate in overlapping EOR but alignment with one or the other of these systems affects both the composition of risks and perceptions about what is risky. Neo-conventional systems appear powerful in terms of managing certain agronomic risks but they generate novel environmental risks that are reallocated in new ways. In Saskatchewan, adoption of NC technologies has been associated with intensified agrichemical use, expansion of monocultures, and clearing and cultivation of wooded land, wetlands, and marginal lands (Gertler, 1999). Yet, many of the ecological effects of NC cropping programmes are not readily observed (Carolan, 2006). Production is sustained by technologies that mask the degradation of natural resources and ecosystems (Buttel, 2006). There is also an inevitable tendency to 'push the system,' for example by taking advantage of improved technologies to spray under windier conditions or to crop erodible land.[14] This is done even in the face of evidence of significant ecological impacts including the widespread effects of pesticides used in NC cropping systems (Pimentel, 2005; Fernandez *et al.*, 2009; Benbrook, 2012; Beckie *et al.*, 2013; Brenzil, 2016; Druille *et al.*, 2016; Loomis *et al.*, 2015; Myers *et al.*, 2016; Eng *et al.*, 2017).

In Saskatchewan, agro-environmental discourses focus primarily on issues purportedly addressed by NC farming systems – e.g. soil erosion, salinization, carbon sequestration, and energy efficiency (Awada *et al.*, 2014). As more issues are revealed, such as air and water pollution, resistant weeds, and declining populations of pollinators, amphibians, and songbirds, the NC agro-food complex moves to mitigate some impacts and to deflect critiques. Resistant weeds, for example, are addressed with schemes for rotating and layering herbicides (Beckie *et al.*, 2013), and with initiatives to produce GM cultivars that are resistant to additional herbicides (Barker, 2014). Efforts are also made to increase the efficiency of fertilizer use and to reduce greenhouse gas emissions by modifying application practices and employing additives that decrease nitrogen loss. Other additives are designed to reduce spray drift and to increase pesticide adherence.

Critics of organic farming can point to risks as well. There are debates about its productivity and what this might imply for land use in the context of growing demand. Despite new research revealing the regenerative powers of living soils – their ability to furnish phosphorus as well as nitrogen, for example – the long-term fertility implications of organic field crop production also remains a topic of debate. Grassland conservationists see organic farming as a threat to long-established pastures and to native prairie given that such lands are eligible for expedited certification as organic. Others allege that reliance on tillage leads to erosion – a point not lost on organic farmers who seek to reduce tillage and add protective cultural practices such as diversified rotations, windbreaks, and forage cover on erosion-prone land. Organic producers also point to green manure crops, tolerance for weeds, and avoidance of burn-off herbicides as soil conserving and building practices.

When organic farms are scrutinized in relation to sustainability, they are typically measured against NC farming systems that have been selectively assessed with respect to what they 'bring to the table.' Yields and controlling soil erosion are emphasized, while potentially negative impacts are ignored because they are more difficult to measure, valorize, or address. The environmental and health advantages of organic agriculture, especially the long-term, diffuse, downstream benefits, are not fully recognized anywhere in the system.[15] Most are not captured by current organic producers or consumers (MacRae *et al.*, 2007); they benefit people and other living things far-removed in time and space.

Though speciality crops and off-farm work are common, Saskatchewan does not seem to encourage diversification of the kind that involves complementary combinations of crops and livestock (Gertler *et al.*, 2002). While there are genuine mixed farms, these are getting rarer. This is not good news for the development of sustainable farming, organic or otherwise. While there are significant differences in terms of risks and trade-offs, it should be acknowledged that neither NC nor organic agriculture has achieved a fool-proof formula for sustainable farming on the Prairies (see Thiessen Martens *et al.*, 2015). Both systems are shaped by neoliberal markets and, while the organic option seems to take such goals seriously and to have more potential for delivering on them (Timmermann and Félix, 2015), neither approach automatically achieves ecological rationality, environmental justice, or distributional equity.

Complex complexities

In rural Saskatchewan, there are marked political and (agri)cultural divides. The sell-off of the CWB, closing of the state-run bus system, privatization of public pastures, adjudication of treaty rights, gun laws, environmental regulation, and organic farming all give rise to polarized debates. Though collective approaches such as production co-ops, marketing co-ops, and community pastures offer advantages, these are deemed impractical by much of the population (Gertler, 2007). Even when organic farming demonstrates viability, many farmers view this as an irreproducible outcome of special circumstances, or they focus on broad allegations that organic production is unsustainable. The mainstream agricultural media provide positive coverage of NC agriculture, which is celebrated as productive and sustainable. When NC crops fail, farms go out of business, or animal or human health problems develop, this is attributed to inclement weather, competitive forces, poor management, or bad luck.

As these two approaches duel for legitimacy and support, deeper reconceptualization of farming to advance sustainability is at once encouraged and hindered. In both the NC and organic camps, additional attention is being paid to complex interactions between soils, crops, and farming practices. For the most part, this has been a long-standing aspect of organic farming and organic research. Due to time, energy, or financial constraints, there are farmers associated with both these orientations who feel that they are not farming as well as they could. The limitations are structural as well as individual. In the neoliberal, productivist climate in which they operate, substantive support for holistic thinking and practices is scarce. On the other hand, this is also a context that makes peer-to-peer mentoring and sharing more important. While some knowledge co-generation occurs in both arenas, it is crucial to organics and more characteristic of this relatively decentralized system of innovation and adaptation.

While practices may be borrowed back-and-forth, the NC and organic agro-food complexes cannot meld. The organic difference is predicated on avoiding central practices of the NC approach. Paradoxically, it is the continued expansion of the NC system and the ongoing industrialization of agriculture that helps to give organics its strong consumer franchize and price premiums. Moreover, the challenge presented by ecological critiques and the success of some organic farmers is a threat that the NC complex cannot ignore. It seems likely that these two complexes will continue to contest each other's productivity and sustainability claims – including the knowledge processes that purport to supply relevant scientific truths. At their core, these are different communities of practice, worlds of production, eco-political cultures – and agro-food complexes. While this does not prevent farmers looking over the fence to learn from and experiment with techniques pioneered by the erstwhile competition, it does prevent easy fraternization and the consummation of any more intimate alliance. Innovative agroecological techniques promising environmental and agronomic benefits represent both possible advantages and potential challenges to both the organic and NC sectors. As long as their technologies and definitions of productivity remain central, agribusiness firms are willing to ally with initiatives to adopt agroecological innovations (see Lang and Heasman, 2004; Wezel *et al.*, 2015). Adding some elements of advanced ecological thinking allows them to assert environmental responsibility and to advance positive-sounding constructs and programmes such as 'sustainable intensification,' and 'climate-smart' and 'conservation agriculture' (Garnett *et al.*, 2013; Climate Smart Agriculture Concerns, 2015).

There is social, cultural, economic, and agronomic diversity in both the NC and organic agro-food complexes. Notwithstanding such heterogeneity and the crossing and blurring of boundaries, the conflict between these epistemologies and ontologies is real and consequential. At stake are regional ecologies, scientific practices, development trajectories, markets, and the moral high ground in terms of food security and sustainability. While the NC complex has deep pockets and more organizational resources, the organic complex draws on broad consumer and NGO support. In terms of production logics, landscapes, and rural economies, the impacts of NC approaches have been significant. The potentials of an organic turn are also evident and its more ecological orientation has already spurred scientific renewal, a new cognitive praxis, and fresh thinking about socially sustainable rural development.

Studying these duelling discourses of agricultural sustainability reveals important things about agriculture and the sociology of agriculture. Agricultural futures are always in flux. They depend on the perspectives of farming households and consumers, agribusiness firms and food purveyors, sector organizations and environmental groups, governments and researchers, and also on unpredictable aspects of globalization and ecological change. Organic farmers have already demonstrated that 'resistance is fertile.' It is up to all of us to try to understand what lies beneath surface manifestations of growth, stagnation, and innovation, and to provide support for deeper thinking about development options and ecologically rational futures.

Notes

1 'Conventional' usually connotes farming systems that depend on a wide array of industrial inputs whereas 'conventionalization' is used to refer to ways that organic farms may come to resemble their non-organic counterparts in terms of enterprise structure and management (Constance *et al.*, 2015). The term 'neo-conventional' has elsewhere been used to describe organic farming that substitutes alternative inputs for products prohibited under organic standards but which disregards certain core principles of agroecology (Lampkin, 2015). This should not be confused with our use of 'neo-conventional' to describe recent tendencies in non-organic crop production.

2 We are indebted to the farmers, marketers, researchers, and others who shared their knowledge and insights. We are grateful to Cathy Holtslander, organic farmer and agricultural policy researcher, who provided a critical review of our penultimate draft. Like the others from whom we learned, she is not responsible for what we have chosen to present.

3 This conceptualization has some affinity with the idea of advocacy coalitions proposed by Sabatier (1988) and used by Whitfield *et al.* (2015) in relation to organizations collaborating to promote conservation agriculture in Africa. Though broader, it also overlaps with the idea of actor-networks formed around specific scientific/societal projects (Busch and Juska, 1997; Gray and Gibson, 2013). This use of agro-food complexes also aligns with Bourdieu's notion of society as a collection of social fields or structured power relationships that simultaneously exist as objective, hierarchical positions and as spaces of symbolic struggle for advantage and legitimation. Drawing on this idea, Roger (2016) shows how the coincidence of economic, political, bureaucratic, and scientific fields works to construct large-scale agriculture as most worthy of state support even where the majority of producers are small-scale family farmers.

4 Compare the Organic Connections Conference and Tradeshow (www.organicconnections.ca) with events such as Canada's Farm Progress Show (www.myfarmshow.com), the Western Canadian Crop Production Show (www.cropproductiononline.com), Canadian Western Agribition (www.agribition.com), Ag in Motion (aginmotion.ca), and FutureFarm Canada Expo (futurefarmexpo.com).

5 Declines in the numbers of mid-size farms were noted throughout North America beginning in the 1980s (Buttel and LaRamee, 1991). Such farms suffer from a 'market access gap' – they are too large for direct marketing and too small to benefit from market advantages of size in the context of increasingly concentrated agribusiness (Stevenson *et al.*, 2014). Organic field crop farms in the Prairie Provinces are typically part of this 'agriculture of the middle.'

6 It is important to acknowledge the complex relationship that organic farmers had with the CWB prior to its dismantling. Some were attracted to organics in part because it allowed them to side-step the delivery quota system that provided equitable access to international markets for non-organic cereals, and many resented the paper transactions and red tape involved in navigating the requirements of the CWB with respect to selling organic grain into export channels.

7 Even though they depend on agrichemicals with problematic environmental impacts, the arrogation of 'conservation tillage' as a moniker associates NC cropping systems with the long history of soil and water conservation innovations such as strip cropping, windbreaks, contour farming, and grassed waterways. Furthermore, conservation tillage is frequently linked to 'conservation agriculture' (Hobbs, 2007; Kassam *et al.*, 2014; Freitas and Lander, 2014; Reicosky, 2015), another (mostly) NC approach with a label that is both evocative and provocative (Giller *et al.*, 2009; Whitfield *et al.*, 2015).

8 The nomenclature can be confusing but we group together no-till and zero-till practices in part because the 2016 Census of Agriculture asked farmers to report land seeded or to be seeded using 'No-till seeding or zero-till seeding (Include direct seeding into undisturbed stubble or sod).' This was to be distinguished from 'Tillage that retains most of the crop residue on the surface' (Include minimum tillage) and 'Tillage that incorporates most of the crop residue into the soil.' It seems likely that some operators would have difficulty deciding if their practices qualified as no-till/zero-till or as minimum tillage.

9 Organic farmers are concerned about the registration and recent introduction of GM alfalfa varieties with herbicide resistance, and they were among those who successfully fought to prevent the introduction of GM wheat. Such introductions threaten markets in importing countries that do not accept GM crops. They could also negatively affect the ability of organic farmers to meet certification requirements for a range of crops (Eaton, 2009).

10 New kinds of investment arrangements are emerging that represent new frontiers in the financialization of agriculture (National Farmers Union, 2014). Input Capital Corporation, an offshoot of the Assiniboia Capital Corporation, a farmland investment company, focuses on buying canola from farmers using multi-year contracts that

provide growers with 'cash upfront for crops yet to be planted' (inputcapital.com). This cash advance could be used to pay down other forms of debt, purchase inputs, or finance land acquisitions.

11 In Canada, organic farmers are the ones who must maintain buffer strips around their fields to try to guard against spray drift from adjoining farms, and this is taken for granted.

12 The 2012 federal budget required government funding to be matched. Corporate proponents of NC-ZT gain economical access to public research institutions using research dollars for which they also get a tax credit. Alternative research agendas are more difficult to fund given that there is less likely to be a wealthy corporate sponsor. This helps to skew the research agenda away from organics.

13 A notable exception is Farmer Direct Co-operative, which brings together several dozen organic field crop farmers and a marketing team that has successfully parlayed the combination of organic production, farmer ownership, and 'Domestic Fair Trade' certification to access U.S. markets such as the upmarket retailer, Whole Foods.

14 On Canada's Atlantic coast, the workings of a similar 'technological treadmill' spurred fishers to adopt techniques that were both productive and destructive, and which concealed the reality of deteriorating fish stocks. Combined with inadequate marine science and flawed policies, this culminated in a multi-decade shutdown of the cod fishery (Neis and Kean, 2003).

15 One should consider the full production system, i.e. factories as well as farms, and second- and third-order effects. The ecological advantages of organic farming include, for example, foregoing the production, transport, and application of agritoxins and fertilizers that have environmental risks, some of which are not fully understood (cf Ascott *et al.*, 2017; Eng *et al.*, 2017).

References

Agriculture and Agri-Food Canada (2014). 'Flexibility of no till and reduced till systems ensures success in the long term,' www.agr.gc.ca/eng/science-and-innovation/agricultural-practices/soil-and-land/soil-management/flexibility-of-no-till-and-reduced-till-systems-ensures-success-in-the-long-term/?id=1219778199286#e1 (Accessed 10 August 2017).

Arnason, R. (2015a). 'Consumers want organic, so why are farmers wary?' *The Western Producer*, www.producer.com/2015/01/consumers-want-organic-so-why-are-farmers-wary/ (Accessed 7 July 2017).

Arnason, R. (2015b). 'Organic group aims to entice growers with price premium info,' *The Western Producer*, www.producer.com/2015/05/organic-group-aims-to-entice-growers-with-price-premium-info/ (Accessed 11 August 2017).

Ascott, M. J., Gooddy, D. S., Wang, L., Stuart, M. E., Lewis, M. A., Ward, R. S. and Binley, A. M. (2017) 'Global patterns of nitrate storage in the vadose zone,' *Nature Communications* 8: 1416 doi:10.1038/s41467–017–01321-w. www.nature.com/articles/s41467-017-01321-w (Accessed 13 November 2017).

Awada, L., Gray, R. and Nagy, C. (2016). 'The benefits and costs of zero tillage RD & E on the Canadian Prairies,' *Canadian Journal of Agricultural Economics*, vol. 64: 417–438.

Awada, L., Lindwall, C. W. and Sonntag, B. (2014). 'The development and adoption of conservation tillage systems on the Canadian Prairies,' *International Soil and Water Conservation Research*, vol. 2, no. 1: 47–65.

Barker, B. (2015). 'Conservation tillage through the decades,' *Top Crop Manager*, www.topcropmanager.com/tillage/conservation-tillage-through-the-decades-17296 (Accessed 11 August 2017).

Barker, D. (2014). 'Debunking popular myths about GE crops portrayed in the media,' Center for Food Safety, www.centerforfoodsafety.org/fact-sheets/3469/debunking-popular-myths-about-ge-crops-portrayed-in-the-media (Accessed 9 July 2017).

Beckie, H. J., Lozinski, C., Shirriff, S. and Brenzil, C. A. (2013). 'Herbicide-resistant weeds in the Canadian Prairies: 2007 to 2011,' *Weed Technology*, vol. 27, no. 1: 171–183.

Beckie, M. A. (2000). 'Zero tillage and organic farming in Saskatchewan: An interdisciplinary study of the development of sustainable agriculture,' PhD thesis, University of Saskatchewan at Saskatoon, Canada.

Bell, S. E., Hullinger, A. and Brislen, L. (2015). 'Manipulated masculinities: Agribusiness, deskilling, and the rise of the businessman-farmer in the United States,' *Rural Sociology*, vol. 80, no. 3: 285–313.

Benbrook, C. M. (2012). 'Impacts of genetically engineered crops on pesticide use in the U.S – the first sixteen years,' *Environmental Sciences Europe*, vol. 24: 24, enveurope.springeropen.com/articles/10.1186/2190–4715–24–24, (Accessed 9 July 2017).

Bouchie, A. (2002). 'Organic Farmers Sue GMO Producers,' *Nature Biotechnology*, vol. 20, no. 3: 210.

Brenzil, C. (2016). 'Herbicide resistance,' Presentation to Saskatchewan Wheat Commission, www.google.ca/search?q=Brenzil+Herbicide+resistance+2016&ie=utf-8&oe=utf-8&gws_rd=cr&ei=_52KWYWxG-rVjwSji6j4Bg (Accessed 9 July 2017).

Bromm, J. and McRae, C. (2009). Feasibility Study 2009: Assessing Saskatoon's Local Organic Production and Potential Institutional Market, Saskatchewan Organic Directorate and CHEP Good Food Inc. http://foodmiles.saskorganic.com/sites/foodmiles.saskorganic.com/resources/SOD_FMC_study.pdf (Accessed 13 November 2017).

Burton, R. J. F. (2004). Seeing through the 'good farmer's' eyes: towards developing an understanding of the social symbolic value of 'productivist' behavior, *Sociologia Ruralis*, vol. 44, no. 2: 95–216.

Burton, R. J. F. (2012). 'Understanding farmers' aesthetic preference for tidy agricultural landscapes: A Bourdieusian perspective,' *Landscape Research*, vol. 37, no. 1: 51–71.

Burton, R. J. F., Kuczera, C. and Schwarz, G. (2008). 'Exploring farmers' cultural resistance to voluntary agri-environmental schemes,' *Sociologia Ruralis*, vol. 48, no. 1: 16–37.

Busch, L. and Juska, A. (1997). 'Beyond political economy: actor networks and the globalization of agriculture,' *Review of International Political Economy*, vol. 4, no. 4: 688–708.

Buttel, F. H. (2006). 'Sustaining the unsustainable: Agro-food systems and environment in the modern world,' in P. Cloke, T. Marsden, and P. Mooney (eds) *Handbook of Rural Studies*, Sage, Thousand Oaks, USA: 213–229.

Buttel, F. H. and LaRamee, P. (1991). 'The disappearing middle: A sociological perspective,' in W. Friedland, L. Busch, F. Buttel, and A. Rudy (eds), *Towards a New Political Economy of Agriculture*, Westview, Boulder, USA: 151–169.

Canada Organic Trade Association (COTA) (2017). 'Organic agriculture by the numbers,' ota.com/sites/default/files/Organic%20Ag%20in%20Canada%20v2.pdf (Accessed 20 May 2017).

Carolan, M. (2005). 'Barriers to the adoption of sustainable agriculture on rented land: An examination of contesting social fields,' *Rural Sociology*, vol. 70, no. 3: 387–413.

Carolan, M. (2006). 'Do you see what I see? Examining the epistemic barriers to sustainable agriculture,' *Rural Sociology*, vol. 71, no. 2: 232–260.

Climate Smart Agriculture Concerns (2015). 'Don't be fooled! Civil society says NO to "Climate Smart Agriculture" and urges decision-makers to support agroecology,' www.climatesmartagconcerns.info/cop21-statement.html (Accessed 8 August 2016).

Constance, D. and Heffernan, W. (1991). 'The global poultry agro/food complex,' *International Journal of Sociology of Agriculture and Food*, vol. 1: 126–141.

Constance, D. H., Choi, J. Y. and Lara, D. (2015). 'Engaging the organic conventionalization debate,' in B. Freyer and J. Bingen (eds) *Re-Thinking Organic Food and Farming in a Changing World, The International Library of Environmental, Agricultural and Food Ethics*, vol. 22, Springer, Dordrecht: 161–185.

Daschuk, J. (2013). *Clearing the Plains: Disease, Politics of Starvation, and the Loss of Aboriginal Life*, University of Regina Press, Regina, Canada.

Deleuze, G. and Guattari, F. (1980). *A Thousand Plateaus: Capitalism and Schizophrenia*, Continuum, London.

Desmarais, A. A., Qualman, D., Magnan, A. and Wiebe, N. (2015). 'Land grabbing and land concentration: mapping changing patterns of farmland ownership in three rural municipalities in Saskatchewan, Canada,' *Canadian Food Studies*, vol. 2, no. 1: 16–47.

Diaz, H. P. and Stirling, R. (2003). 'Degradation of Work in the Canadian Prairies,' in H. P. Diaz, J. Jaffe and R. Stirling (eds) *Farm Communities at the Crossroads*, CPRC, Regina, Canada: 31–44.

Druille, M., García-Parisi, P. A., Golluscio, R. A., Cavagnaro, F. P. and Omacini, M. (2016). 'Repeated annual glyphosate applications may impair beneficial soil microorganisms in temperate grassland,' *Agriculture, Ecosystems & Environment*, vol. 230: 184–190.

Eaton, E. (2009). 'Getting behind the grain: the politics of genetic modification on the Canadian Prairies,' *Antipode*, vol. 41, no. 2: 256–281.

Eng, M. L., Stutchbury, B. J. M. and Morrissey, C. A. (2017). 'Imidacloprid and chlorpyrifos insecticides impair migratory ability in a seed-eating songbird,' *Scientific Reports* 7: 15176. DOI:10.1038/s41598–017–15446-x. www.nature.com/articles/s41598-017-15446-x (Accessed 13 November 2017).

Fairbairn, M. (2014). '"Like Gold with Yield": Evolving Intersections Between Farmland and Finance.' *The Journal of Peasant Studies*, vol. 41, no. 5: 777–795.

Farm Credit Canada (FCC) (2017a). 2016 Farmland Values Report, www.fcc-fac.ca/en/about-fcc/governance/reports/farmland-values-report-2016.pdf (Accessed 22 July 2017).

Farm Credit Canada (FCC) (2017b). Historic Farmland Values, 1985–2016, www.fcc-fac.ca/fcc/about-fcc/corporate-profile/reports/farmland-values/farmland-values-historic-2016.pdf (Accessed 22 July 2017).

Fernandez, M. R., Zentner, R. P., Gehl, D. T., Selles, F., Huber, D. and Basnyat, P. (2009). 'Glyphosate associations with cereal diseases caused by Fusarium spp. In the Canadian Prairies,' *European Journal of Agronomy*, vol. 31, no. 3: 133–143.

Freitas, P. L. de and Landers, J. N. (2014). 'The transformation of agriculture in Brazil through development & adoption of zero tillage conservation agriculture,' *International Soil and Water Conservation Research*, vol. 2, no. 1: 35–46.

Frick, B. (ed.) (2013). *Organic Farming on the Prairies*, 2nd edn, Saskatchewan Organic Directorate, Regina, Canada.

Garnett, T., Appleby, M. C., Balmford, A., Bateman, I. J., Benton, T. G., Bloomer, P., Burlingame, B., Dawkins, M., Dolan, L., Fraser, D., Herrero, M., Hoffmann, I., Smith, P., Thornton, P. K., Toulmin, C., Vermeulen, S. J. and Godfray, H. C. J. (2013). 'Sustainable intensification in agriculture: premises and policies,' *Science*, vol. 341, no. 6141: 33–34.

Gertler, M. (1999). 'Sustainable communities and sustainable agriculture on the Prairies,' in J. Pierce and A. Dale (eds) *Communities, Development, and Sustainability across Canada*, University of British Columbia Press, Vancouver.

Gertler, M. (2007). 'Production co-operatives and co-operative production: legacy and promise,' in M. Knuttila and R. Stirling (eds) *The Prairie Agrarian Movement Revisited*, Canadian Plains Research Centre, University of Regina Press, Regina, Canada.

Gertler, M., Jaffe, J. and Swystun, L. (2002). *Beyond Beef and Barley: Organizational Innovation and Social Factors in Farm Diversification and Sustainability*, Research Report 02.01. Centre for the Study of Co-operatives, University of Saskatchewan, Saskatoon, Canada.

Giller, K. E., Witter, E., Corbeels, M. and Tittonell, P. (2009). 'Conservation agriculture and smallholder farming in Africa: the heretics' view,' *Field Crops Research*, vol. 114, no. 1: 23–34.

Gray, B. J. and Gibson, J. W. (2013). 'Actor–networks, farmer decisions, and identity,' *Culture, Agriculture, Food and Environment*, vol. 35, no. 2: 82–101.

Guerra, J. (2017). *Organic Agriculture in the Prairies: 2015 Statistics*, Canada Organic Trade Association, Ottawa, Canada, www.ota.com/canada-ota (Accessed 2 August 2017).

Guerra, J. and Martin, L. (2017). *The State of Organics*, Canada Organic Trade Association, Ottawa, Canada, www.ota.com/sites/default/files/StateofOrganics_2017-FINAL 2.pdf (Accessed 20 July 2017).

Hall, A. and Mogyorody, V. (2007). 'Organic farming, gender, and the labour process.' *Rural Sociology*, vol. 72, no. 2: 289–316.

Hedley, D., Karunagoda, K., Dyack, B. and Mussell, A. (2017). *Understanding Agricultural Support: Summary Report*, Canadian Agri-Food Policy Institute, Ottawa, Canada www.capi-icpa.ca/news/2017/subsidies-paper_170731.html (Accessed 1 August 2017).

Heit, J. and Gertler, M. (2009). *Co-operative Marketing Options for Organic Agriculture*, Centre for the Study of Co-operatives, Saskatoon, Canada.

Henry, L. (2017). 90 Years of Saskatchewan Farmland Prices. *Grainews*, www.grainews.ca/2017/02/03/90-years-of-saskatchewan-farmland-prices-2/ (Accessed 3 July 2017).

Hobbs, P. R. (2007). 'Conservation agriculture: what is it and why is it important for future sustainable food production?' *Journal of Agricultural Science*, vol. 145: 127–137.

Holm, F. A. and Johnson, E. N. (2009). 'The history of herbicide use for weed management on the prairies,' *Prairie Soils and Crops*, vol. 2: 1–10, www.prairiesoilsandcrops.ca (Accessed 9 July 2017).

Holtslander, C. (2015). 'Losing Our Grip, 2015 Update: How Corporate Farmland Buy-Up, Rising Farm Debt, and Agribusiness Financing of Inputs Threaten Family Farms and Food Sovereignty,' National Farmers Union, Saskatoon, Canada, www.nfu.ca/issues/losing-our-grip-2015-update (Accessed 30 May 2016).

Howard, P. H. (2016). *Concentration and Power in the Food System: Who Controls What We Eat?* Bloomsbury Academic, New York.

Jaffe, J. and Gertler, M. (2008). 'Sustainable development in the new economy: risk, vulnerability and eco-social justice.' *Currents: New Scholarship in the Human Services*, vol. 7, no. 2: 18.

Jaffe, J. and Quark. A. A. (2006). 'Social cohesion, neoliberalism, and the entrepreneurial community in rural Saskatchewan,' *The American Behavioral Scientist*, vol. 50, no. 2: 206–225.

Johnson, L. B. and Schnakenberg, G. (2017). 'Gendering strategies for civic agriculture: the case of Blue Ridge Women in Agriculture and the High Country Farm Tour,' *Journal of Rural Studies* vol. 55: 181–192.

Kassam, A., Derpsch, R. and Friedrich, T. (2014). 'Global achievements in soil and water conservation: The case of conservation agriculture,' *International Soil and Water Conservation Research*, vol. 2, no. 1: 5–13.

Keller, J. C. (2014). '"I wanna have my own damn dairy farm!": Women farmers, legibility, and femininities in rural Wisconsin, U.S.' *Journal of Rural Social Sciences*, vol. 29, no. 1: 75–102.

Khaledi, M. with Gray, R., Weseen, S. and Sawyer, E. (2007). 'Assessing the Barriers to Conversion to Organic Farming: An Institutional Analysis,' Report submitted to Advancing Canadian Agriculture and Agri-Food Saskatchewan (ACAAFS), Department of Agricultural Economics, University of Saskatchewan, Saskatoon, Canada.

Khaledi, M., Weseen, S., Sawyer, E., Ferguson, S. and Gray, R. (2010). 'Factors influencing partial and complete adoption of organic farming practices in Saskatchewan, Canada,' *Canadian Journal of Agricultural Economics/Revue canadienne d'agroeconomie*, vol. 58: 37–56.

Lamine, C. (2015). 'Sustainability and resilience in agrifood systems: Reconnecting agriculture, food and the environment,' *Sociologia Ruralis*, vol. 55, no. 1: 41–61.

Lamine, C. and Bellon, S. (2009). 'Conversion to organic farming: a multidimensional research object at the crossroads of agricultural and social sciences. A review,' *Agronomy for Sustainable Development*, vol. 29, no. 1: 97–112.

Lampkin, N. (2015). 'Organic food systems – an agro-ecological perspective,' in A. Meybeck, S. Redfern, F. Paoletti, and C. Strassner (eds) *Proceedings of an International Workshop: Assessing sustainable diets within the sustainability of food systems*, FAO, Rome: 49–53 www.fao.org/3/a-i4806e.pdf (Accessed 2 July 2017).

Lang, T. and Heasman, M. (2004). *Food Wars: The Global Battle for Mouths, Minds, and Markets*, Earthscan, London.

Légère, A., Shirtliffe, S. J., Vanasse, A. and Gulden, R. H. (2013) 'Extreme Grain-Based Cropping Systems: When Herbicide-Free Weed Management Meets Conservation Tillage in Northern Climates,' *Weed Technology*, vol. 27, no. 1: 204–211.

Lindwall, W. (2009). 'Engineering Breakthroughs in Conservation Tillage/Direct Seeding Technology,' *Proceedings of Landscapes Transformed: The Quiet Triumph of Conservation Tillage and Direct Seeding*. www.kis.usask.ca/CTConf/proceedings/Lindwall.pdf (Accessed 8 July 2017).

Lockeretz, W. (2007). 'What explains the rise of organic farming?' in W. Lockeretz (ed.) *Organic Farming: An International History*, CAB International, Cambridge, USA.

Loomis, D., Guyton K., Grosse, Y., El Ghissasi, F., Bouvard, V., Benbrahim-Tallaa, L., Guha, N., Mattock, H. and Straif, K. (2015). 'Carcinogenicity of lindane, DDT, and 2,4-dichlorophenoxyacetic acid,' *The Lancet Oncology*, vol. 16, no. 8: 891–892.

MacRae, R. J., Frick, B. and Martin, R. C. (2007). 'Economic and social impacts of organic production systems,' *Canadian Journal of Plant Science*, vol. 87: 1037–1044.

McManus, C. R. (2011). *Happyland: A History of the "Dirty Thirties" in Saskatchewan, 1914–1937*, University of Calgary, Calgary, Canada.

Milligan, P. (2015). 'Transitioning to organic farming,' *Grainews*, www.grainews.ca/2015/03/30/transitioning-to-organic-farming-3/ (Accessed 11 August 2017).

Morgan, K., Marsden, T. and Murdoch, J. (2008). *Worlds of Food: Place, Power, and Provenance in the Food Chain*, Oxford University Press, Oxford.

Müller, B. (2008). 'Still feeding the World? The political ecology of Canadian Prairie farmers,' *Anthropologica*, vol. 50, no. 2: 389–407.

Murray, W. E. (1998). 'The global agro-food complex, neoliberalism and small famers in Chile: Lessons for the Pacific Islands?' *The Journal of Pacific Studies*, vol. 22: 27–59.

Myers, J. P., Antoniou, M. N., Blumberg, B., Carroll, L., Colborn, T., Everett, L. G., Hansen, M., Landrigan, P. J., Lanphear, B. P., Mesnage, R., Vandenberg, L. N., Saal, F. S. von, Welshons, W. V. and Benbrook, C. M. (2016). 'Concerns over use of glyphosate-based herbicides and risks associated with exposures: a consensus statement,' *Environmental Health*, vol. 15: 19–32.

National Farmers Union (2014). 'Using farm debt to spin farmers' canola into investors' gold,' *Union Farmer Monthly*, vol. 62, issue 3: 1–3.

Neis, B. and Kean, R. (2003). Why fish stocks collapse: An interdisciplinary approach to understanding the dynamics of 'fishing up' in R. Byron (ed.), *Retrenchment and Regeneration in Rural Newfoundland*, University of Toronto Press: Toronto: 65-102.

Padel, S. (2001). 'Conversion to organic farming: A typical example of the diffusion of an innovation?' *Sociologia Ruralis*, vol. 41, no. 1: 40–61.

Pimentel, D. (2005). 'Environmental and economic costs of the application of pesticides primarily in the United States,' *Environment, Development and Sustainability*, vol. 7, no. 2: 229–252.

Ploeg, J. D. van der (2009). *The New Peasantries: Struggles for Autonomy and Sustainability in an Era of Empire and Globalization*, Earthscan, London.

Reicosky, D. C. (2015). Conservation tillage is not conservation agriculture. *Journal of Soil and Water Conservation*, vol. 70, no. 5: 103A–108A.

Roger, A. (2016). 'Power in the field. Explaining the legitimisation of large-scale farming in Romania,' *Sociologia Ruralis*, vol. 56, no. 2: 311–328.

Rud, H. (2013). 'Reconciling the Divide: An Analysis of Farmers' Land Strategies Within the Corporate-Environmental Food Regime,' MA Thesis, University of Regina at Regina, Canada.

Rutherford, A., Gimby, M., Smith, G., Mair, A. and Coxworth, E. (1992). 'Viability of Organic Farm Practices,' Saskatchewan Research Council, Saskatoon.

Sabatier, P. A. (1988). An advocacy coalition framework of policy change and the role of policy-oriented learning therein, *Policy Science*, vol. 21, no. 2: 129–168.

Sachs, C. E., Barbercheck, M. E., Brasier, K. J., Kiernan, N. E. and Terman, A. R. (2016). *The Rise of Women Farmers and Sustainable Agriculture*, University of Iowa Press, Iowa City, USA.

Saskatchewan Crop Insurance Corporation (SCIC) (nd). 'Crop insurance: the organic option,' www.saskcropinsurance.com/ci/organic/ (Accessed 9 July 2017).

Selfa, T., Laroi, A., and Burnham, M. (2015). Promoting ethanol in rural Kansas: Local framings and cultural politics. *Journal of Rural Studies*, vol. 39: 63–73.

Statistics Canada (2017a). 2016 Census of Agriculture: Farm and Farm Operator Data. www.statcan.gc.ca/pub/95-640-x/95-640-x2016001-eng.htm (Accessed 20 May 2017).

Statistics Canada (2017b). Saskatchewan remains the bread basket of Canada. Catalogue no. 95–640-X www.statcan.gc.ca/pub/95-640-x/2016001/article/14807-eng.htm (Accessed 30 July 2017).

Stevenson, G. W., Clancy, K., Kirschenmann, F. and Ruhf, K. (2014). 'Agriculture of the middle,' in P. B. Thompson and K. M. Kaplan (eds) *Encyclopedia of Food and Agricultural Ethics*, Springer, Dordrecht, Netherlands.

Storper, M. and Salais, R. (1997). *Worlds of Production: The Action Frameworks of the Economy*, Harvard University Press, Cambridge, USA.

Thevenot, L. (2001). 'Pragmatic regimes governing the engagement with the world,' in K. R. Schatzki, K. Knorr-Cetina, and E. von Savigny, (eds) *The Practice Turn in Contemporary Theory*, Routledge, London.

Thiessen Martens, J. R., Entz, M. H. and Wonneck, M. D. (2015). Review: Redesigning Canadian prairie cropping systems for profitability, sustainability, and resilience. *Canadian Journal of Plant Science*, vol. 95: 1049–1072.

Timmermann, C. and Félix, G. F. (2015). 'Agroecology as a vehicle for contributive justice,' *Agriculture and Human Values*, vol. 3, no. 32: 523–538.

Waiser, W. (2016). *A World We Have Lost: Saskatchewan Before 1905*, Fifth House, Calgary, Canada.

Warick, J. (2017). '"It's a strong and vibrant industry right now": $368M farm subsidies unnecessary, says professor,' www.cbc.ca/news/canada/saskatoon/farm-subsidies-agriculture-saskatchewan-budget-1.4013441 (Accessed 30 July 2017).

Weiss, S. J. (2008). 'Canola Farmers' Responses to the Introduction of Genetically Modified and Commoditized Agricultural Seed: Understanding Discursive Foundations for Action,' MA Thesis, University of Regina at Regina, Canada.

Wezel, A., Soboksa, G., McClelland, S., Delespesse, F. and Boissau, A. (2015). 'The blurred boundaries of ecological, sustainable, and agroecological intensification: a review,' *Agronomy for Sustainable Development*, vol. 35, no. 4: 1283–1295.

Whitfield, S., Dougill, A. J., Dyer, J. C., Kalaba, F. K., Leventon, J. and Stringer, L. C. (2015). 'Critical reflection on knowledge and narratives of conservation agriculture,' *Geoforum*, vol. 60: 133–142.

Willer, H. and Lernoud, J. (2017). 'The world of organic agriculture: Summary,' in H. Willer and J. Lernoud (eds) *The World of Organic Agriculture: Statistics and Emerging Trends*, FiBL & IFOAM-Organics International, Bonn: 25–33, www.organic-world.net/yearbook/yearbook-2017.html (Accessed 7 July 2017).

Winson, A. (1993). *The Intimate Commodity: Food and the Development of the Agro-Industrial Complex in Canada*, Garamond, Toronto.

Youngberg, G. and DeMuth, S. P. (2013). 'Organic agriculture in the United States: A 30-year retrospective,' *Renewable Agriculture and Food Systems*, vol. 28, no. 4: 294–328.

11 Contested sustainability discourses as lived experience

Conflicted feelings towards meat in consumers' narratives and life stories

Robert Magneson Chiles

Introduction

Any conversation about global sustainability in the twenty-first century would be remiss without an in-depth examination of meat production and consumption. Fisheries are collapsing, livestock are a major contributor to anthropogenic climate change, and massive quantities of land, fertilizers, pesticides, and water are being devoted to feed crops. Intertwined with these questions about modern meat's *ecological* sustainability are ongoing concerns over the *social* sustainability of meat production consumption, namely with respect to horizontal and vertical integration in the meat processing industry, the health risks of meat consumption, the rights and wellbeing of meatpackers and agricultural workers, bacteriological contamination, and humanity's ethical obligations to animals. These critiques stand in glaring contrast to the widespread understanding that meat is a nutritional, economic, and cultural necessity. Studying meat provides a singular opportunity to examine contested sustainability discourses across multiple dimensions of modern social life.

When stripped of the bark of academic jargon, discourses are essentially *stories* that we tell ourselves and each other. Those discourses which circulate the most widely and achieve dominance are first and foremost *narratives* that speak to broadly shared meanings and understandings. Czarniawska (2004) observes that on the one hand, narrative is a mode of communication, as people use storytelling to interpret, teach, and learn. On the other hand, narrative is also a mode of knowing – it organizes the world according to contestable plot lines and human intentionality rather than absolute/scientific truth. It is for this reason that Vance (1995) describes storytelling as an act of theorizing. Unlike scientific theorizing, however, the theorizing that emerges from narrative is based not on hypothetical-deductive reasoning but rather on those stories that most resonate with people's lived experiences, personal truths, and deepest convictions. We cannot, therefore, contest sustainability discourses – as implied by the title of this volume – without also contesting people's life stories. As noted by Adams (2008: 177–179),

> If narratives are tools and if the crafting and sharing of stories involve morals, then a discussion of ethics is a necessary component of narrative

> inquiry ... if stories teach us how to act, we must critically evaluate these stories to observe hidden and problematic politics.

In this chapter, I use narrative analysis to examine the contested sustainability discourses that surround meat consumption in the life stories of three people: Mark, Gabby, and Patty. I met each of them while conducting a focus group study on the resilience of meat culture in the aftermath of controversy (Chiles, 2016a). Upon asking six small groups of consumers what they liked and didn't like about meat, I situated their responses in a broader social context through the use of secondary historical materials (see Chiles and Fitzgerald, 2017) and a content analysis of meat-related news coverage in the *New York Times* (1981–2013). I found that the meat-related stories which appeared most frequently in the *New York Times* – i.e. taste, restaurants, cooking, and affordability – reflected the everyday concerns of many of my focus group participants (Chiles, 2016b). Conducting the focus groups provided an opportunity to temporarily puncture the dominant discourses surrounding meat production and consumption by providing consumers with the opportunity to openly share the anxieties, concerns, and worries about meat that they usually kept to themselves.

Of the 38 participants that I met over the course of the study, the stories of Mark, Gabby, and Patty were particularly evocative in their representation of the desires and anxieties that animate contemporary meat culture. Narrative analysis is an ideal method by which to understand these conflicting forces. As noted by Creswell and Poth (2017), narrative analysis showcases participants' life experiences as reconstructed through the sharing of epiphanies, plot twists, key characters, and causal explanations. A core element of any narrative is the 'moral of the story,' i.e. the ethical implications of what has just been shared. Polletta *et al.* (2011: 111) poignantly observe that 'events in a story project a desirable or undesirable future. They make a normative point ... Insofar as stories draw on a cultural stock of plots, they communicate the normative values that are associated with those plots.'

The narratives that are shared in this chapter collectively illuminate the key discourses which have shaped normative understandings about meat throughout American history. Their stories are expressions of scarcity, abundance, comfort, guilt, pride, tradition, and apprehension. The purpose of this chapter is to unpack these tensions and unravel the different threads which tug on the hearts, minds, and stomachs of consumers like Mark, Gabby, and Patty.

Mark

Mark is a 32-year-old single black man with a tall build and broad shoulders. He is currently unemployed, lives with his sister, attends community college full-time, and has seven children from previous relationships. During construction season, he does subcontracting work – power washing, siding, fixing decks, and related tasks. I met Mark when I was recruiting for a focus group discussion on meat in America at a local community college. He had a real knack for breaking the ice in the group with his offbeat commentary:

> Oh, bologna ain't meat ... No it ain't, man. They grind up eyeballs and all type of stuff and throw it in there. Throw some soy in there.
>
> Know what's funny? People eat hot dogs. Hot dogs the same thing ... For real. That's pig intestine with a bunch of grinded up everything.
>
> A lot of people say McDonald's, Taco Bell type of things is like drunk food. You gotta have some liquor and alcohol fighting with it in your stomach.

I have so much fun listening to Mark's quips that I ask him to participate in a follow-up interview in his home, and he agrees. This gives me the opportunity to learn more about his life story and how it affects his current beliefs and behaviours about meat.

Mark grew up in the Harold Ickes Homes in Chicago, a public housing project that was located next to the Robert Taylor Homes. 'When you're in the projects, it's like a whole city,' he tells me. He could buy 'penny candy' for a few cents from a woman who lived on the seventh floor, his parents could go to an adjacent building to buy clothes, and local residents would 'hit the tracks' to steal and resell goods from parked cargo trains. His mother ran a free child care service at a local neighbourhood centre, and his father made drums for a living until losing a finger in a work-related accident. They split up when Mark was nine. The family struggled financially, and this had a direct impact on the quality of food that was available:

> We was eating chicken in the can. It was bad ... We was eating sugar sandwiches and stuff like that. You know ... Yeah, whatever we could get our hands on. Made us pretty creative ... It made you want the thing you didn't have the most of. Meat was the thing. Meat was a big issue. We had chicken, pork in the can, stuff like that. You couldn't even keep it long enough. It wasn't good at all. You looked at it; it looked like what it was. You know, feet, beaks and all type of stuff they just throw in there. You know.

He also tells me that his family pushed him to eat vegetables and home-cooked meals, but he would beg and cry for his parents to take him to McDonald's instead. In hindsight, he attributes his rebellious attitude around food to advertising, popular culture, and a desire to fit in. 'I just wanted what I thought everybody was supposed to want,' he told me. He believes that these experiences contributed to making him a 'die hard meat eater:'

> I grew up really poor. Nobody in my family's even seen an ocean or mountain or anything. We got put where we at and that's where we at. I know guys ain't never leave my neighborhood, still today. Where I'm from in

> Chicago. I think I became a meat eater because it was the thing we didn't have. We didn't have it in abundance. And when we did have it, you know what I'm sayin', it was like, dig in. And we ended up fighting for the last piece every time. Now I'm an adult, I can have it at my will.

At the same time, Mark expresses deep ambivalence about meat. He felt that his meat consumption was unhealthy for him, and that as an African American man, he was at risk for diabetes and premature death. 'I'm probably the most meat-eatingest person,' Mark tells me. 'I made a decision a long time ago that I'm gonna die before 66 because of this reason.' He likes the meat that he sees in advertisements, and yet he feels uncomfortable with where the meat comes from. Still, he wants it:

> The advertising, I'm looking at the pictures, they make it really appealing. It's kinda hard to turn down. And I know that chicken is a cubicle chicken. I know the chicken on top is pooping on the chicken on the bottom. And he's only three weeks old and this chicken is … You know what I'm sayin? But when I go to KFC, boy, I'm putting the meat in one side and spittin' the bones out the other. I be getting' down. I like meat. I can't eat just noodles or something. It's gotta be meat in the noodles, know what I'm sayin'? You know.

Later in the interview, Mark admits to me that he's scared of changing his lifestyle and eating differently. He feels that his desire to eat more healthfully is being overwhelmed and overridden by the large meat companies, and he laments his apparent lack of self-efficacy:

> I don't think it's a choice anymore … The larger companies, it's to the point where they could almost sell anything … I think that when you get to the point where you talk to a lot of people and they're saying we don't agree with the things they feed us and stuff like that, you know, there's got to be some element of control from up top. Towards these people. They might not actually be forcing. But I think they hand it to you … I figure I'm a stupid guy. That knows better and does anyway. That's what stupid means to me. Dumb is just can't comprehend. But I can comprehend and I do it anyway.

In many ways, Mark's coming-of-age, from the meat-deprived days of his youth to his carnivorous present parallels the coming-of-age of meat in America. Between 1909 and 2007, per capita meat consumption in the United States nearly doubled, and prized cuts became increasingly accessible and affordable to more and more people (Daniel *et al.*, 2011; Ogle, 2013). The industrialization of meat production allowed middle-class Americans to have meat, as Mark put it, at their will. With this boon, however, came a steep price. As cheap meat proliferated, so too did discourses about food safety, public health, workers' rights, the environment, and ethical obligations towards animals. Food justice advocates have also argued that fast food restaurants and other elements of the 'cheap meat for all' bargain actually contribute to socioeconomic and racial disparities rather than alleviating them, as poor people and racial minorities are more likely to suffer from meat-related health problems.

Many people, like Mark, argue that large companies have used massive advertising and lobbying budgets to disseminate discourses that shape consumers' attitudes and desires. Still, Mark also downplays the depth and breadth of corporations' power by commenting that he 'knows better' but 'does anyway.'

Gabby

Gabby is an 18-year-old single white woman. Originally from Phoenix, Arizona, she is currently attending college at a large out-of-state university. She has a slight build, long blonde hair, and is shy yet friendly. Unlike Mark, Gabby says that she had 'a privileged upbringing.' Her father is a commercial real estate broker and her mother earns a 'meager' salary as a fourth grade teacher at a private school. Gabby says that her family 'was probably lower middle class' during her early childhood but has been 'steadily upper-middle-class for the past probably 5–10 years' with occasional financial downturns. Regardless of the family's economic situation, however, Gabby tells me that her family never scrimped on getting quality meat:

> Meat is a big part of my family. My dad is a big time cook. And so every Sunday night, we have a family dinner. We try to do it most nights. But Sunday is kind of like the grill night. And he cooks for hours and hours. And we sit down and have a big meat dinner. And he loves meat. He loves meat. So that's something I associate with my dad ... Eating meat was always part of my life. I didn't question it for a long time.

Gabby's relationship with meat was disrupted in seventh grade, when she participated in a pig dissection at school. She enjoyed learning about the biology, but was disgusted when she made the connection between the pig cadaver and the foods she ate:

> My teacher through the whole process would give us these facts about what part of the pig is what you eat in a hot dog and stuff like that. So that kind of grossed me out. And the fact that my mom would like, make bacon for me in the morning. And I would have to go to school and dissect the pig. And I started relating those things and thinking about it more ... And then around the same time, I guess like a year later, I had a teacher in history who started reading us bits of Upton Sinclair's *Jungle* during class, where they talk about meat production and all of the horrors of that that were going on at the time that the book was written. So that really grossed me out for a while too.

From time to time, she would ask her parents if she could become a vegetarian, but her parents said no. Gabby said that she wouldn't eat other forms of protein besides meat, and so her parents were concerned about her nutrition. Perhaps more significantly, vegetarianism also contradicted the centrepiece of the family meal as well as the family's Christian values. Her parents argued that God gave humans dominion over animals, that there were more important problems in the world to worry about, and that there were a lot of deprived people in the world who would love to have what she was turning down. 'So the only thing I would have in defense would be to say that meat production, I'm not okay with the process of that,' Gabby explains to me. 'So my family started doing local meat.' Her father agreed to make the switch after being convinced that local meat tasted better and was higher quality.

Despite her success in nudging the family towards local meat, Gabby still feels unsettled with her meat consumption. She recounts that the meat in the dorms doesn't taste as good as what she had at home, so she eats less of it and has a less balanced diet overall, which diminishes her energy level and mood. She feels that it would be very difficult for her to get adequate nutrition without any meat in her diet. At the same time, however, she commented that the meat consumed in the house was about 50 per cent locally sourced and 50 per cent conventionally sourced, and her feelings of discomfort towards eating meat have not completely gone away:

> I would still have problems with meat every once in a while. I would hear some horrible fact about meat processing. I mean we do not always eat local meat. But most of the time we do … It does seem silly not to eat something that I do already have and that my family is already going to make. And also, I enjoy meat … I know that it's good for me, and I do like the flavor … And I do feel, a lot of times guilty, because I don't think about it as much as I should, and I will eat meat that in the back of my head, I think like, I don't know anything about this, this is very bad. But I'll still eat it. Um – so I think that I should be thinking about it more than I do, and things like this [focus group] help but at the same time, I don't know that I'll change anything.

Mark and Gabby come from very different backgrounds. Nonetheless, they both express positive and conflicted feelings about meat. While such feelings are rarely shared openly over the course of everyday life, Mark's and Gabby's anxieties about meat are not in the least bit unusual.

Patty

Patty is a 53-year-old divorced white female with five children. She performs home health care work for her live-in, disabled boyfriend, and she also works part-time at a national retailer as a stock clerk. She reports her household income at $19,000-$36,000. When I arrive at her apartment complex to interview her, I quickly learn that Patty's current household situation is quite chaotic. Three generations live in a very modestly sized apartment unit, seven people in total: Patty, her boyfriend, her two oldest daughters, and three grandkids. When I walk in the door to conduct the interview, I see two toddlers running around, two television sets on at the same time (one daughter was watching TV and two grandkids were playing video games), degraded carpet, and clutter gathered everywhere – clothes, debris, and trash. The TVs are loud and there is lots of chatter and activity, so we conduct the interview in her daughter's bedroom.

I ask Patty how she grew up, and she tells me that she was raised in the city with her siblings. Her father worked for 42 years as a meat cutter for a large processing company, and her mother was a nursing assistant. Her family spent a lot of time, however, on her grandparents' rural homestead outside the city. Patty explains that when her mother was growing up in the country, the family lived in extreme poverty and would use whatever they could find to make a meal. The grandparents raised pigs and chickens, and her grandpa hunted for wild game and fished. When visiting her grandparents as a kid, Patty would chase and catch the chickens and give them to her grandma. Her grandma would then snap their necks, de-feather them, and gut them. Patty has fond memories of the experience:

> Mm-mm. I'd watch 'em hang the deer, after you hang it and de-blood it or whatever. And watch 'em whack off the different parts and all that. And that was kind of cool … Yeah. Same with ducks, chickens, rabbits. Squirrels. All that kind of stuff. Go hunting with grandpa, catch the animals, come home and watch him skin 'em and all that and clean 'em up and cook 'em up. Never bothered me.

She was disappointed that her kids and grandkids hadn't been exposed to rural life as she had:

> Unfortunately my children haven't been exposed to anything like that. But that was all a part of growing up. I think some of it was the bonding. To be able to go out with grandpa and hunt them. Even though we didn't have a lot of luck when us kids were along. You know, 'cause we're noisy. But it just, it just was like wow. We caught this and now we're gonna eat it and it tastes so good. And I don't know. It just seems like that's how it should be.

Within a single generation, the family had shifted from self-provisioning to wage-labour and mass consumption. This had a direct impact on the frequency and type of meat that was bought and prepared. Her grandparents were financially worse off than her parents, so once her parents were able to afford meat and have chicken every night, 'It was like, well, this is really special … And it became something you just did,' Patty told me. She went on to say that both of her parents were excellent cooks. Her parents prepared lasagna with ground beef, pork hocks with sauerkraut and potatoes, blood sausage, and 'plate steaks'

– steaks that were almost as large as a dinner plate. Her father would also make head cheese by boiling pig's head, feet, and tails. 'And that's how I grew up eating meat,' Patty explains. When Patty and her now ex-husband were raising their own family, they bought chuck 'dollar steaks' for between $.75 and $1. Patty also has fond memories of making meals out of chicken necks, gizzards, and hearts, and she says that it was really inexpensive to buy them. Chicken wings used to be the cheapest – she would buy them with her first husband, and save the tips in the freezer for soup.

The family foodways would shift again when Patty began raising children and then grandchildren. Given the current challenges of living with her boyfriend, daughters, and grandkids, the primary discourses that shape Patty's current food philosophy are accessibility, affordability, and domestic tranquility. They buy most of their food at Walmart because they can walk to it easily and the prices are low. Patty moreover comments to me that 'Walmart has excellent brands' in terms of its meat quality. They also get meat and other foods from a local community food pantry, and sometimes they'll get beef from her sister's farm.

Patty says that she no longer cooks the exotic meats or the different types of cuts that she grew up with. 'Mostly because the kids would go, oh gross,' she laments. 'But that also comes from their friends and a lot of their friends are from even bigger cities.' She adds that her grandkids are even pickier. For example, one of her grandsons doesn't eat fish. Patty says that she had a duck in the freezer, and one of the kids refused to eat it even though they'd never tried it. One of the grandkids 'could live on chicken wings and hot dogs. What is it with kids and hot dogs?' she wonders aloud. The grandkids also want to eat before it's time to sit down for dinner. Patty tells me that they were used to being indulged by their mothers, because it's easier for the mothers to just feed them, and Patty herself has resorted to this strategy. 'It's like here, eat this and shut up. Go away and let me finish what I'm doing,' she says. Patty adds that, 'I know that at this point I can still be an example. But I kind of feel like moms have more control. And my daughters don't always want to listen to what I have to say.'

While Patty feels that living off the land is a better way to produce and consume meat ('that's how it should be'), she also defends the legitimacy of the industrial meat system. Towards the end of the interview, I ask her to talk more about mass production and whether or not the meat that goes to Walmart, McDonald's, and Burger King is cruelly produced. She says that she is uncomfortable with the process, but still defends the legitimacy of their business practices:

> That's really hard. Part of me says I do think in some ways they are [cruel] … I don't think that they intend to go out there and be mean to them. It's not like they beat 'em or anything like that. They certainly wouldn't starve 'em. But I think they turn a blind eye to what we're doing may not be the right thing and we're trying to do too much and make too much money. But hey, you know, next week, we have to put out thirty thousand chickens

> so what choice do we have? So I think a lot of it comes down to dollar bills. It really does.

Tellingly, she uses the pronouns 'they' and 'we' interchangeably when referring to these companies, which implies a level of self-identification with them. She also argues that meat production isn't cruel on the basis that 'God made animals for us to consume … We are not animals. We're human beings. God created us in his image.' When I ask her about her religious background, Patty explains me that she was born and raised Catholic and currently attends an Apostolic Evangelical church.

Patty's religious views illustrate how the discourses of tradition and religion can maintain the legitimacy of a consumer good, even when the material process of producing it has been desacralized. Companies like Walmart and the large meat processor that her father worked for have contributed to the decline in the rural livelihoods that Patty yearns for. Nonetheless, she steadfastly resolved that meat would always be a part of her life, regardless of where it came from. Without meat, she tells me that she feels unsatisfied after a meal. Most important for her is that her family wants it. Through cooking meat, she symbolically transforms industrially produced meat from a profane commodity into a sacred gift for her family. These bonds and connections powerfully legitimate modern apparatuses of production, and companies seize upon these discourses to defend their business practices. The discourse of family values – which asserts that the American family comes together when it enjoys meat together – can be found in 'Feeding you like family,' a 1989 umbrella concept ad campaign that Saatchi & Saatchi developed for Tyson Foods:

> Recent Tyson television ads contain multiple vignettes, quick cuts on the idealized state of the family meal in real situations of modern family eating. The ads reflect modern families – single-parent households, friends, teens at a microwave – and they show Tyson convenience products being used to fulfill traditional needs. The Saatchie campaign is centered around sentimental association to the family and family meals. While we might toss convenience products into the microwave with little emotional thought, in our minds there may remain a subliminal link to an older-style kitchen, perhaps a traditional American mother wearing an apron and standing before a stove. Saatchi's promotional campaign takes advantage of these cultural memories, emphasizing product quality and Tyson name awareness.
> (Schwartz, 1991: 114–115)

Historicizing contested sustainability discourses

Meat production has quite often come at a great ecological and social cost: erosion, polluted waterways, slain wildlife, eradicated indigenous communities, and epidemic diseases. We therefore must avoid the temptation to romanticize the pre-industrial method of meat production as a kind of Holy Grail for the land and the community. Nutritionally, for the vast majority of people living in modern societies, meat is simply unnecessary. While this has not always been the case over the course of human history, when it comes to the consumption of domesticated livestock, meat has traditionally provided a very small proportion of dietary calories. From a political-economic standpoint, there can be no doubt that meat producers have reaped tremendous profits from their business, from the large Roman estates to Tyson Foods. At the same time, the discourse of anthropocentrism – another cornerstone of Western history – can be traced back to ancient Greece, and it was amplified in later years by Romans, Hebrews, Christians, scientists, cowboys, and many others. Although the specific ideologies have changed over the years, the core tenet of human superiority over animals and nature persists, as it has with Patty and also with Gabby's family. Gabby, for example, shared with her focus group that

> My parents kept pushing that I wouldn't be a vegetarian. And my dad would use things like 'It's against our religion. Because God created animals for us to eat. It's silly to think of it in an animal rights way.' He said we don't think like that … My family's always said 'No, we eat meat, animals aren't equal to us.'

Dating back to the late Middle Ages, meat has also been understood as essential to human health. Despite the fact that this proposition has been thoroughly rejected by numerous leading dietary organizations, many continue to subscribe to it, and evolutionary nutrition is only the most recent version of this thesis. All the while, from antiquity through to the present day, those who refrained from meat consumption have been subject to ridicule (Chiles and Fitzgerald, 2017).

Despite the continuities in discourses that have dominated Western history, there has also been change. Gradually, as Western societies grew more stratified and occupations became more specialized, the dirty work of slaughtering animals was removed from the daily sight of the urban consumers who demanded it. Through spatial-temporal distancing, producers have helped to keep meat production out of sight and out of mind. While the generational shift from meat as subsistence good to a mass produced commodity is often discussed in terms of broader shifts in the global economy, this transition has been facilitated and legitimated by deeply ingrained consumer habits, values, and practices. Evidence of these legitimating beliefs and behaviours can be found in the story of Patty's family.

For consumers like Gabby, who feel uncomfortable eating meat off of the bone, today's marketplace offers a nearly infinite array of boneless, skinless, and ready-to-eat options. Nonetheless, to the extent that there is continued ignorance about how meat is produced, it is a willful ignorance. In order for the enjoyment of meat eating to continue, consumers may feel the need to exercise some degree of cognitive-affective distance from the animals who are transformed into food (Chiles, 2016b). This was evident in my conversation with Mark and many other consumers, even those who professed to eat 'nose-to-tail.'

MARK: If I had a chicken just sitting in a coop and I expected to make food that would be easier to do if I didn't name him … I couldn't eat a guy – or eat a chicken, that I had named. You know. Took time to name it, watch it grow up. Couldn't do that.

RC: What if the chicken was treated as well, as humanely as possible.

MARK: That would make it even harder.

Reconciling contested discourses

At the beginning of this chapter, I introduced Mark and Gabby: people from radically different backgrounds who were struggling to come to terms with their meat consumption. They both enjoyed the taste and flavour of meat, but this enjoyment had been disrupted by the knowledge that meat involves the death and, quite often, the suffering of animals. At the same time, both of them were unwilling to stop eating meat. While Patty feels more secure in her meat eating, she remains troubled by the conditions under which it is produced. The discourses of taste, flavour, and abundance that Mark, Gabby, and Patty are contesting in these stories are the very discourses that have given their lives meaning. Indeed, their negative and positive associations with meat are intertwined with one another.

For decades, and particularly over the past ten years, small-scale, local, and organic meat advocates and producers have sought to assuage consumers' conflicting feelings by changing the way that meat is produced, and they have achieved many noteworthy successes. Even for those who consume local products, however, lingering doubts about the efficacy and ethics of these products often remain. Moreover, there are numerous structural and environmental obstacles facing the local food movement. Small-scale slaughter plants are steadily disappearing, land resources are dwindling, and the political economic system favours large-scale, grain fed livestock operations. Many consumers still struggle with the ethics of killing animals, and this is a philosophical question that ultimately cannot be resolved through animal welfare reforms alone. The persistence of this dilemma is evident in my conversation with Mark. At first, he tells me that it is okay to kill animals for food, and that it isn't cruel to do so. Upon further reflection, however, he comments that 'it's pointless to let life express itself just to slaughter it and kill it. I think that's kind of sick. You know. There's something sick about it.' Moreover, while he feels that it would be better to treat animals more humanely, he also believes that would make it even more difficult for society to kill animals for food:

> You couldn't have slavery if the masses respected slaves. Eventually that led to not being slavery … You gotta keep your mind where it is. You don't teach a soldier to respect the Germans and go fight Germans. You know … You do the opposite. You've got to keep them demonistic. If you treat the animal like shit, guess what, it's easier. Way more easier to slaughter something that never, you know, never one second of his life has been treated

humanely. You know what I mean? If I'm putting a little bow tie on it and stuff like that, I'm not gonna be slapping no seasoning on you in two weeks.

The distance between meat production and consumption is further maintained by mass media coverage, which overwhelmingly discusses meat-related issues from the standpoint of cuisine and economy: the two topics that are most relevant to the everyday lives of consumers. Nutrition and food safety receive less coverage, and the broader ethical issues pertaining to workers, animals, and the environment receive less coverage still.

Every now and again, a meat-related crisis will capture the public's attention. This occurs when meat production or consumption suddenly appears to people as hazardous, disgusting, or unethical. The most disruptive events usually involve some kind of combination of these three elements, as was the case with the lean finely textured beef (LFTB) or 'pink slime' crisis of 2012 (see Greene, 2012). Disruptions can also occur at a deeply personal level, as evidenced by Gabby's refusal to eat pork ribs after participating in a pig dissection (Chiles, 2016a).

There have been many books and articles in recent years that have extensively documented the many concerns and grievances associated with meat production and consumption. Nonetheless, most of the discourses that are circulated in these texts pay insufficient attention to how these controversies have deeply perplexed ordinary consumers. If we are to cultivate a more comprehensive understanding of ecological and social sustainability, we must look beyond the familiar and highly visible debates that take place between the Global North and the Global South, corporations and activists, democrats and republicans. We must also examine how sustainability discourses are actively contested within the hearts, minds, and narratives of individual people.

References

Adams, T. E. (2008). A review of narrative ethics. *Qualitative Inquiry*, 14(2): 175–194.

Chiles, R. M. (2016a). Food system fragility and resilience in the aftermath of disruption and controversy. *Journal of Agricultural and Environmental Ethics*, 29(6): 1021–1042.

Chiles, R. M. (2016b). Hidden in plain sight: How industry, mass media, and consumers' everyday habits suppress food controversies. *Sociologia Ruralis*. Available from: http://onlinelibrary.wiley.com/doi/10.1111/soru.12152/abstract (Accessed 17 November 2016).

Chiles, R. M. and Fitzgerald, A. J. (2017). Why is meat so important in Western history and culture? A genealogical critique of biophysical and political-economic explanations. *Agriculture and Human Values*. Available from: https://link.springer.com/article/10.1007/s10460-017-9787-7 (Accessed 7 June 2017).

Creswell, J. W. and Poth, C. N. (2017). *Qualitative inquiry and research design: Choosing among five approaches*. Sage Publications, Thousand Oaks.

Czarniawska, B. (2004). *Narratives in social science research*. Sage Publications, Thousand Oaks.

Daniel, C. R., Cross, A. J., Koebnick, C. and Sinha, R. (2011). Trends in meat consumption in the USA. *Public Health Nutrition*, 14(4): 575–583.

Greene, J. L. (2012). Lean Finely Textured Beef: The "Pink Slime" Controversy. Congressional Research Service, www.foodpoisonjournal.com/uploads/image/R42473.pdf. (Accessed 18 April 2018).

Ogle, M. (2013). *In meat we trust: an unexpected history of carnivore America.* Houghton Mifflan Harcourt Publishing: New York.

Polletta, F., Chen, P. C. B., Gardner, B. G. and Motes, A. (2011). The sociology of storytelling. *Annual Review of Sociology*, 37: 109–130.

Schwartz, M. (1991). *Tyson: From Farm to Market (c)*. University of Arkansas Press, Fayetteville.

Vance, L. (1995). Beyond just-so stories: Narrative, animals, and ethics. *Animals and Women: Feminist Theoretical Explorations*. In C. Adams and J. Donovan (eds), *Animals and Women: Feminist Theoretical Explorations*: 163–191. Duke University Press: Durham.

Part IV

Contested agrifood governance

12 Shifting visions of sustainability in United States agriculture

A case study of the role of multi-stakeholder governance

Jason Konefal and Maki Hatanaka

Today, it is widely acknowledged that accelerating the sustainability of agriculture is a pressing need. However, while there is widespread agreement that agriculture needs to become more sustainable, what sustainable agriculture entails is widely contested (Constance, 2009; National Research Council, 2010; Thompson, 2010). In its most recent report on agricultural sustainability, the U.S. National Research Council (2010: 5) states: 'agricultural sustainability is a complex, dynamic, and political concept that is inherently subjective ...' More recently, the USDA has stated: 'some terms defy definition. Sustainable agriculture has become one of them' (Gold, 2015).

Given the lack of consensus as to what counts as *sustainable* agriculture, a first step in transitioning agricultural systems is defining sustainability. This tends to include, first, developing general principles. It then entails translating such principles into usable metrics, standards, benchmarks, and/or measurements – i.e. tools that can be used to foster and measure changes in agricultural practices (Busch, 2011; Loconto and Fouilleux, 2014). Research finds that this process of defining sustainability and developing implementation and measurement tools is highly important. Not only does it significantly influence the implementation of sustainability on farms, it also has the potential to affect agrifood markets and supply changes (Loconto, 2010). Consequently, the process of defining sustainability and developing metrics and standards is quite politicized (Hatanka, Konefal and Constance, 2012; Cheyns and Riisgaard, 2014). Unruh and Ettenson (2010: 212) observe that the world is in the midst of a 'green frenzy' in which there is 'a tooth-and-claw-competition among a growing pack of stakeholders, including environmental activists, think tanks, bloggers, industry associations, consultants, and your rivals, all clamoring to establish and impose their will on green standards.'

Much of the process of codifying sustainability generally, and agricultural sustainability in particular, is taking place in multi-stakeholder initiatives (MSIs) (Newell, Pattberg and Schroeder, 2012). MSIs are a collective form of governance in which stakeholders self-govern activities and things that they have a vested interest in (Tamm Hallstrom and Bostrom, 2010). In the U.S., MSIs devoted to sustainable agriculture began to emerge in 2007. The analytical focus of this chapter is on the two MSIs that have become the most prominent and are at the forefront of trying to transition U.S. agriculture towards greater sustainability: Field to Market and the Stewardship Index for Specialty Crops (SISC). Field to Market has developed environmental sustainability metrics for commodity crops, and SISC environmental metrics for speciality crops (i.e. fruits and vegetables). Both initiatives are made up of different combinations of stakeholders from agrifood supply chains, civil society organizations, and university researchers. Importantly, both include many of the lead market actors in the U.S. agrifood system, such as Walmart, Unilever, Monsanto, Cargill, Del Monte Foods, and Kellogg's.

This chapter examines how MSI governance is affecting the meaning of sustainable agriculture in the case of the United States. While sustainable agriculture has largely been uncodified in U.S. agriculture until fairly recently, there have been social movements focused on the sustainability of agriculture for quite some time. Commonly referred to as alternative agrifood movements, these movements have advocated for an alternative system to the industrial agriculture system, which they view as largely unsustainable. Specifically, in place of industrial agriculture, they advocate for a food and agriculture system that is localized, utilizes agroecological practices, and is democratic (Constance *et al.*, 2014). While this vision of sustainable agriculture continues to have widespread support among organic, local food, and urban food advocates, this alternative vision is largely not incorporated into the MSI principles and metrics for sustainable agriculture in the US.

With the emergence of MSIs for sustainable agriculture, we find the development of a different definition of sustainable agriculture that competes with the alternative vision. Specifically, Field to Market and SISC are advancing a view of agricultural sustainability that largely corresponds with the emerging notion of sustainable intensification (SI). SI is the idea that accelerating sustainability in agriculture is possible with increases in productivity and eco-efficiencies (Garnett *et al.*, 2013). It is grounded in the belief that faced with growing population and resource scarcities, agriculture will need to undergo significant intensification to both 'feed the world' and be environmentally sustainable (Godfray *et al.* 2010; Garnett *et al.*, 2013; Godfray, and Garnett, 2014).

We contend that the shift from the alternative vision of sustainable agriculture to that of SI has implications not only for agricultural practices, but also for what agricultural sustainability means more generally and how to achieve it. Specifically, Field to Market and SISC are contributing to a more general shift from understanding sustainability as a political and normative challenge to seeing it as a technical one (Freidberg 2013, 2014). The result is a reframing of agricultural sustainability transitions from requiring changes in the structure of the U.S. agrifood system to a process of fostering innovation. While both of these visions of sustainability are likely to be implemented in U.S. agriculture going forward, the market power associated with Field to Market and SISC indicate that the SI vision is likely to be much more widely adopted in U.S. agriculture.

The findings in this chapter are based on two data sources. The first set of data is reports and webpages by Field to Market and SISC that explain their metrics. Field to Market has published multiple national reports on their metrics (Field to Market, 2018), and SISC has a metric webpage, metric documents, and webinars on their metrics (SISC, 2018). Analysis of these documents provided data on the metrics and their criteria, justifications undergirding the metrics, and information on the objectives of each of the initiatives. Second, we conducted in-depth interviews between 2011 and 2013 with participants and facilitators in Field to Market and SISC, as well as with professionals and activists working on issues related to agricultural sustainability who were not formally part of these projects. Interviews focused on two primary topics: the ways that the metric development process works and understandings of sustainable agriculture.

The remaining sections of the chapter are organized accordingly. First, we discuss the alternative agriculture movement and its vision of agricultural sustainability. Second, we provide an overview of SI and its key tenets are introduced. Third, we introduce Field to Market and SISC and discuss each of their metrics. Fourth, using the alternative and SI visions of sustainability as guides, we analyse the ways that Field to Market and SISC codify sustainability. We conclude with a discussion of the implications of the rapid development and diffusion of SI for sustainable agriculture in the United States. Specifically, we highlight the ways that sustainability governance is increasingly becoming a form of 'technopolitics.'

Alternative agriculture and agricultural sustainability

The industrial agriculture system has long been critiqued as being environmentally, socially, and economically unsustainable. Critics argue that industrial agriculture contributes to a multitude of environmental problems, including global climate change, water and air pollution, biodiversity loss, and soil degradation (Khan and Hanjra, 2009). Socially, industrial agriculture has been criticized for contributing to declining social capital in rural communities and a loss of food sovereignty for many people (Lobao and Sharp, 2013). Economically, it is criticized for forcing farmers onto technology treadmills that threaten their livelihoods (Cochrane, 1958).

In response to the problems associated with industrial agriculture, there has been a diverse set of movements, initiatives, and research focused on transitioning agriculture towards more sustainable structures and practices. Among the most prominent movements and initiatives are organics, agroecology, food sovereignty, Fair Trade, food justice, and localism. While these efforts are quite diverse, and not always in agreement, collectively they constitute what is commonly referred to as the alternative agrifood movement (Constance *et al.*, 2014; Hinrichs and Eshleman, 2014).

Given the diversity of the alternative agrifood movement, there is not an agreed upon collective definition of sustainable agriculture. In fact, while the various groups that constitute the alternative agrifood movement claim to be committed to sustainable agriculture, sustainability is a vague ideal for most of the movement (Forssell and Lanoski, 2015). Nevertheless, in reviewing the considerable research on the alternative agrifood movement, a set of shared characteristics that constitute a sustainable agriculture system can be identified. The first is the notion that a sustainable food and agriculture system entails the (re)localization of food and agriculture (Hinrichs, 2000). Localization entails a shortening of relations between farmers and consumers. Often this means spatially in terms of developing geographically bounded agrifood systems. However, it has also been used to denote a shortening of supply chains through 'constructing transparent chains in which products reach the consumer with a significant degree of value-laden information' (Renting, Marsden and Banks, 2003: 398). As localization aims to socially embed food and agriculture, it allows for the inclusion of values that are often seen as congruent with sustainability, such as fairness, environmental conservation, and community development, to be incorporated into food and agricultural transactions and decision-making (Hinrichs, 2000).

A second characteristic of the alternative agrifood movement is a commitment to agroecological farming practices (Gliessman, 2013). Agroecologists reject input intensive industrialized forms of agriculture, in favour of agricultural practices and technologies that utilize and build on natural processes. While the implementation of agrocecological farming practices on the ground can be quite diverse, they are built around the following agro-ecological processes: 'organic matter accumulation and nutrient cycling, soil biological activity, natural control mechanisms, and resource conservation and regeneration' (Altieri, 2002: 8). Thus, much of the alternative agrifood movement rejects industrial agricultural practices, such as monocropping, synthetic inputs, and genetic modification, as unsustainable.

A third characteristic of the alternative agrifood movement is a commitment to democracy and social justice (Allen, 2004). This includes initiatives that seek to empower farmers and support farmer livelihoods, such as Fair Trade. Perhaps the strongest iteration of this commitment is the movement for food sovereignty, which seeks to ensure the 'right of peoples to define their own food and agriculture' (Holt Giménez and Shattuck 2011). The commitment to people having a voice in the food system is also evident in the notion of citizen-consumers that many alternative agrifood organizations advance. Citizen-consumers are people for whom what they eat is informed by their ethical values and political commitments (Johnston, 2008). Hence, for the alternative agrifood movement, a sustainable food and agriculture system is one that is democratic, fair, and equitable.

Marsden's (Marsden, 2012, Marsden and Farioli, 2015) notion of an 'eco-economy' is among the most comprehensive theorizations of a sustainable food and agriculture system that builds on the principles of alternative agrifood movements. An agricultural eco-economy is 'place-based,' based on agroecological principles, and characterized by sovereignty and 'reflexive governance' (Marsden, 2012). Such a food and agriculture system, if constituted, would be very different from the current industrial agriculture system. This means, from the perspective of alternative agrifood movements, a sustainable agrifood system entails, not modifications, but significant transformations to the industrial agrifood system.

Sustainable intensification

A more recent vision of sustainable agriculture is sustainable intensification (SI). In the SI vision of sustainability, the food and agriculture system is seen as facing two critical challenges in the coming decades. The first is food security. Since the food crisis of 2007, there is a renewed awareness of food insecurities by policymakers, academics, and agribusiness. For example, Morley, McEntee, and Marsden (2014) observe that there has been a boom of supranational and government reports on food security over the past decade. While food insecurity is a pressing problem currently, many envision it becoming an even greater challenge in the future given projected global population growth (Godfray *et al.*, 2010; Garnett *et al.*, 2013; Godfray, and Garnett, 2014). Further compounding food insecurity challenges are ongoing dietary transitions to higher caloric foods in many parts of the world (Godfray *et al.*, 2010; Garnett *et al.*, 2013; Godfray, and Garnett, 2014).

The second is environmental challenges. These include resources scarcities, global climate change, and environmental degradation. Competition for scarce resources, particularly land and water, is viewed as likely to increase in the near future (Garnett *et al.*, 2013). Consequently, extensification is not viewed as a viable strategy to meet future food demands. There is also growing recognition that agriculture contributes significantly to global climate change, and that global climate change will impact agriculture. Lastly, there is pressure on agriculture to limit its negative environmental externalities, such as biodiversity loss and pollution.

Proponents of SI advocate for the (eco)intensification of agriculture as the most effective approach for addressing the twin environmental and food security challenges. Proponents of SI have developed a general four-part agricultural sustainability roadmap:

> (i) that increased production must play at least some role in meeting the food security challenge of the next fifty years; (ii) that the vast majority of this increase must come from existing agricultural land; (iii) that increasing the sustainability of food production is of equal importance; and (iv) that we must consider a broad range of tools and production methods to achieve these goals.
>
> (Godfray and Garnett, 2014: 4)

In other words, SI entails increasing both yields and eco-efficiencies via technological innovation and improvements in management (Garnett *et al.*, 2013; Godfray *et al.*, 2010). In short, in an SI approach, a key component of sustainable agriculture is growing more food with fewer resources.

A defining characteristic of SI is its openness to all forms of agriculture. Proponents argue that SI is not an 'ideologically driven' approach to agriculture sustainability in that there are no judgements as to whether or not specific forms of agriculture are sustainable. For example, Godfray and Garnett (2014: 6), leading academic proponents of SI, argue that 'ideas from biotechnology, conventional farming, agroecology, and organic farming may all be used to achieve this goal of sustainable yield increases.' While SI purports to be inclusive of all forms of agriculture, technological innovation is the area that is most emphasized. For example, proponents of SI tend to view precision agriculture – 'the information-based management of agricultural production systems' (Gebbers and Adamchuk, 2010: 828) – as a key tool for increasing the sustainability of agriculture. Precision agriculture entails the use of technologies and associated management practices to increase the efficiency of all aspects of production. Examples of precision agriculture technologies include crop management software programs, soil and water sensors, and geographic information systems for farm machinery.

Given SI's emphasis on technology, for proponents of SI, sustainability is largely an issue of innovation (Garnett *et al.*, 2013; Godfray, and Garnett, 2014). This means that accelerating the sustainability of agriculture is primarily a techno-scientific challenge. Specifically, it is a challenge of developing technologies and management practices that increase the eco-efficiency of agriculture. Thus, from this perspective of SI, the primary challenge in transitioning agriculture towards greater sustainability is developing incentive structures that foster the development and adoption of eco-efficient technologies and management practices.

Codifying sustainability: multi-stakeholder governance in the United States

Since the start of the millennium, MSIs have become a leading form of sustainability governance (Newell, Pattberg and Schroeder, 2012). Both business interests and civil-society organizations viewed MSIs as an opportunity to advance their interests. Namely, they offered companies a mechanism of self-governance and, thus, would allow them to avoid potentially more onerous governmental regulations (Gunningham and Kagan, 2005). At the same time, MSIs provided civil-society organizations with an alternative to relying on governmental action to advance their agenda (Bartley, 2007; King and Pearce 2010). Initially, MSIs were understood to have great potential as a democratic form of governance. This is because MSIs purport to bring together representatives of all potentially affected actors, and use participatory, consensus-based, and transparent practices (Beisheim and Dingwerth, 2008; Tamm Hallstrom and Bostrom, 2010).

However, recent research indicates that MSIs may favour some actors and interests, as a result of their embeddedness in the market (Cheyns and Riisgaard, 2014; Ponte, 2014). As Cashore notes (2002: 503), non-state governance systems 'derive their policy-making authority not from the state, but from the manipulation of global markets and attention to customer preferences.' Given MSIs' embeddedness in the marketplace, this means that, at least to a degree, their legitimacy is derived from the market. Consequently, MSIs may favour economic interests over other ones (e.g. environmental and social), and they may favour the positions of lead market actors (Konefal, Hatanaka, and Constance, forthcoming).

In the sections below, two MSIs that have developed sustainability metrics for U.S. agriculture are examined: Field to Market and SISC. First, we examine the general contours of each initiative, including their membership and governance procedures. Second, we provide an overview of the metrics that both Field to Market and SISC have developed.

Field to Market

Field to Market was convened by the Keystone Center in 2006 to explore the question of sustainability in U.S. agriculture. The initial meeting consisted of 12 stakeholders from agribusiness and civil society organizations. At these early meetings, a set of general principles and objectives was developed to guide the initiative. They included, first, a commitment to 'technology neutral' approaches to sustainability. Second, the process was to be 'science-based.' That is, the sustainability of U.S. agriculture would be scientifically assessed and the resultant metrics would be based on scientific research. Third, the initiative would be collaborative and seek to include actors throughout supply chains. Fourth, the initiative would seek to develop outcome-based metrics. Fifth, the initiative decided to focus on on-farm sustainability, as this is where the most immediate sustainability concerns were thought to be. Lastly, the focus of Field to Market would be commodity crops. Initially, this included corn, soy, cotton, and wheat. Potatoes, rice, barley, peanuts, and sugar beets have since been added.

From the initial 12 members, Field to Market's membership expanded to 28 members in 2009 when the first metrics were released. In 2012, when Field to Market released its second national report (Field to Market, 2012), membership had further expanded to 44 members. By 2018, its membership had further expanded to 126 members (Field to Market, 2018). Historically, the membership of Field to Market has been skewed towards lead actors in the conventional U.S. agrifood system (Konefal, Hatanaka, and Constance, forthcoming). For example, when Field to Market released its second national report in 2012, 31 of the 44 members were from grower associations and companies, as well as organizations associated with the input, processing, and retailing industries. Seven members were from mainstream environmental advocacy organizations, and there were no stakeholders representing labour or community development, or who were associated with alternative forms of agriculture (see Table 12.1).

In March 2013, Field to Market became an independent non-profit organization, which has led to a reorganization and formalization of its governance structure. First, a formal membership structure was created. It includes four full

Table 12.1 Field to Market Membership (Field to Market, 2018)

Agribusiness	*Brands & Retailers*	*Civil Society*	*Grower Organizations*	*Affiliate*
Agrible	American Bakers Association	American Farmland Trust	Farm Bureau	Alabama A&M and Auburn Universities Extension
Agricultural Retailers Association	American Crystal Sugar Company	C-AGG	American Peanut Council	Colorado State University
Certified Crop Advisor	General Mills	Conservation Technology Information Center	American Soybean Association	Columbia University School of Professional Studies
ADM	GMA	Ducks Unlimited	California Rice	Context
Bartlett	Ingredion	Environmental Defense Fund	Cotton Incorporated	Dartmouth
BASF	Kellogg's	Environmental Initiative	Edge	Farm Journal Media
Bayer CropScience	Land O'Lakes	National Association for Conservation Districts	Iowa Agriculture Water Alliance	Fleishman Hillard
Biotechnology Innovation Organization	Mars	Pheasants Forever	Illinois Corn Growers Association	Houston Engineering Inc.
Bunge	McDonald's	The Conservation Fund	Illinois Soybean Association	International Plant Nutrition Institute
Cargill	Nestle Purina	The Nature Conservancy	Indiana Soybean Alliance	Iowa State University
CF	Pepsico	World Wildlife Federation	Innovation Center for U.S. Dairy	Kansas State University
CHS	Riceland		National Alfalfa & Forage Alliance	Keystone Policy Center
Coolplanet	Coca-Cola		National Association of Wheat Farmers	Michigan State University AgBioResearch
Corn Refiners Association	Unilever		National Barley Growers Association	MIT Joint Program on the Science and Policy of Global Change

CropLife America	Walmart	National Corn Growers Association	National Association of State Conservation Agencies
Dow		National Cotton Council of America	National Association of State Departments of Agriculture
DuPont Pioneer		National Council of Farmer Cooperatives	North Carolina State University
FMC		National Farmers Union	Penton Agriculture
GMO Renewable Resources		National Sorghum Producers	Texas A&M Agrilife Research Extension
Healthy Food Ingredients		Nebraska Corn Board	Texas Tech University
Hortau		United Soybean Board	The Sustainability Consortium
IntelliFarms		USA Rice Federation	Thompson Coburn LLP
John Deere		U.S. Grains Council	University of Arkansas Division of Agriculture
K Coe Isom		U.S. Soybean Export Council	The University of Georgia
Leaf			The University of Nebraska Extension
LifeLine Foods			The University of Tennessee Extension
Marrone Bio Innovations			The University of Wisconsin-Madison College of Agricultural & Life Sciences
Monsanto			United States Department of Agriculture Natural Resources Conservation Service

continued

Table 12.1 Continued

Agribusiness	*Brands & Retailers*	*Civil Society*	*Grower Organizations*	*Affiliate*
MyFarms				
Nachurs				
National Milk Producers Federation				
New Leader				
Nutrien				
Redox				
Simplot				
Syngenta				
Tate & Lyle				
The Fertilizer Institute				
Mosaic				

membership categories (agribusiness, brands and retailers, civil society, and grower organizations). Full members have voting rights. There are also affiliate members, who mostly consist of university research centres, government agencies, and consultants. Affiliate members do not have voting rights, but participate in the general assembly and serve on standing committees. There are also associate members. Associate members are largely agrifood supply chain actors who participate in Field to Market, but do not have voting rights. Second, a formal governance structure was established. It consists of a general assembly, board of directors, executive committee, and several standing committees. Standing committees include metrics, verification, education and outreach, and awards and recognition. In addition, there are both science and technology advisory councils (Field to Market, 2018).

Metrics

In its most recent report, Field to Market defines

> sustainable agriculture as a system that meets the needs of the present while improving the ability of future generations to meet their own needs by increasing productivity to satisfy future food, fiber, and fuel demands and improving the environment, human health, and the socioeconomic well-being of agricultural communities.
>
> (Field to Market, 2016: 6)

Congruent with Field to Market's conceptualization of sustainability, Field to Market's five original metrics focus on 'overall efficiency as well as total resource use' (Field to Market, 2016: 10).[1] Its five original metrics, which were presented in its 2012 report, are (1) land use, (2) soil conservation, (3) irrigation water use, (4) energy use, and (5) greenhouse gas emissions.

The land use metric measures the efficiency of agricultural land by calculating planted area per unit of production. In other words, it is primarily a measure of crop yields. The soil conservation metric measures soil loss per acre. The irrigation water metric measures acre-inches of water applied per additional unit of production. That is, it is a 'measure of the overall efficiency of irrigation water applied in terms of the incremental improvement it produces in crop yield' (Field to Market, 2016: 9). The energy use metric measures both direct and indirect energy use. More specifically, it assesses the 'BTU of energy use per unit of production' (Field to Market, 2016: 17). The greenhouse gas emissions metric gauges both the direct and indirect production of carbon dioxide per unit of production.

In its 2016 report, Field to Market has developed three new environmental metrics: soil carbon, water quality, and biodiversity. In introducing these metrics, Field to Market noted that 'sustainable agriculture is broader than efficiency' (Field to Market, 2016: 10). Hence, these metrics extend the focus of Field to Market beyond agricultural productivity and eco-efficiencies. For the biodiversity metric, Field to Market has developed the Habitat Potential Index (HPI), which 'quantifies farm-level habitat and landscape change from year to year' (Field to Market, 2016: 70). The HPI will measure both production and non-production land cover types, such as forest, grasslands and savannas, wetlands, surface waters, and buffer strips. For its soil carbon metric, Field to Market is using a government soil conditioning metric developed by the USDA Natural Resource Conservation Service (NRCS). The soil conditioning metric measures soil organic matter, field operations, and soil erosion. It provides farmers with an indication of whether soil carbon is increasing or decreasing. For its water quality metric, Field to Market has adopted another USDA NCRS metric.[2] The water quality measures 'field characteristics, soil physical factors that influence erosion, nutrient management factors, tillage management factors, pest management factors, irrigation management, and additional conservation practices' (Field to Market, 2016: 87). It provides farmers with a score that indicates the water quality of their runoff.

In addition to developing metrics, Field to Market has also developed what it refers to as a Fieldprint® Calculator (Field to Market, 2013). The FieldPrint® Calculator is an online software program that farmers can use to measure their performance. Specifically, farmers enter their performance data on Field to Market's eight environmental metrics, and using Field to Market's algorithms the calculator analyses their sustainability performance. Field to Market also makes the farmers' aggregated data available to all of its members through its Fieldwork® Platform. This allows members to assess collective changes in U.S. agricultural sustainability.

Stewardship Index for Specialty Crops (SISC)

SISC was started in 2008 to develop sustainability metrics for speciality crops (i.e. fruits and vegetables) in the United States. It was founded by the Natural Resource Defense Council, Western Growers Association, and Sure Harvest, who then recruited additional producers, processors, buyers, and advocacy organizations to be part of the initiative.

SISC is organized into a three-part membership structure: (1) environmental and public interest groups; (2) growers, suppliers, and trade associations, and (3) buyers and trade associations. Membership is roughly balanced across the three groups (see Table 12.2). Currently, SISC's membership consists of 25 members across the three groups, and three additional experts (SISC, 2018). Membership includes lead market actors in food and agriculture, such as Unilever and Campbell's Soup, as well as major growers' associations (e.g. Western Growers). SISC also has several environmental advocacy organizations, and a small number of representatives that are more closely associated with alternative agriculture, such as the Community Alliance with Family Farmers and Jacobs Farm. Unlike Field to Market, which has undergone tremendous growth in membership, SISC has kept its membership size relatively constant. There is some turnover in SISC's membership, as some stakeholders rotate out and new ones join. For example, Walmart was previously a member, but this is no longer the case.

Table 12.2 SISC Membership (SISC 2018)

Environmental and Public Interest Groups	*Growers, Suppliers and Trade Associations*	*Buyers and Trade Associations*	*Additional Experts*
American Farmland Trust	Community Alliance with Family Farmers	California Sustainable Winegrowing Alliance	SurHarvest
Willamette Alliance	Jacobs Farm/DelCabo	Campbells Soup	Sustainable Food Lab
Natural Resource Defense Council	Farm Fresh Direct	Compass Group	University of Arkansas
The Nature Conservancy	Georgia Fruit and Vegetable Association	Del Monte Foods	
The Sustainability Consortium	National Potato Council	Produce Marketing Association	
Washington State University	Western Growers	Sodexo	
Sustainable Conservation	Agriculture Capital Management	Unilever	
The Freshwater Trust	Lipman Produce	Smuckers	
World Wildlife Fund			

These three groups, along with three independent expert members, make up the coordinating council. The coordinating council is responsible for the development and approval of metrics. To be approved, a metric must have the support of a majority of each of the three member groups. This means that a single membership group can block a metric, even if a majority of the other two membership groups support a potential metric. In addition to the coordinating council, there is a steering committee that oversees the initiative's daily activities. Two representatives from each of the three stakeholder groups constitute the steering committee. There is also a series of metric review committees made up of external stakeholders and experts that provide advisory input and feedback on proposed metrics (SISC, 2018).

Metrics

SISC does not define sustainable agriculture. Rather, it states that it 'regards sustainability as an ongoing commitment to advancing environmental, social and economic values' (SISC, 2018). It further states that it does not intend to 'identify who or what is "sustainable," but instead to provide common measurement tools to help all participants implement and communicate clearly about sustainable activities' (SISC, 2018). SISC's primary objective is to promote sustainable practices and not denounce those who are not engaged with becoming more sustainable. Congruent with this stated objective, SISC has focused on developing metrics through which farmers can measure inputs and outputs, and in so doing improve their sustainability.

In 2013, SISC released its first five metrics: applied water use efficiency, energy use, nitrogen use, phosphorous use, and soil organic matter. They have since added one additional metric: habitat and biodiversity (SISC, 2018a). The applied water metric measures the amounts of applied water needed to produce a crop. Specifically, farmer irrigation efficiency is calculated by dividing the acre-inches of applied water by the ton of product harvested. The energy use metric measures fuel and electricity consumption, and energy used in the production of fertilizers. Farmer energy efficiency is calculated by measuring total BTU by ton of product harvested. The nitrogen use metric measures the pounds of nitrogen applied per ton of product harvested. Nitrogen inputs include synthetic, organic, applied in irrigation water, and fixed nitrogen in leguminous crops. The phosphorous use metric measures the pounds of phosphorous added minus the pounds of phosphorous recommended per ton of product harvested. The soil organic matter metric gauges soil quality according to the amount of total organic carbon in the soil. Specifically, it measures current soil organic matter per soil organic matter potential.

The biodiversity and habitat metric 'documents and tracks the extent of on farm habitat and practices associated with positive biodiversity results' (SISC, 2018b: 1). This metric differs from the others in that it is both quantitative and qualitative, and assesses practices in addition to outcomes. Farmers need to document their acreage in production as well as non-productive acreage (e.g.

forest, riparian, and wetland). Additionally, they need to document their farming and conservation practices. Using their survey data, farmers can then calculate production and developed areas as a percentage of total farmland, natural habitats as a percentage of total farmland, and the number of management practices they implement for both cropped and non-cropped areas. Currently, SISC is working on developing two additional metrics: greenhouse gas emissions and simple irrigation efficiency. SISC has also developed a free demonstration metric calculator farmers can use to test their performance using SISC's metrics.

Field to Market, Stewardship Index for Specialty Crops, and sustainability

Field to Market and SISC have become powerful actors in agricultural sustainability governance in the United States. As each of their metrics has the potential to be widely implemented, Field to Market and SISC may substantially impact the form and content of sustainability transitions in U.S. agriculture. Thus, it is important to analyse how each envisions sustainable agriculture. Using the alternative agrifood movement and SI conceptualizations of sustainability, the vision of sustainability that Field to Market and SI are respectively advancing is assessed. The focus is on four components of sustainability: environmental impacts, measuring sustainability, farming practices, and the role of technology.

A first difference between the alternative agrifood and SI visions is how they conceptualize the environmental dimension of sustainability. The alternative agrifood vision of sustainability takes a systems perspective in that the focus is on maintaining the functional integrity of agro-ecological systems (Thompson, 2010). Hence, from the alternative agrifood perspective, environmentally sustainable agriculture is equated with the use of agroecological practices that integrate agricultural and ecological processes. Practically, this means that environmentally sustainable agriculture entails such things as 'organic matter accumulation and nutrient cycling, soil biological activity, natural control mechanisms, and resource conservation and regeneration.' (Altieri, 2002: 8). In contrast, the SI approach conceptualizes environmental sustainability more in terms of resource sufficiency (Thompson, 2010). Thus, from the perspective of SI, environmental sustainability is a question of ensuring sufficient natural resources and limiting the harmful externalities associated with agriculture (e.g. greenhouse gases).

Applying the alternative and SI notions of environmental sustainability to Field to Market and SISC, we find that the metrics of both initiatives are more closely aligned with SI. For Field to Market, its five original metrics are focused on productivity and eco-efficiencies: land use, soil conservation, irrigation water use, energy use, and greenhouse gas efficiencies. Similarly, four of SISC's six metrics – applied water use efficiency, energy use, nitrogen use, and phosphorous use – also focus on resource efficiencies. Hence, congruent with SI, both approaches tend to take an accounting approach to environmental sustainability in that there is a significant emphasis on calculating resource use. While the most recent metrics developed by Field to Market on biodiversity and soil carbon, and SISC on biodiversity, indicates a broadening of both initiatives to begin to incorporate environmental dimensions that are associated with agroecology, the focus of both initiatives continues to tilt towards resource sufficiency. For example, neither initiative has metrics focused on key agro-ecological processes, such as pest and weed control mechanisms.

A second difference between the alternative agrifood movement and SI vision is how they gauge agricultural sustainability. From the perspective of the alternative vision, the sustainability of agriculture is typically assessed according to farming and market practices. For example, are organic or agroecological processes being used?; is it local scaled?; is it embedded in the community?; and is it socially just? In contrast, in the SI approach, sustainability is gauged in terms of inputs and outputs. Thus, the emphasis is on the development and use of quantitative measurements that assess such things as resource use and greenhouse gas emissions. Both Field to Market and SISC take an SI approach in how they gauge sustainability. The metrics of both Field to Market and SISC focus on measuring farmer performance. Specifically, SISC has developed 'quantitative performance metrics' (SISC, 2018), while Field to Market is working to 'facilitate quantification and identification of key impact areas and trends' in agriculture (Field to Market, 2012: 5). Thus, congruent with SI, the focus is identifying, quantifying, and measuring resource inputs and their impacts on the environment.

A third difference between the alternative agrifood vision and SI is the relationship between agricultural practices and sustainability. As noted above, from the position of alternative agrifood movements, sustainable agriculture is characterized by specific practices. For example, for agroecologists there is a link between environmental sustainability and farming practices. Practices such as 'cover crops, green manures, intercropping, agroforestry, and crop-livestock mixtures' (Altieri, 2002: 8) are key to environmental sustainability. At the same time, other practices are rejected as environmentally unsustainable, such as monocropping, the use of synthetic pesticides and herbicides, and the use of genetically-modified organisms. In contrast, SI does not advance specific practices or kinds of agriculture as more sustainable than others. Rather, proponents of SI advance it as a neutral approach that views different farming styles and agricultural approaches as having the potential to contribute to environmental sustainability.

Both Field to Market and SISC strongly align with SI in terms of how they conceptualize the relationship between farming practices and sustainability. Neither initiative advances a particular form of agriculture as sustainable, nor makes any judgements as to the sustainability of specific agricultural practices or technologies. For example, neither Field to Market nor SISC has developed a metric for pesticides, herbicides or genetically-modified organisms. In interviews, stakeholder participants from both initiatives expressed the position that if a technology or practice can increase the sustainability of agriculture, its use should be considered. For example, a Field to Market committee member, who is with an environmental advocacy organization, commented: 'if we become convinced that GMOs [genetically-modified organisms] are a really critical part of getting to that vision of hyper productivity and environmental sustainability, then yes, GMOs are on the table.' This approach fits with SI, in that it is open to all forms of agriculture, *as long as* they contribute to sustainability. In contrast, for proponents of alternative agrifood sustainability, the use of chemical inputs and genetically-modified organisms tends to be viewed as unsustainable.

A fourth difference between the alternative and SI vision is the relationship between technologies and sustainability. Although alternative agrifood movements are not anti-technology, they are critical of many agricultural technologies. On the one hand, technologies, such as synthetic pesticides and herbicides, are criticized for their negative environmental impacts. On the other hand, technological innovations in agriculture are criticized for locking farmers on 'technology treadmills' that undermine their economic viability, particularly for small farmers (Cochrane, 1958). For these reasons, many proponents of alternative agrifood tend to conceptualize technological innovations as a key component of current agricultural unsustainabilities. In contrast, proponents of SI view technological innovation as one of the most important components for further increasing agricultural sustainability. While it purports to be ideologically neutral, a key component of nearly all conceptualizations of SI is the need for technological innovations focused on eco-efficiencies.

We find that there is some divergence between Field to Market and SISC in terms of the degree to which they stress the relationship between agricultural sustainability and technological innovation. While SISC's metrics incentivise technological innovation, SISC does not advocate for technological innovations as the key driver of a sustainability transition in U.S. agriculture. However, Field to Market positions technological innovation as key to accelerating the sustainability of agriculture. Field to Market takes the position that increases in the sustainability of agriculture (i.e. gains in productivity and eco-efficiencies) since 1980 have largely resulted from technological improvements (Field to Market 2016). They contend that a new round of technological innovations, particularly oriented towards eco-efficiencies, is needed to further accelerate the sustainability of U.S. agriculture. Thus, in terms of technological innovation, we find that Field to Market's vision of sustainability largely fits with SI given their strong advocacy of technological innovations.

In sum, the analysis of Field to Market and SISC and each of their metrics indicates that the form of agricultural sustainability that they are advancing more closely aligns with the SI approach. This represents a clear shift in the approach to agricultural sustainability in the United States, which has historically been associated with alternative agrifood movements. Hence, at least in the case of the United States, the codification of sustainability and the development of sustainability metrics using MSIs are affecting the meaning of agricultural sustainability. In particular, Field to Market and SISC have promulgated an approach to agricultural sustainability that focuses on the quantification and measurement of resource use and environmental impacts. It is an ideologically neutral approach in that it does not make any judgements as to the sustainability of different farming systems, practices, or technologies. As such, the sustainability metrics of both Field to Market and SISC are accommodative of the current structures and practices of the U.S. food and agriculture system.

Discussion

This chapter examines two initiatives that have positioned themselves at the forefront of transitioning U.S. agriculture towards greater sustainability. The absence of many aspects of the alternative vision of agricultural sustainability from the metrics developed by Field to Market and SISC indicates that much of the alternative vision of sustainability has little support among lead actors in the food and agriculture system. Rather, an SI vision of sustainable agriculture is being advanced through MSI governance for U.S. agriculture.

Given the widespread support of Field to Market and SISC by growers' associations, agribusiness, processors and retailers, an SI approach to agricultural sustainability has the potential to be widely implemented in U.S. agriculture. To date, there have already been 60 pilot projects using Field to Market metrics. These projects have included roughly 1,500 farmers and approximately two million acres of farmland, and many leading market actors as well, such as ADM, Cargill, General Mills, Unilever, and Kellogg's. While the widespread implementation of these metrics is not yet in place, given the involvement of lead market actors in these MSIs, the likelihood of imminent implementation appears to be quite high.

What do these findings mean regarding the notion and practices of 'sustainable agriculture' in the U.S.? We contend that the emergence of these MSIs is generating at least, a partial reframing of sustainable agriculture in the U.S. Historically, with its connections to alternative agrifood movements, sustainable agriculture is thought of in terms of a set of specific interconnected values and practices. Most notably, it has been associated with such values as environmentalism, sovereignty, localism, justice, and democracy. 'Sustainable' forms of agriculture, such as local and organic agriculture, are then understood as reflecting these values. In contrast, the approaches of Field to Market and SISC, and SI more generally, represent a more technical and calculative approach to sustainability. They fit with the increasing turn of sustainability governance towards 'technopolitics' (Freidberg, 2014). In this approach, sustainability is the purview of experts, science-based techniques and calculations, and is addressed through technical interventions. The result is a reframing of sustainability from 'norms to numbers' (Freidberg, 2013; 582).

While our analysis demonstrates that SI is on the rise in U.S. agriculture, it needs to be highlighted that at this point in time the trajectory of a sustainability transition in U.S. agriculture remains contested. On the one hand, it appears that the metrics of Field to Market and SISC are likely to be widely implemented across large swaths of U.S. agriculture. This means that mainstream conventional agriculture (i.e. not farmers engaged in alternative forms of agriculture, such as organic and local) is likely to be affected by efforts to implement SI. On the other hand, a wide array of alternative agrifood movements and initiatives continue to expand and gain niche market share (USDA ERS, 2017). Given these two trajectories, it is quite likely that different parts of U.S. agriculture will implement different visions of sustainable agriculture moving forward. While there may be room for both approaches, questions remain regarding the viability of both approaches to sustainability. Alternative agriculture faces questions as to its ability to feed the world, in terms of both productivity and affordability. At the same time, critics question whether SI is capable of generating significant environmental gains, as well as SI's viability for small farmers, given the potential costs of the technological innovations often associated with it.

Notes

1 While Field to Market has been more focused on environmental dimensions of sustainability, it has also developed five socioeconomic sustainability indicators: farm financial health, farm profitability, generation of economic value, worker safety, and labour productivity (Field to Market, 2016). However, these have yet to be incorporated into the Fieldprint® Calculator.

2 Field to Market states that the use of the USDA NCRS metric is temporary, while it finishes developing its own.

References

Allen, P. (2004). *Together at the Table: Sustainability and Sustenance in the American Agrifood System*, The Pennsylvania State University Press, University Park, PA.

Altieri, M. (2002). 'Agroecology: The science of natural resource management for poor farmers in marginal environments,' *Agriculture, Ecosystems and Environment*, vol. 193: 1–24.

Bartley, T. (2007). 'Institutional Emergence in an Era of Globalization: The Rise of Transnational Private Regulation of Labor and Environmental Conditions,' *American Journal of Sociology*, vol. 113, no. 2: 297–351.

Beisheim, M. and Dingwerth, K. (2008). 'Procedural Legitmacy and Private Transnational Governance: Are the Good Ones Doing Better,' *SFB-Governance Working Paper Series*, 14.

Busch, L. (2011). *Standards: Recipes for Reality*, MIT Press, Cambridge, MA.

Cashore, B. (2002). 'Legitimacy and the Privatization of Environmental Governance: How Non-State Market Driven (NSMD) Governance Systems Gain Rule Making Authority,' *Governance: An International Journal of Policy and Administration*, vol. 15, no. 4: 503–529.

Cheyns, E. and Riisgaard, L. (2014). 'Introduction to the symposium,' *Agriculture and Human Values*, vol. 31, no. 3: 409–423.

Cochrane, W. W. (1958). *Farm Prices: Myth and Reality*, University of Minnesota Press, St. Paul, MN.

Constance, D. H. (2009). 'Sustainable Agriculture in the United States: A Critical Examination of a Contested Process,' *Sustainability*, vol. 2, no. 1: 48–72.

Constance, D. H., Friedland, W. H., Renard, C. and Rivera-Ferre, M. G. (2014). 'The Discourse on Alternative Agrifood Movements,' in D. H. Constance, C. Renard and

M. G. Rivera-Ferre (eds), *Alternative Agrifood Movements: Patterns of Convergence and Divergence*, Emerald, New York: 3–46.

Field to Market (2012). *Environmental and Socioeconomic Indicators for Measuring Outcomes of On-Farm Agricultural Production in the United States: Second Report*, www.fieldtomarket.org (Accessed 18 September 2013).

Field to Market (2013). 'Fieldprint calculator,' https://calculator.fieldtomarket.org/fieldprint-calculator (Accessed 18 September 2013).

Field to Market (2016). *Environmental and Socioeconomic Indicators for Measuring Outcomes of On-Farm Agricultural Production in the United States Third Edition*, ISBN: 978–0-692–81902–9.

Field to Market (2018). 'Home,' https://fieldtomarket.org/ (Accessed 31 January 2018).

Forssell, S. and Lankoski, L. (2015). 'The sustainability promise of alternative food networks: an examination through "alternative" characteristics,' *Agriculture and Human Values*, vol. 32, no. 1: 63–75.

Freidberg, S. (2013). Calculating sustainability in supply chain capitalism, *Economy and Society*, 42(4): 571–596.

Freidberg, S. (2014). 'Footprint Technopolitics,' *Geoforum*, vol. 55, no. 2: 178–189.

Garnett, T., Appleby, M. C., Balmford, A., Bateman, I. J., Benton, T. G., Bloomer, P., Burlingame, B., Dawkins, M., Dolan, L., Fraser, D., Herrero, M., Hoffmann, I., Smith, P., Thornton, P. K., Toulmin, C., Vermeulen, S. J. and Godfray, H. C. (2013). 'Agriculture. Sustainable intensification in agriculture: premises and policies,' *Science*, vol. 341, no. 6141: 33–34.

Gebbers, R. and Adamchuk, V. I. (2010). 'Precision agriculture and food security,' *Science*, vol. 327, no. 5967: 828–831.

Gliessman, S. (2013). 'Agroecology: Growing the roots of resistance,' *Agroecology and Sustainable Food Systems*, vol. 37, no. 1: 19–31.

Godfray, H. C. J. and Garnett, T. (2014). 'Food security and sustainable intensification,' *Philosophical Transactions of the Royal Society*, vol. 369: 1–10.

Godfray, H. C. J., Beddington, J. R., Crute, I. R., Haddad, L., Lawrence, D., Muir, J. F., Pretty, J., Robinson, S., Thomas, S. M. and Toulmin, C. (2010). 'Food security: the challenge of feeding 9 billion people, *Science*, vol. 327, no. 5967: 812–818.

Gold, M. V. (2015). 'Sustainable Agriculture: Definitions and Terms,' Special Reference Briefs Series no. SRB 99–02, United States Department of Agriculture, National Agricultural Library, Beltsville, MD. www.nal.USDa.gov/afsic/sustainable-agriculture-definitions-and-terms#toc1 (Accessed 18 April 2018).

Gunningham, N. and Kagan, R. A. (2005). 'Regulation and Business Behavior,' *Law & Policy*, vol. 27, no. 2: 213–218.

Hatanka, M., Konefal, J. and Constance, D. H. (2012). 'A tripartite standards regime analysis of the contested development of a sustainable agriculture standard,' *Agriculture and Human Values*, vol. 29, no. 1: 65–78.

Hinrichs, C. (2000). 'Embeddedness and local food systems: notes on two types of direct agricultural market,' *Journal of Rural Studies*, vol. 16, no. 3: 295–303.

Hinrichs, C and Eshleman, J. (2014). 'Agrifood movements: Diversity, aims, and limits,' in C. Bailey, L. Jensen and E. Ransom (eds) *Rural America in a Globalizing World: Problems and Prospects for the 2010s*, West Virginia University Press, Morgantown, WV: 138–155.

Holt Giménez, E. and Shattuck, A. (2011). 'Food crises, food regimes and food movements: rumblings of reform or tides of transformation?' *The Journal of Peasant Studies*, vol. 38, no. 1: 109–144.

Johnston, J. (2008). 'The citizen-consumer hybrid: ideological tensions and the case of Whole Foods Market,' *Theory and Society*, vol. 37, no. 3: 229–270.

Khan, S. and Hanjra, M. A. (2009). 'Footprints of water and energy inputs in food production – Global perspectives,' *Food Policy*, vol. 34, no. 2: 130–140.

King, B. G. and Pearce, N. A. (2010). 'The Contentiousness of Markets: Politics, Social Movements, and Institutional Change in Markets,' *Annual Review of Sociology*, vol. 36: 249–267.

Konefal, J. Hatanaka, M. and Constance, D. H. (forthcoming). 'Multi-stakeholder initiatives and the divergent construction and implementation of sustainable agriculture in the USA,' *Renewable Agriculture and Food Systems*.

Lobao, L. and Sharp, J. (2013). 'Agriculture and Rural Development,' in G. Green (ed.), *Handbook of Rural Development*, Edward Elgar, North Hampton, MA: 115–138.

Loconto, A. (2010). 'Sustainably Performed: Reconciling Global Value Chain Governance and Performativity,' *Journal of Rural Social Sciences*, vol. 25, no. 2: 193–225.

Loconto, A. and Fouilleux, E. (2014). 'Politics of private regulation: ISEAL and the shaping of transnational sustainability governance,' *Regulation & Governance*, vol. 8, no. 2: 166–185.

Marsden, T. (2012). 'Towards a Real Sustainable Agri-food Security and Food Policy: Beyond the Ecological Fallacies?' *The Political Quarterly*, vol. 83, no. 1: 139–145.

Marsden, T. and Farioli, F. (2015). 'Natural powers: from the bio-economy to the eco-economy and sustainable place-making,' *Sustainability Science*, vol. 10, no. 2: 331–344.

Morley, A., McEntee, J. and Marsden, T. (2014). 'Food futures: framing the crisis,' in T. Marsden and A. Morley (eds), *Sustainable Food Systems: Building a New Paradigm*, Routledge, New York: 30–36.

National Research Council (2010). *Toward Sustainable Agriculture Systems in the 21st century*. The National Academy Press, Washington, DC.

Newell, P., Pattberg, P. and Schroeder, H. (2012). 'Multiactor governance and the environment,' *Annual Review of Environment and Resources*, vol. 37: 365–387.

Ponte, S. (2014). 'Roundtabling sustainability: lessons from the biofuel industry,' *Geoforum*, vol. 54: 261–271.

Renting, H., Marsden, T. and Banks, J. (2003). 'Understanding alternative food networks: exploring the role of short food supply chains in rural development,' *Environment and Planning* A, vol. 35, no. 3: 393–411.

Stewardship Index for Specialty Crops (SISC) (2018). 'Home,' www.stewardshipindex.org/ (Accessed 31 January 2018).

Stewardship Index for Specialty Crops (SISC) (2018a). 'Working Metrics 1.0,' www.stewardshipindex.org/working_metrics.php (Accessed 31 January 2018).

Stewardship Index for Specialty Crops (SISC) (2018b). 'Habitat and Biodiversity,' www.stewardshipindex.org/amass/documents/document/16/SISC-Habitat-and-Biodiversity_final.pdf (Accessed 31 January 2018).

Tamm Hallstrom, K. and Bostrom, M. (2010). *Transnational Multi-Stakeholder Standardization*, Edward Elgar, Northampton, MA.

Thompson, P. B. (2010). *The Agrarian Vision: Sustainability and Environmental Ethics*, University of Kentucky Press, Lexington, KY.

United States Department of Agriculture, Economic Research Service (2017). *Organic Market Overview*, www.ers.USDa.gov/topics/natural-resources-environment/organic-agriculture/organic-market-overview/ (Accessed 31 January 2018).

Unruh, G. and Ettenson, G. (2010). 'Winning in the Green Frenzy,' *Harvard Business Review*, November: 110–116.

13 Understanding the challenge of problem definition in multistakeholder initiatives

Lessons from sustainability policy frames in Canadian non-state food strategies

Margaret Bancerz

Stubborn policy controversies (Rein and Schön, 1993), or wicked policy issues (Churchman, 1967) are becoming more prevalent as society becomes more complex. These are policy issues that are not well defined and are extremely difficult to solve (Churchman, 1967; Head, 2008; Rittel and Webber, 1973). A good example of a wicked policy issue is food policy (Peters and Pierre, 2014). It is complex and multidimensional, touching social, economic, environmental, and cultural spheres, and thus reaching into several policy domains (Barling, Lang and Caraher, 2002). Lang (2009) argues that food policy is about

> how food is produced and grown to how it is processed, distributed and consumed; from the structures that shape food supply to those that determine health and environment; from the sciences and processes that unlock food's potential to the formal governance and lobbies that seek to control it; from the impact the food system's dynamics have on society to the way its demands are factored into policy-making itself.
>
> (2009: 157)

From this definition, it is obvious that food policy is extremely complex and very political, requiring a comprehensive approach to the different policy domains it touches. As a result, food policy space is always contested (Lang, 2009). Traditionally, Canada has only developed agricultural and food safety policies (Hedley, 2006; MacRae, 2011), but has never had an integrated policy reflecting the complexities of food. It was not until 2015 that the Canadian federal government mandated the creation of a comprehensive food policy for Canada. In 2017, the government launched a multistakeholder consultation process to develop this policy.

Collaboration between multiple actors has become a widespread method used to address complex policy issues like food. One of the most popular methods employed is multistakeholder initiatives (MSIs). MSIs are considered fitting governance mechanisms applied to wicked policy issues (Calton and Payne, 2003; Bevir and Richards, 2009), because they include many actors affected by the issue who are searching for a solution. However, MSIs are also very challenging and come with their own set of governance issues. A good example of such a challenge is developing a shared understanding of a policy issue (Rühli *et al.*, 2015).

This chapter serves as a warning to policymakers to be aware of the common challenges found in MSIs, such as the difficulties in forging a shared understanding of issues. To show this, I have chosen to look at the issue of sustainability in all four non-state food strategies in Canada. Sustainability was chosen because it is perceived as a common consensus frame. This is a frame based on a concept that enjoys widespread acceptance in terms of its shared values and objectives (Candel *et al.*, 2014). In this study, the sustainability frame therefore functions as 'low-hanging fruit' to show that even with a consensus frame like sustainability, it proves difficult to forge a common understanding of not only the issue, but a solution as well. I demonstrate this by analysing the sustainability frame in each non-state food strategy using critical frame analysis.

This chapter begins with a brief overview of sustainability framing and how it relates to problem definitions and policy solutions. Next, I provide a short literature review of MSIs, and a brief description of each strategy examined. The methods, results, discussion and conclusion follow.

Policy framing and sustainability

The manner in which problems are framed influences and shapes the way issues are thought about and how issues are acted upon (Shaw, 2010). Policy is inherently ideological and policymaking is a strategic use of discourse through words, language, and framing (Entman, 1993). Framing highlights specific features of reality promoted by ideology to support 'a particular problem definition, causal interpretation, moral evaluation, and/or treatment recommendation' (1999: 52). Policy framing begins with the assumption that policy is not objective and can be interpreted in multiple ways (Verloo, 2005). However, this is challenging because it produces multiple social realities and can promote diverse calls to action (Rein and Schön, 1993). Verloo (2005) defines a policy frame as 'an organising principle that transforms fragmentary or incidental information into a structure and meaningful policy problem, in which a solution is implicitly or explicitly enclosed' (2005: 20). This means that a frame is a specific lens used to interpret and understand a particular problem that is then structured into a policy solution. Policy frames are not seen as intentional ways in which actors represent reality (Verloo and Lombardo, 2007). Therefore, they do not describe an objective reality, but rather shape reality and give it meaning through the construction of policy (Verloo, 2005).

Bacchi (2000) argues that framed policy discourses reveal obvious solutions within the parameters of their defined policy issue. This is because the problem that is constructed matches the solutions proposed. Frame analysis therefore allows the researcher to deconstruct and expose implicit and explicit dimensions of an argument that can encourage collective policy action (Fletcher, 2009). To explore policy framing and its impacts, this chapter looks at sustainability and how it has been framed in food policy documents developed by non-state actors in Canada. Sustainability is perceived as a consensus frame. This is a frame based on a concept that enjoys widespread acceptance in terms of its shared values and objectives. It would be difficult to find someone against 'sustainability' (Candel *et al.*, 2014). Sustainability is a frequently used concept, and is often used in opposing ways. While, as a broad term, it may resonate with everyone, when used by groups subscribing to different worldviews, it becomes very contested (Candel *et al.*, 2014). Disagreement likely ensues when more precise action strategies are developed out of sustainability.

The general idea of sustainability originated from the United Nation's document, *Our Common Future* or *The Brundtland Report*, and its definition of sustainable development: 'development that meets the needs of the present without compromising the ability of future generations to meet their own needs' (World Commission on Environment and Development, 1987: 43). From this, the idea of sustainability was derived as a three-legged stool approach which included environmental, economic, and social sustainability. Nonetheless, this was a very broad definition. As such, it has since been moulded to fit different policy goals and interests, placing an emphasis on specific pillars (environmental, economic, and social).

Aiking and de Boer (2004) argue that the relationship between food and sustainability was formed alongside the global discussions on sustainability in the 1980s. The modern or industrial food system has had many adverse environmental and health effects on humans and the planet (Horrigan, Lawrence and Walker, 2002). Many, from non-governmental organizations, to global institutions, to large food companies have realized that the current direction of the global food system is unsustainable and are calling and attempting to develop a more sustainable one. However, these concerted efforts are very often contradictory. These inconsistencies stem from different framings of sustainability and how they are interpreted and applied to the food system. Gillespie Jr. (2010) summarizes this issue very well:

> I think the biggest challenge that we face in achieving a sustainable food system is neither determining what we want nor developing the knowledge needed to get us there, but rather it is how do we actually get there, that is, 'the steering question' of collective action.
>
> (2010: 3)

Multistakeholder Initiatives (MSIs)

MSIs represent a novel governing tool that became popular in the 1990s in the international arena[1] (Pattberg and Widerberg, 2016). Martens (2007) argues that MSIs originated from the general frustration of governments, international organizations, and non-governmental organizations because of the slow pace at which global negotiation processes proceeded, and the absence of will and capacity of a number of players to commit to global agreements. Bäckstrand (2006), on the other hand, argues that MSIs arose because of the inability of intergovernmental diplomacy to confront increasingly multidimensional and complex issues. MSIs are voluntary partnerships that are made up of both state and non-state actors from different sectors (Pinkse and Kolk, 2012; Powers and Jablonski, 2015). MSIs include stakeholders from business, civil society, and governments (Roloff, 2008) and have played vital roles in forming international sustainable development policies (Pattberg and Widerberg, 2016). These initiatives often form non-hierarchal processes that create level-playing fields requiring shared responsibility and cooperation between those involved. The objective of MSIs is to bring different stakeholders together who have been affected by a shared and complex issue in order to produce a policy solution. Multistakeholder collaborations tend to be issue driven and are created when different stakeholders are compelled to cooperate around a pressing and multifaceted issue (Roloff, 2008), combining their expertise and skills to solve it (Fransen, 2012). MSIs can vary in format, organization, and agenda (Epstein and Nonnecke, 2016), and so there is no ideal design of an MSI.

There are many benefits when using MSIs to solve complex policy issues. They often present opportunities to learn from other stakeholders through interaction, which can also overcome learning gaps in the group (Pinkse and Kolk, 2012). MSIs also develop opportunities for capacity-building between governments, industry, and civil society which can produce valuable social bonds and trust between stakeholders (Antonova, 2011). Over time, common values, norms, and goals are developed that can prevail over stereotypes, prejudices, and discrimination in relationships between stakeholders (Rühli *et al.*, 2015). Rühli *et al.* (2015) find that MSIs are especially constructive when solving wicked policy issues because there is a lot of dialogue, collaboration, and inclusivity in the group to dissect the issue. MSIs may also address deficits in governance, implementation, and participation that often appear in traditional governance formats (Bäckstrand, 2006).

Even though there are many benefits to MSIs, they also carry a lot of risk. While at times MSIs may address governance and participation deficits (Bäckstrand, 2006), they can also conceal inequities between stakeholders in the negotiation (Edmunds and Wollenburg, 2001) and decision-making processes (Almeida, Getschko and Afonso, 2015). Other false assumptions about MSIs include the idea that: they are neutral and objective spaces of negotiation; that all stakeholders must be involved for the process to be effective; and that any failures of MSIs can be contributed to poor communication (Edmunds and Wollenburg, 2001). In fact, one of the major reasons as to why MSIs fail is because of the absence of a common vision. Problem definitions and measurable goals must be agreed upon at the beginning of the partnership to ensure the success of the MSI (Pattberg and Widerberg, 2016). It is vital that the stakeholders involved in the MSI produce a shared understanding of the common policy issue in order to determine how to proceed in the partnership (Rühli *et al.*,

2015). Problem definition alignment must be developed with all participants involved in the MSI. Detailed and clearly defined goals also have a stabilizing impact on the partnership, especially when those involved are investing their own time and resources (Rühli *et al*., 2015).

What makes this more difficult is that language, especially scientific language, is never neutral. Words can be interpreted differently by various actors depending on experiences, and previous perceptions of issues (Edmunds and Wollenberg, 2001). A different frame may be used to examine or understand the same issue. It is often assumed that out of every MSI, consensus emerges (or must emerge), however this may not always be possible. Often, even if it seems as if consensus has been achieved, the 'consensus' itself may continue to connote different meanings for different actors. Accordingly, clarity is key and should always be sought (Edmunds and Wollenberg, 2001).

The organizations behind Canada's non-state food strategies

In total, there have only been four national level food strategies or policies developed in Canada. All four of these were developed by non-governmental organizations described below.

The Canadian Federation of Agriculture

The Canadian Federation of Agriculture (CFA), founded in 1935, is Canada's largest farm organization made up of provincial farm groups and commodity groups. It represents over 200,000 Canadian farmers in Ottawa and speaks on their behalf with government. The goal of the organization is to enable the success of Canadian farmers (CFA, n.d.), and to bring forward a policy consensus to the Canadian government from the farming community (personal communication, 19 November 2015). The CFA responds to policies created by government affecting the farming community, but also suggests policy options to the government (personal communication, 19 November 2015).

The CFA came out with its National Food Strategy in 2011 and largely drew on the participation of farmers, input suppliers, and processors (personal communication, 13 November 2015). Sponsors included Syngenta, RBC Royal Bank, Keystone Agricultural Producers, the Ontario Federation of Agriculture, Dairy Farmers of Canada, Egg Farmers of Canada, Canadian Pork Council, Kubota, Chicken Farmers of Canada, and Canadian Hatching Egg Producers (Currie and Etsell, 2011).

Canadian Agri-food Policy Institute

The Canadian Agri-food Policy Institute (CAPI) is an independent, non-partisan policy research institute that focuses on the agri-food sector. It was established in 2004 by the federal government. Its stakeholders are broadly defined as agri-food players (CAPI, n.d.a). CAPI is funded by Agriculture and

Agri-Food Canada, some provinces, and Farm Credit Canada. Projects are also funded by different organizations and the private sector. Its mandate is to take a mid to long term perspective on emerging issues in agriculture, and to create a space for research and balanced discussion relating to current policy and the food system (personal communication, 1 June 2015).

CAPI's strategy, *Canada's Agri-food Destination*, was developed in 2011 by a mix of agri-food value chain actors, government, academia, and non-state organizations (CAPI, 2011). It was sponsored by a number of organizations: Alberta Agriculture and Rural Development, Campbell Company of Canada, Canadian Federation of Independent Grocers, Canadian Pork Council, Canola Council of Canada, Dairy Farmers of Canada, Dieticians of Canada, Environment Canada, Health Canada, International Institute for Sustainable Development, Maple Leaf Foods Inc., Manitoba Agriculture, Food & Rural Initiatives, McGill World Platform for Health and Economic Convergence (McGill University), Meyers Norris Penny LLP, Ministère de L'Agriculture, des Pêcheries et de l'Alimentation du Québec, Nestlé Nutrition Canada, North America Strategic Agriculture Institute, Nova Scotia Agricultural College, Nova Scotia Department of Agriculture, Public Health Agency of Canada, Pulse Canada, Research and Development Institute for the Agri-Environment, Richard Ivey School of Business (University of Western Ontario), Royal Bank of Canada, Sustain Ontario, Spur Ventures, L'union des producteurs agricoles du Québec, University of British Columbia, University of Toronto, University of Waterloo, Veg Pro International Inc., Vineland Research and Innovation Centre, and Y U Ranch (CAPI, n.d.b).

Food Secure Canada (FSC)

FSC is a non-governmental organization and leader of the Canadian food movement. It is a national member-based organization made up of individuals and organizations that come together to advance food security and food sovereignty through three objectives: zero hunger, healthy and safe food, and sustainable food systems (FSC, n.d.).

Around 2010, FSC began engaging in campaigns to develop a national food policy where it worked to bring peoples' voices to the federal government (personal communication, 26 August 2015). The People's Food Policy Platform (PFPP) was developed by Food Secure Canada (FSC) in 2011. The PFPP was formed out of cross-national talks with over 3,500 Canadians. The PFPP had core funding from Heifer International Canada, with additional sponsorship from Inter Pares, the Assembly of First Nations and USC Canada. It was instigated by members of FSC (FSC, 2011).

Conference Board of Canada

The Canadian Food Strategy (CFS) was developed by the Centre for Food in Canada, housed within the think tank, Conference Board of Canada (CBoC). The CBoC identifies itself as an independent, not-for-profit, objective, non-partisan, and non-state organization with expertise in economic trends, organizational performance, and public policy issues. It has defined itself as leading the way in shaping national dialogue on important policy issues (CBoC, 2010: 5).

The CBoC's Strategy was completed in 2014. It was based on 20 research reports, and benefitted from three annual Food Summits that brought together various state and non-state stakeholders, consultations, and a board of investors including industry, government, foundations, and academia (Bloom, 2014). The CFS had a number of investors: Loblaw Companies Ltd., Maple Leaf Foods, Heinz Canada, Nestlé Canada Inc., Parmalat Canada, Ontario Ministry of Agriculture, Food and Rural Affairs, Ontario Ministry of Health and Long-Term Care, Agriculture and Agri-Food Canada, Public Health Agency of Canada, Cargill Canada, Cott Beverages Canada, McCain Foods Canada, Weston Foods, Farm Credit Canada, Cavendish Farms (Irving Group), Olymel L.P., Saputo Inc., Deloitte & Touche LLP, IBM Canada, PepsiCo Canada, the Government of Prince Edward Island, KPMG LLP, The J.W. McConnell Family Foundation, Metcalf Foundation, The Canadian Federation of Agriculture, The Heart and Stroke Foundation, and the University of Guelph.[2]

Methodology – critical frame analysis

Four Canadian non-state food strategies were analysed using critical frame analysis, focusing specifically on 'sustainability' in policy documents. To capture the nuances of sustainability framing within a food strategy, sections in the strategies with explicit reference to sustainability were targeted. In addition, the introductions of each document, contextualizing the strategy and the purpose behind it were also used to gain a better understanding of the organization's approach to food sustainability.

According to Verloo (2005), policy frames have standard formats which include an implicit or explicit depiction of a diagnosis (problem), and an implicit or explicit prognosis (solution). Critical frame analysis was used to map the way in which an issue was framed (Verloo and Lombardo, 2007). Data was analysed using sensitizing questions (Verloo, 2005), looking for 'crucial dimensions of frames' rather than developing 'hierarchical sets of codes' as in content analysis, or 'typologies of frames' as in some discourse analysis (van der Haar and Verloo, 2016). I used a modified version of Verloo's (2005) framework methodology of critical frame analysis to analyse all four food strategies. There were two general elements to the framework: the policy problem constructed (diagnosis), and the policy solution constructed (prognosis). Both the problem and solution sections had a subsection identifying attribution of roles in the policy problem and the policy solution. 'Sensitizing questions' were used under each element to identify themes in each policy document:

- Policy problem (diagnosis)
    - What is represented as the problem?
    - What is seen as a problem?

- Roles attributed to various actors in the policy problem (attribution of responsibility)
    - Causality (who is seen to have made the problem?)
    - Responsibility (who is seen as responsible for the problem?)
    - Problem holders (whose problem is it seen to be?)
- Policy solution (prognosis)
    - What to do?
    - How to achieve goals (strategy/means/instrument)?
    - Hierarchy/priority in goals
- Roles attributed to various actors in the policy solution
    - Call for action and non-action (who should [not] do what?)
    - Who is acted upon (targeted groups)?

These questions constructed a template for coding policy documents to develop a summarized 'super-text' which made implicit meanings of the policy document more explicit (Verloo, 2005).

Results

Below, I present a summary analysis of each document based on the answers to the sensitizing questions listed above.

Canadian Federation of Agriculture's (CFA) 'National Food Strategy'

The CFA's Strategy interspersed sustainability throughout the document. The problem was seen as the food system not functioning 'as effectively as it should' and not being 'sustainable in its current form' (Currie and Etsell, 2011: 3). The CFA presented several issues in the Canadian food system: a rising global population requiring a 'sustainable food supply' (CFA, n.d.: 2), poor eating habits and lifestyles of Canadians which have contributed to poor health (Ibid.: 2), the nominal and negative income of food producers and processors across Canada (Ibid.: 2), the exit of the food processing industry in Canada (Ibid.: 2), complex regulatory infrastructure harming the food producing, processing, and distribution sectors (Ibid.: 2), and the importation of foods that do not align with Canadian food and input safety requirements (Ibid.: 3). No clear blame was attributed to any actor or institution throughout the document, however, it was insinuated that the reason for these issues is ineffective governance and policies. The government and the private sector appeared to be responsible for the problems (Ibid.: 12), and the problem is described as affecting Canada as a whole.

Solutions were outlined in nine strategic objectives and their relevant key outcomes. There did not appear to be a hierarchy or priority in goals, although there was a clear emphasis on economic and environmental goals. Most policy solutions were assumed to be achieved through policy. Out of the document's nine strategic objectives (or main drivers of the Strategy), three included sustainability as part of their objectives, and four included sustainability in their key outcomes. The first strategic objective, 'The Canadian food chain is driven by its diverse, sustainable innovative and profitable farm and food supply sectors' (Currie and Etsell, 2011: 9), discussed investment in innovation, research and agricultural career promotion, competitiveness within Canada and abroad, more efficient food chains, and the rewarding of Canadian producers and processors for providing ecosystem goods and services. The second strategic objective, 'Canada's food chain capacity meets future demand in a sustainable manner' (Ibid.: 10) emphasized sustainably increasing both production and processing of food, succession planning, balancing energy and food crops, growing a sustainable workforce, decreasing food waste throughout the food chain, conserving and using more renewable resources, expanding research for increasing sustainable production, renewable inputs and outputs, and resiliency of the sector in the face of climate change (Ibid.: 10). The final strategic objective, 'Canada is a global leader in ensuring international governance agreements impacting food are based on science, sustainability, fairness, consistency and enforceability' (Ibid.: 11) focused on fair international agreements that do not damage the domestic market and promote environmental standards in food processing, production and inspection, increasing food security globally, and Canada becoming a leader in reducing greenhouse gas emissions and increased use of renewable resources in the food sector (Currie and Etsell, 2011: 11).

Other key outcomes peppered in strategic objectives largely focused on sustainable production (Currie and Etsell, 2011: 6–8), sustainable processing (Ibid.: 8), sustainable systems (Ibid.), and sustainable energy and the growing of food and energy crops (Ibid.: 11). Roles for fulfilling policy solutions throughout the document were largely given to government, producers, and processors. Occasionally, researchers and consumers were called upon to assist with policy solutions. Generally, governments, producers, and processors were targeted groups of action in policy solutions.

Canadian Agri-Food Policy Institute's (CAPI) 'Canada's Agri-Food Destination'

CAPI's Strategy generally encompassed sustainability in one area of its document. In its introduction, *Canada's Agri-Food Destination* outlined key food system issues. The general problem was represented as Canada not realizing its full potential as a major agricultural producer (CAPI, 2011: 5). The causes of the issues outlined in the document were a direct result of current policies and practices in the food sector (Ibid.). There were several problems described at the outset of the document: subpar agrifood performance, unsustainable healthcare and poor diets and lifestyles, a growing global population and subsequent need for increased food production and the consequential rise in environmental impact, as well as falling research and development investment in Canada, inadequate responses to the issues in the food sector, and a generally deficient Canadian economy (Ibid.). The document also had five 'enablers of change,' one of which was the 'leadership in sustainability' which encompassed the large part of discussions on sustainability. While it did not explicitly identify issues in this section, it did repeat the necessity to produce more food because of a growing global population, the desire of consumers to be more informed about where their food comes from, and the need to set domestically developed sustainability standards (Ibid.: 66). The government was blamed for these issues resulting from inadequate and ineffective policies. Problem holders were not made evident, but it is assumed from the document that everyone in Canada suffers because of this policy inadequacy.

The section on sustainability provided clear solutions to issues specifically related to sustainability in the sector: the development of a domestic sustainability standard, the creation of sustainability farm plans, developing national programmes supporting actors in the food system that produce ecological goods and services, coordination of public and private science and technology research, and finally the development of a national climate change agrifood strategy. While there was no hierarchy between the handfuls of solutions provided, there was a clear propensity for a strong Canadian agri-food economy. The call to action was made towards the government, and to an extent the private sector, to improve the policies surrounding the sector. Producers and processors were targeted groups in policy solutions.

Food Secure Canada's (FSC) 'Resetting the Table'

FSC integrated sustainability throughout its document. As such, there were no distinct sections in the document where sustainability was discussed exclusively. Food system problems were described at the beginning of the document which included high rates of obesity, food insecurity, farmers and fishers going out of business, the environment being pushed to its limits, no national school meal plan, and no citizen participation in the food system (FSC, 2011: 1, 5). The current industrial food system was represented as the problem: the document ascribed fault to major food companies, financial speculators, and a patchwork of government policies that do not address the complexities of the food system. This problem was described as affecting major groups in the sector including farmers, fishers, Indigenous peoples, and citizens. Interspersed references to sustainability focused on sustainable management of the food ecosystem including fisheries (Ibid.: 12, 18), encouraging sustainable practices that do not harm the ecosystem, the fostering of strong and sustainable economic development of rural economies, and education promoting environmental sustainability (Ibid.: 18). There did not appear to be a hierarchy between goals, although there was a strong ecological and social focus throughout the document. Policy was used to developed solutions which were meant to benefit producers, fishers, and citizens. Government was called to action and targeted groups were Indigenous peoples, farmers and fishers, and citizens.

Conference Board of Canada's (CBoC) 'Canadian Food Strategy'

The CBoC's Strategy had a clear section in the document entitled 'Environmental Sustainability' that was one of five key elements of the Strategy. Under each element were goals, desired outcomes, action strategies, and specific actions. The introduction of the document gave a brief overview of some issues in the food system. These included the need to grow Canada's agri-food sector, the rise in global food demand and the subsequent requirement to produce more food, the need to develop more domestic opportunities for industry and to create new markets, consumer concerns which include food safety, health, environment, and household security, the issue of Canada's food industry not being prosperous and viable in the long-term, and the need to increase agri-food exports sustainably (Bloom, 2014: 4). The problem was represented as a missed opportunity to grow Canada's food industry and export market. However, no blame was attached to the current sector's situation. The environmental sustainability section did not expand on specific issues, but emphasized the need to increase production and export levels while reducing environmental impact, the sustainable management of natural resources, waste reduction, and climate change adaptation (Bloom, 2014: 15). Action strategies focused on decreasing food waste by improving harvest and production methods and supply chain efficiency. Education was emphasized to encourage consumers to purchase environmentally friendly products and become food waste literate. The section also discussed sustainable management of natural resources, including a more efficient use of energy and low-carbon food systems, the protection and maintenance of land and biodiversity, improving air quality and decreasing greenhouse gas emissions, the joint development of environmental standards and labels, and a national metric system assessing the food system's environmental impact (Bloom, 2014: 25). There did not appear to be a hierarchy in desired outcomes or action strategies. Most solutions required policy from both public and private sectors; consumers played a small part in policy solutions. Industry, including producers and fisheries, as well as consumers were targeted groups in solution formulation (Bloom, 2014: 25).

Discussion

Each policy document was examined using the modified version of Verloo's (2005) critical frame analysis framework explained above. However, analysis proved to be more challenging with certain documents. While CAPI and CBoC had special chapters devoted to sustainability, CFA, and especially FSC, integrated sustainability throughout their documents. In these cases, and especially with the FSC, the problem definitions and contextualization at the beginning of policy documents were used as areas to identify policy problems.

All documents included sustainability as part of their food strategies for Canada. While there were some overlaps between documents in terms of problem definition of sustainability in the food system, solutions tended to vary more. The CFA, CAPI, and CBoC all began by contextualizing the food problem as one of growing global populations, increased demand for food, and therefore a required increased growth in food production, which would increase agriculture's environmental impact. The CFA, CAPI, and FSC all emphasized the problem with current policies that were either inadequate or harmful to the food sector. All documents mentioned the dire situation of the food industry in Canada. Where the CFA and FSC specifically focused on producers, CAPI mentioned the 'sector' in general, and CBoC indicated 'industry' in general. Although there was some overlap, there were also several distinct problem definitions. For example, the CFA mentioned the import of food products not meeting Canadian food and input safety requirements; CAPI mentioned decreased investment research and development; FSC focused on the Canadian environment being pushed to its limits, and CBoC on the need to create more markets and increase exports more sustainably.

Every strategy implied that the problem was one concerning all Canadians, with FSC focusing specifically on farmers, fishers, Indigenous peoples, and citizens. Both the CFA and CAPI regarded ineffective policies as the causes of sustainability problems in the sector. The FSC also blamed policy, but particularly faulted major food companies and financial speculators. The CBoC did not blame anyone or anything in particular.

The 'policy problem' in each document was also represented slightly differently between strategies. The CFA represented the issue as the food system not functioning as effectively as it should; CAPI represented it as Canada not realizing its full potential as a major agri-food producer; FSC as the current industrial agri-food system; and the CBoC as a missed opportunity for the sector. The major difference between these problem representations is that the FSC identified food system issues as developing out of *structural* issues, while the other organizations deemed this a *policy failure*. Nonetheless, every organization sought policy change in their document.

The variations in problem definition or solution formulation can be attributed to the different framings of sustainability in the documents. While some major ideas of sustainability problems in the Canadian food system may have overlapped between strategies, there were far fewer overlaps in solutions and fewer strategies overlapping each other. As such, even though some problem definitions intersected between strategies, they were interpreted slightly differently and for this reason resulted in diverse policy solutions. It is also important not to ignore who the strategies were formulated by and who sponsored or invested in the endeavour. For example, CBoC is considered as a more right-leaning think tank, often supported by large industry and interested in markets and economic forecasting. For this reason, it is not surprising that CBoC's Strategy supported large industry and the expansion of export markets. On the other hand, FSC is a civil society organization based on the idea of food sovereignty. It is funded by foundations and its member base. As a result, it seems sensible that *The People's Food Policy* is very critical of the current system and focuses very heavily on the social aspects of sustainability in the food system.

These comparisons between policy documents were not meant to show detailed similarities and differences. What they were intended to show was the complexity in settling on a clear vision of what sustainability is and what should be done about it in the food system. Food is a wicked policy issue (Peters and Pierre, 2014) and is multifaceted and complex in nature. Sustainability is just one dimension of food policy and can have many elements and dimensions itself. It is however a consensus frame (Candel *et al.*, 2014), something that resonates with everyone. Every strategy included sustainability in their document and emphasized its importance. However given its often broad definition,[3,4] and its frequent use in everyday language, its meaning is often twisted and contorted to fit different ideologies and perspectives. For this reason, it is difficult to have a common understanding of sustainability, and even more challenging to develop common policy solutions for it.

Recall that framing features specific elements of reality promoted by ideology and supported by particular problem definitions and interpretations of issues (Entman, 1993). Sustainability was framed differently in each policy document, although the framing by the CFA, CAPI, and CBoC may have had similar elements. When contemplating a food policy for Canada within an MSI, it is crucial to be attentive to the different frames used and understood in food policy, especially when contending with consensus frames like sustainability. Without a clear and common vision of the problem between participants in the MSI, the process is doomed to fail (Pattberg and Widerberg, 2016). If the MSI lacks a common understanding of the issues the Canadian food system faces, it will be very difficult to proceed in a multistakeholder partnership for food policy development (Rühli *et al.*, 2015).

Sustainability is also a good example to consider when reflecting on forging consensus within an MSI. As Edmunds and Wollenberg emphasize (2001), language is never neutral. Language becomes important when expressing ideas because it does not mirror reality, but rather is grounded in context and practices of society (Fischer, 2003). As Fischer (2003) notes, 'given that the languages of politics inscribe the meanings of a policy problem, public policy is not only expressed in words, it is literally "constructed" through the language(s) in which it is described' (2003: 43). For this reason, discourse becomes critical to the recognition and definition of policy problems (Ridgway, Lawrence and Woods, 2015). This is especially important for a concept like sustainability which can be interpreted to mean different things for different actors even when consensus appears to have been achieved (Edmunds and Wollenberg, 2001).

Conclusion

Problem definition can be one of the most important stages in policy development (Howlett, Ramesh and Perl, 2009). This stage of the policy cycle deals with the recognition of issues as policy problems, and their definition. Problem definition is viewed by some as a socially constructed process where the framing of ideas and normalcy plays a significant role in policy development (Howlett, Ramesh and Perl, 2009). As Fischer (2003) notes,

> policymaking is a constant discursive struggle over the definitions of problems, the boundaries of categories used to describe them, the criteria for their classification and assessment, and the meanings of ideals that guide particular actions.
>
> (2003: 60)

In 2017, the Canadian federal government began a multistakeholder process to develop a food policy for Canada. This chapter serves as a cautionary example for policymakers and demonstrates the potential challenges experienced in multistakeholder processes. One of the biggest contributors to the collapse of such processes is the failure to develop a clear problem definition of the policy issue being considered. To demonstrate this, I analysed four non-state food strategies and how they problematized sustainability in the food system. Sustainability as a consensus frame is a concept that generally resonates with everyone (Candel *et al.* 2014), but can also be framed differently by actors leading to various policy solutions. In order to reach a common policy solution for the Canadian food system, a clear and common understanding of the food policy problem is vital.

Often, conflicting frames of an issue can appear unsolvable because the same body of evidence is used to come up with different problem definitions and policy solutions. It is even difficult to compare different perspectives of the same policy issue, because often under different perspectives the problem itself has changed (Rein and Schön, 1993). However, this does not mean that food policy development in Canada, or anywhere else, is hopeless. MSIs are important to use as policymaking mechanisms, especially since this is a policy area where multiple actors and values are involved. However, to be a successful MSI, a number of precautions must be considered.

MSIs are not neutral collaborative zones and should by no means be considered as apolitical processes. It is advised that rather than attempting to glaze over or ignore political differences between actors during the process, these distinctions should be used to construct 'alliances,' to learn and 'test ideas strategically' with the goal of empowering often disadvantaged groups like civil society through increased decision-making power (Edmunds and Wollenberg, 2001). Thus, dialogue between actors about political context and power relations should begin at the outset of the process (Edmunds and Wollenberg, 2001). Some tension between actors within the process will always linger (Rühli *et al.*, 2015), given that diverse interests and identities are under one space (Calton and Payne, 2003). Second, outcome evaluation should be transparent, and be cooperatively developed to arrange for the identification of successes and recording of progress to encourage positive performance (Pattberg and Widerberg, 2016). Last, conflict resolution procedures should also be developed and integrated into the multistakeholder governance structure and processes to ease conflicts between actors (Rühli *et al.*, 2015).

There are clear potential synergies between different perspectives of food policy as evidenced in the four non-state food strategies. MSIs may help some food actors transcend stalemates that often dominate binary food policy debates (Clapp, 2016). These debates are frequently polarized by different frames and interpretations of policy problems. Stakeholders often do not realize they are communicating past each other and speaking different policy languages. This prevents progress towards unearthing a middle ground necessary to advance in the development of complex policies like food. Forming and nurturing genuine partnerships through MSIs and engaging in open discussions may assist food actors with different understandings of policy issues to embrace food policy complexity, find common ground, and be more open to different policy frames. However, synergies between actors should be nuanced and clarified. The most testing moment within food policy development in Canada will occur once broad ideals developed by diverse stakeholders are translated into concrete policies.

Further research can focus on how stakeholders from different perspectives and ideologies can begin to find common ground, common language, and genuine collaboration for a more just, healthier, and sustainable food system.

Notes

1 Almeida, Getschko and Afonso (2015) argue that the International Labour Organization was the first to identify MSIs' usefulness when engaging in discussions with governments, employers, and unions in 1919.

2 The policy document itself lists fewer investors, because some opted to leave their names out of the document.

3 As found in the United Nation's document, *Our Common Future* or *The Brundtland Report*.

4 Only two out of the four strategies defined 'sustainability.'

References

Aiking, H. and de Boer, J. (2004). 'Food Sustainability: Diverging Interpretations.' *British Food Journal* vol. 105, no. 5: 359–365.

Almeida, V., Getschko, D. and Afonso, C. (2015). 'The Origin and Evolution of Multistakeholder Models.' *IEEE Internet Computing* vol. 19, no. 1: 74–79.

Antonova, S. (2011). '"Capacity-building" in Global Internet Governance: The Long-term Outcomes of "Multistakeholderism".' *Regulation & Governance* vol. 5, no. 4: 425–445.

Bacchi, C. (2000). 'Policy as discourse: What does it mean? Where does it get us?' *Discourse: studies in the cultural politics of education* vol. 21, no. 1: 45–57.

Bäckstrand, K. (2006). 'Multi-Stakeholder Partnerships for Sustainable Development: Rethinking Legitimacy, Accountability and Effectiveness.' *European Environment* vol. 16, no. 5: 290–306.

Barling, D., Lang, T. and Caraher, M. (2002). 'Joined-up Food Policy? The Trials of Governance, Public Policy and the Food System.' *Social Policy & Administration* vol. 36, no. 6: 556–574.

Bevir, M. and Richards, D. (2009). 'Decentring Policy Networks: A Theoretical Agenda.' *Public Administration* vol. 87, no. 1: 3–14.

Bloom, M. (2014). 'From opportunity to achievement: Canadian food strategy,' www.conferenceboard.ca/e-library/abstract.aspx?did=6091 (Accessed 27 February 2017).

Calton, J. and Payne, S. (2003). 'Coping with Paradox: Multistakeholder Learning Dialogue as a Pluralist Sensemaking Process for Addressing Messy Problems.' *Business & Society* vol. 42, no. 1: 7–42.

Candel, J. J. L., Breeman, G. E., Stiller, S. J. and Termeer, C. J. A. M. (2014). 'Disentangling the Consensus Frame of Food Security: The Case of the EU Common Agricultural Policy Reform Debate.' *Food Policy* vol. 44: 47–58.

CAPI (n.d.a). 'About the institute,' www.capi-icpa.ca/about.html (Accessed 27 February 2017).

CAPI (n.d.b). 'Supporters.' www.capi-icpa.ca/supporters.html (Accessed 24 July 2017).

CAPI (2011). 'Canada's agri-food destination,' www.capi-icpa.ca/destinations/CAPI-Agri-Food_Destination_FULL.pdf (Accessed 27 February 2017).

CBoC. (2010). 'Proposal July 2010: Centre for Food in Canada,' Conference Board of Canada, Ottawa.

CFA (n.d.). 'About us,' www.cfa-fca.ca/about-cfa/ (Accessed 27 February 2017).

Churchman, W. (1967). 'Wicked Problems.' *Management Science* vol. 14, no. 4: B141–B142.

Clapp, J. (2016). 'Big thinking lecture: Navigating the global food fight: Trade, food security and the battle for policy space' [video]. http://congress2016.ca/program/events/clapp (Accessed 27 February 2017).

Currie, N. and Etsell, G. (2011). 'Towards a national food strategy: A framework for securing the future of food,' Canadian Federation of Agriculture, Ottawa.

Edmunds, D. and Wollenberg, E. (2001). 'A Strategic Approach to Multistakeholder Negotiations.' *Development & Change* vol. 32, no. 2: 231–253.

Entman, R. (1993). 'Framing: Toward clarification of a fractured paradigm,' *Journal of Communication*, vol. 43, no. 4: 51–58.

Epstein, D. and Nonnecke, B. (2016). 'Multistakeholderism in Praxis: The Case of the Regional and National Internet Governance Forum (IGF) Initiatives.' *Policy & Internet* vol. 8, no. 2: 148–173.

Fischer, F. (2003). *Reframing Public Policy: Discursive Politics and Deliberative Practices*, Oxford University Press, Toronto.

Fletcher, A. L. (2009). 'Clearing the Air: the Contribution of Frame Analysis to Understanding Climate Policy in the United States.' *Environmental Politics* vol. 18, no. 5: 800–816.

Fransen, L. (2012). 'Multi-stakeholder Governance and Voluntary Programme Interactions: Legitimation Politics in the Institutional Design of Corporate Social Responsibility.' *Socio-Economic Review* vol. 10, no. 1: 163–192.

FSC (n.d.). 'Who We Are.' https://foodsecurecanada.org/who-we-are (Accessed 27 February 2017).

FSC (2011). 'Resetting the table: A people's food policy for Canada,' https://foodsecurecanada.org/sites/foodsecurecanada.org/files/fsc-resetting-2015_web.pdf (Accessed 27 February 2017).

Gillespie Jr., G. W. (2010). '2009 AFHVS Presidential Address: The Steering Question: Challenges to Achieving Food System Sustainability.' *Agricultural Human Values* vol. 27: 3–12.

Head, B. (2008). 'Wicked Problems in Public Policy.' *Public Policy* vol. 3, no. 2: 101–118.

Hedley, D. (2006). 'Why Is There No Canadian Food Policy in Place? Canadian Issues/Thèmes Canadiens,' www.mcgill.ca/files/mwp/2749__CITC _Magazine_Food_v6.pdf (Accessed 27 February 2017).

Horrigan, L., Lawrence, R. S. and Walker, P. (2002). 'How Sustainable Agriculture can Address the Environmental Human Health Harms of Industry Agriculture.' *Environmental Health Perspectives* vol. 110, no. 5: 445–456.

Howlett, M., Ramesh, M. and Perl, A. (2009). *Studying Public Policy: Policy Cycles & Policy Subsystems*, Oxford University Press, Toronto.

Lang. T. (2009). 'Food Policy.' *Perspectives in Public Health* vol. 129, no. 4: 157.

MacRae, R. (2011). 'A Joined-Up Food Policy for Canada.' *Journal of Hunger and Environmental Nutrition* vol. 6, no. 4: 424–457.

Martens, J. (2007). 'Multistakeholder Partnerships – Future Models of Multilaterlism?' *Dialogue on Globalization Occasional Papers Berlin* no. 29: 1–75.

Pattberg, P. and Widerberg, O. (2016). 'Transnational Multistakeholder Partnerships for Sustainable Development: Conditions for Success.' *Ambio* vol. 45: 42–51.

Peters, B. G. and Pierre, J. (2014). 'Food Policy as a Wicked Problem: Contending with Multiple Demands and Actors.' *World Food Policy* vol. 1, no. 1: 2–9.

Pinkse, J. and Kolk, A. (2012). 'Addressing the Climate Change Sustainable Development Nexus: The Role of Multi-stakeholder Partnerships.' *Business & Society* vol. 51, no. 1: 176–210.

Powers, S. and Jablonski, M. (2015). *The Real Cyber War: The Political Economy of Internet Freedom*, University of Illinois Press, Urbana, US.

Rein, M. and Schön, D. (1993). 'Reframing Policy Discourse,' in F. Fischer and J. Forester (eds) *The Argumentative Turn in Policy Analysis and Planning*, Duke University Press, Durham, NC.

Ridgway, E. M., Lawrence, M. A., and Woods, J. (2015). 'Integrating Environmental Sustainability Considerations into Food and Nutrition Policies: Insights from Australia's National Food Plan.' *Frontiers in Nutrition* vol. 2, no. 29: 1–13.

Rittel, H. and Webber, M. (1973). 'Dilemmas in a General Theory of Planning.' *Policy Sciences* vol. 4, no. 2: 155–169.

Roloff, J. (2008). 'Learning from Multi-Stakeholder Networks: Issue-Focussed Stakeholder Management.' *Journal of Business Ethics* vol. 82, no. 1: 233–250.

Rühli, E., Sachs, S., Schmitt, R. and Schneider, T. (2015). 'Innovation in Multistakeholder Settings: The Case of a Wicked Issue in Health Care.' *Journal of Business Ethics*. doi:10.1007/s10551–015–2589–1.

Shaw, S. E. (2010). 'Reaching the parts that other theories and methods can't reach: How and why a policy-as-discourse approach can inform health-related policy. *Health*, vol. 14, no. 2: 196–212.

van der Haar, M. and Verloo, M. (2016). 'Starting a Conversation about Critical Frame Analysis: Reflections on Dealing with Methodology in Feminist Research.' *Politics & Gender* vol. 12, no. e9: 1–7.

Verloo, M. (2005). 'Mainstreaming Gender Equality in Europe: A Critical Frame Analysis Approach.' *The Greek Review of Social Research* vol. 11, no. B: 11–34.

Verloo, M. and Lombardo, D. (2007). 'Contested Gender Equality and Policy Variety in Europe: Introducing a Critical Frame Analysis Approach,' in M. Verloo (ed.) *Multiple Meanings of Gender Equality. A Critical Frame Analysis of Gender Policies in Europe*. CEU Press, New York.

World Commission on Environment and Development. (1987). *Our Common Future*, Oxford University Press, Toronto.

14 Standardizing 'unused' land

The politics of indicators in land classification

Daniel Bornstein

Introduction: sustainability roundtables

Multi-stakeholder sustainability roundtables are a form of private regulation to address the adverse environmental and social impacts of globalized agriculture. Typically organized around a particular commodity, they develop standards for sustainable production, and grant certification to agribusiness companies in compliance with them. Roundtables' standard-setting processes bring together private industry players with transnational and local civil society organizations, including influential environmental groups and smaller advocacy groups concerned with social issues.

The roundtables' decision-making process claims to have balanced representation and participation by all stakeholders, but I follow critical perspectives in challenging such a premise of inclusivity. Numerical equivalence among different actors does little to resolve the vast power asymmetry between corporate actors and NGOs (Cheyns and Riisgaard, 2014). The need to keep industry players satisfied with the outcome – so that they decide to apply for certification – places limitations on the accepted discourses of sustainability. Groups pushing an agenda highly critical of large-scale agriculture, for example, may be sidelined from the debate (Cheyns, 2012). In addition, participants capable of framing their views in technical terms may be perceived by other stakeholders to have more credibility than those who rely upon other forms of narrative (Cheyns, 2012; Djama *et al.*, 2011).

Third-party auditors assess and verify compliance by companies seeking certification. The audit process provides assurance that the agricultural production process did in fact meet the roundtable's specified social and environmental objectives (Bain and Hatanaka, 2010). Auditors gain their credibility from their appeal to 'techno-scientific' values of independence and objectivity (ibid). This presumed value-neutrality of the auditor is problematic, however. Auditors' concern with technical measurement closes down debate on the contested meanings of sustainability (Merry, 2011). Second, auditors' evidentiary standards for what constitutes rigorous proof excludes local communities whose claims are rooted in other ways of knowing (Silva-Castaneda, 2012). Altogether, then, auditors' decisions have potential implications for uneven control over resources (Selfa *et al.*, 2014).

Objective and argument

This chapter's objective is to interrogate the role of private sustainability standards in constructing land classification systems. It provides a case of the Roundtable on Sustainable Biomaterials (RSB), which has developed a standard for identifying 'unused' farmland for biofuel production. Shifting production onto such empty land helps to decouple agricultural expansion from forest loss. I trace the process through which standards construct land as 'unused.'

I argue that throughout the standards-development process, the shifting power relations among technical experts, standard-setters, and private industry produce different ways of categorizing farmland. These contrasting classifications correspond to two directly opposed models of agriculture: one rendering land amenable to export-oriented biofuel production, the other granting recognition to local food production needs.

I advance scholarly understanding of the role of technical indicators in 'rendering land investable' (Goldstein and Yates, 2017). That is, indicators do the 'work' of making the land valuable for investment, favouring certain land uses to the exclusion of others (Goldstein and Yates, 2017; Li, 2014). My objective is to illuminate the highly contested process of producing those indicators that render land 'investable.'

Initially, the proposed technical indicators appeared to be in conflict with the land's value for export biofuel production. Technical experts advised that in order for land to qualify as 'unused,' it must not interfere with future food production needs. This approach classifies land in a way that is attentive to local subsistence concerns. A pilot project drawing upon this indicator thwarted the designation of the land as 'unused.' Then, however, the standard-setting body omitted the indicator as a basis for defining 'unused' land – in turn rendering land investable for biofuel companies and foreclosing consideration of local food production priorities.

The standard-setting body's imperative to attract private industry to sign up for certification explains this shift in how to classify farmland. To maintain their relevance in global governance, roundtables must entice agribusiness companies to voluntarily submit themselves to regulation. But companies are unlikely to do so if they face a risk of being non-compliant. In the context of land classification, this meant the RSB had to axe an indicator that would stifle the ability of biofuel companies to successfully demonstrate the unused status of their land.

This chapter has several parts. I first review critical perspectives on land classification. I then explain how sustainability roundtables are entwined with public regulation, and specifically explain the RSB's standard to deal with carbon emissions from indirect changes in land use. I then describe the RSB's use of satellite imagery and interview data as acceptable forms of proof for defining 'unused' land. I proceed to discuss the outcome of a pilot project intended to demonstrate the viability of shifting production onto unused land. Next, I show how the roundtable deliberately abandoned the indicator responsible for the pilot project's collapse, and how this re-working of technical measurement transforms what sustainability means on the ground. I conclude by exploring the potential for land classification schemes to advance a just, locally autonomous vision of sustainability.

Critical perspectives on land classification schemes

Growing demand for farmland, coupled with climate mitigation concerns, has led to land use competition between agricultural production and forest conservation. To resolve this trade-off, land classification schemes have become salient for their ability to designate certain lands as most desirable for agriculture and others as valuable for forest carbon storage. Constructing land as 'empty' or 'unused' has been one particular attempt to expand agricultural production while keeping intact tropical forests valued for their carbon storage (Borras and Franco, 2012; Goldstein, 2014). This rationale for classification seeks to reconcile conflict between differing land uses, amounting to a sort of land use planning strategy. A second rationale has more to do with promoting economic development. Framing land as empty and available has been part of the narrative employed by agribusiness firms, development agencies, and governments in their push to modernize agricultural systems in developing countries (Geisler, 2012; Borras and Franco, 2012). Through this lens, the land is an untapped resource and must be integrated into the global market to render it productive. Of course, these two rationales often overlap, but their distinction makes clear the wide range of actors and legitimacy claims tied to discourses of unoccupied land.

Land classification constitutes a form of standardization. It creates a common metric by which to categorize farmland across a diverse range of contexts, 'constructing uniformities across time and space' to 'make things work together over distance' (Timmermans and Epstein, 2010: 71). In turn, a wide range of farming systems, with their distinct social and ecological contexts, are rendered comparable. Comparability lays the groundwork for imposing a single land use strategy across all these spaces; that is, they are rendered governable. This uniformity underlies the World Bank's determination that:

> The currently non-cultivated area suitable for cropping that is non-forested, non-protected, and populated with less than 25 persons/km2 (or 20 ha/household) amounts to 446 million ha. This is equivalent to almost a third of globally cropped land which is 1.5 billion ha.
>
> (World Bank, 2010: xxxiv–xxxv)

Yet, land classification functions as a form of 'simplification' to render legible distant landscapes (Scott, 1998). Constructing equivalent, comparable entities across space involves erasing their local differences (Timmermans and Epstein, 2010). In turn, the construction of categories to unify different landscapes renders invisible the practices and values of local communities (Borras and Franco, 2010). The procedures of measurement that underlie classification require an abstraction of place-based social and ecological dimensions. Accordingly, critical perspectives have equated the construction of distant farmland as 'available' with the dispossession of smallholder farmers (Goldstein, 2014; White *et al.*, 2012). Under this framing, systems of classification are a form of social exclusion. Yet, importantly, Baird (2014) reminds us that such categorizations are also capable of *preventing* accumulation in certain spatial and political contexts.

Hybrid governance and the mutual dependence of public and private regulation

Sustainability roundtables, a form of private regulation, are deeply interconnected with European Union public regulation, constituting 'hybrid governance' (Ponte and Daugbjerg, 2015). The RSB and the EU are 'mutually dependent' (ibid.). On the one hand, the EU Renewable Energy Directive's sustainability criteria require verification at the site of production, because they encompass impacts that cannot simply be detected in the final product. These criteria pertain to reducing greenhouse gas emissions and prohibiting the conversion of high-carbon stock and high-biodiversity areas. By carrying out such verification, transnational private schemes like the RSB enable the EU to 'extend its authority and capacity to implement the environmental sustainability criteria beyond its territorial borders' (Ponte and Daugbjerg, 2015: 104). The EU has approved 19 certification schemes through which biofuel companies can demonstrate their compliance with the EU biofuel sustainability criteria. These private certification schemes are the dominant mechanism through which biofuel producers comply with the EU regulations.

On the other hand, the roundtables rely upon the EU's sustainability requirements for their own legitimacy in global governance. Roundtables' standards are influenced in part by the rules set in the EU. The RSB developed its standard specifically to focus on compliance with the EU's sustainability criteria, which enables the RSB to attract fuel companies interested in tapping the EU market. For example, the RSB's EU-specific standard contains a list of 'no-go' and 'no conversion' areas to align with the EU criteria's emphasis on carbon and biodiversity conservation. Importantly, however, many of the private standard schemes also contain social criteria, going well beyond the EU's limited concern with ecological impacts.

RSB and the problem of carbon emissions from indirect land use change

The Roundtable on Sustainable Biomaterials (RSB) is a global multistakeholder standards-setting body that provides certification to individual companies. Unlike the commodity-specific roundtables, RSB applies to all sources of biofuel. RSB governance is based on dividing stakeholders into seven different 'chambers.' Business interests (i.e. those embedded in the commodity chain) and civil society groups each comprise three chambers.[1] The representation within civil society is quite wide-ranging: international environmental groups, rights-based groups, and smallholder farmers' associations, to name a few. As noted before, however, the numerical equivalence between business and civil society chambers does not reflect any equality in their respective capacities to exert influence in standards-setting (Cheyns and Riisgaard, 2014). A seventh chamber is devoted to international agencies, governments, consultants, and auditors, but they have no voting power and instead serve in an advisory capacity.

The RSB's Assembly of Delegates, the roundtable's highest decision-making body, provides a glimpse of the RSB's most influential members. Members of the Assembly function as representatives for their entire chamber. The Delegates for the civil society chambers include World Wildlife Fund USA; the Associated Labor Unions-Trade Union Congress of the Philippines; the France-based Ethical Sugar; the Philippines-based Rural Development Institute of Sultan Kudarat; the Centre for Empowerment and Development in Nepal; the conservation NGO Applied Environmental Research Foundation based in India; and The Energy and Resources Institute also in India. Among the Delegates representing business interests are big players in the aviation industry such as Boeing and South African Airways.

I focus on the RSB's development of a standard to promote biofuel cultivation on 'unused land.' The standard's objective is to reduce the carbon emissions associated with crop-based biofuels. In particular, it seeks to resolve the problem of indirect land use change (ILUC). ILUC occurs when biofuel production replaces another crop, leading to the clearance of forests elsewhere in the world to grow this crop (Searchinger *et al.*, 2008). This deforestation constitutes the loss of a carbon 'sink' considered valuable for climate mitigation. The biofuel production site is culpable for these carbon emissions because it sets in motion the cascade effect that generated deforestation. The logic of targeting unused land is that if biofuels are cultivated on land without current production, then there is no previous land use displaced into carbon-rich forests (Ecofys, EPFL, WWF, 2012).

Upon producing biofuel on unused land, agribusiness companies obtain a 'credit' that represents the extent of avoided ILUC-induced carbon emissions. The amount of biomass certified as ensuring avoided ILUC emissions is based on the land area under cultivation and the crop yield. For example, a grower who produces a yield of 4.2 tons per hectare on 100 hectares of previously unused land would have 420 tons of biofuel certified as minimizing ILUC.

The ILUC effect reflects the tension between agricultural land expansion and the protection of forests crucial for climate mitigation. In particular, the ILUC issue threatens to undermine the viability of biofuel production as a climate mitigation strategy. Biofuels' legitimacy is based on their ability to reduce carbon emissions relative to fossil fuels. Yet, the ILUC effect suggests that biofuel production may be associated with more carbon emissions than previously anticipated. Seen in this light, the RSB's effort to reduce ILUC-based emissions reflects an attempt to boost the credibility of the biofuel sector.

The RSB standard is titled 'RSB Low ILUC Risk Biomass Criteria and Compliance Indicators.' Its content is based on a report co-authored by technical experts from Ecofys, the RSB secretariat, and World Wildlife Fund International. The report, 'Low Indirect Impact Biofuel (LIIB) Methodology,' functioned as the 'scientific basis' for the RSB standard (RSB, personal communication, 2015). Yet, a vitally important piece of guidance from the report was omitted from the RSB's low ILUC standard, as I explore in this chapter. The LIIB methodology contains six authors from Ecofys, and one from each of the two other organizations. Ecofys is an international energy and climate consultancy based in the Netherlands with a team of over 600 experts, and has advised governments, corporations, and NGOs.

The expert-authored LIIB methodology was intended as a guide for policymakers and certification schemes interested in addressing indirect land use change emissions. The RSB had no binding commitment to utilize the exact methodology; rather, the report was seen as a starting point to stimulate consideration of how best to tackle indirect land use change. The experts who authored the guidelines, then, can be seen as having an advisory role, rather than direct authorship of the standard. This explains why it was possible for the RSB standard to differ from that suggested by the LIIB methodology.

The issue of indirect land use change shows clearly how private standard-setters seek legitimacy by claiming their relevance to public regulations. The RSB's decision to develop an approach for combating ILUC emissions reflected its interest in contributing to EU policy debates (Sapp, 2013). In this sense, the RSB process functioned as a 'test-bed' where relevant industry players and scientists could deliberate on controversial issues that they anticipate being enshrined in government regulations (Fortin and Richardson, 2013: 19). The RSB's executive director framed the significance of the ILUC standard on the basis of the EU's amendment of its sustainability criteria to address ILUC (RSB press release, 2015).

The RSB's standard for reducing ILUC was approved in June 2015. The roundtable has not yet issued any certifications for compliance with the standard.

Methodology

This analysis of sustainability standards draws upon content analysis of technical documents and reports. Accessed through the roundtable's website, I examined the RSB's standard for indirect land use change emissions, with the purpose of investigating how the standard defined 'unused' farmland. I was particularly interested in what indicators counted as sufficient evidence to demonstrate land's unused status. The RSB standard cited the LIIB methodology as its basis. In turn, I then examined the report outlining this methodology – drafted by Ecofys, WWF, and RSB – in order to explore its approach to classifying unused farmland. I compared the LIIB methodology's indicators for unused land with those listed in the RSB standard. One of the most valuable sections of the LIIB methodology report was its summary of the WWF-led pilot project for unused land in Mozambique. I located the WWF report on the pilot project to gain further insight. Upon recognizing a discrepancy between the LIIB methodology's and the RSB standard's indicators for identifying unused land, I interviewed a high-level member of the RSB secretariat to understand the reasoning behind the difference.

The three documents analysed here enable me to depict the evolution of the standards-setting process, with each document corresponding to a different juncture of the process. Together, these documents offer insights into the evolving power relations among technical experts, the standard-setting body, and private industry throughout the process of standards development. The earliest document, the LIIB methodology, reflects the influence of experts who endeavoured to create a rigorous definition of unused land. The report on the WWF pilot project illuminated how those experts' approach accounted for complex local realities that were antagonistic to agribusiness companies' endeavour to identify land for sustainable biofuel production. The published RSB standard, when juxtaposed with the previous two, shed light on the power of industry in private standards, diminishing the voice of experts whose guidance had proven at odds with industry's interests.

RSB's approach to identify unused land: criteria and indicators

Sustainability standards consist of two parts, criteria and indicators. The criteria represent the approach to production and land use. The indicators specify the forms of evidence needed to demonstrate compliance with the criteria. In this section, I provide an overview of the indicators adopted by the RSB to show proof of unused land. These were the indicators published at the culmination of the standards-setting process. Another crucial indicator, concerned with future food demand, was recommended and tested during the standard's development, but was excised from the standard's final version, as I show later.

According to the RSB standard's criteria, companies must demonstrate that the land they have identified is unused at the time of initiating biofuel cultivation, as well as three years beforehand. This approach contains several indicators that are accepted as evidence of compliance. The first is satellite imagery data, which provides a portrait of land cover. Second, interviews with local communities are necessary to provide on-the-ground verification. Importantly, the companies interested in acquiring the land are the ones carrying out the interviews. The RSB specifies that interviews are to take place with 'the land owner or tenant and with other local people/ local authorities.' In addition, the RSB standard includes an indicator mandating that shifting cultivation must not be taking place on the land in question. The standard notes, 'this indicator refers to practices in some countries (outside Europe, North-America and Australia) to have very long crop rotation systems, in which fields can be left fallow for ten years before being planted again' (RSB, 2015). Interviews are the required method to demonstrate the absence of this practice.

The evolution of indicators: from spotlighting local priorities to re-working representations of distant land

Experts advising the RSB developed an indicator to account for future food production on the land, but in the end this approach was excluded from the standard. Its omission by the RSB reflects the need for the roundtable to remain attractive to industry stakeholders interested in seeking certification. The indicator would complicate the efforts of biofuel companies to gain certification, by imposing on them a more rigorous burden for proving the land's unused condition. I now turn to tracing the decision-making process to show how this indicator features prominently in the initial testing phase, but is ultimately eliminated.

While only briefly discussing the satellite imagery and interview approaches above, the crux of my analysis focuses on the indicator for avoiding competition with future food production. This is because the latter indicator featured as the main point of divergence between the technical experts' guidance and the RSB standard. The experts had recommended the use of satellite imagery and interviews, and the RSB heeded their advice in this particular regard. But because the RSB chose to eliminate the requirement to anticipate future food production needs, it is this indicator that forms the basis of my chapter's assessment of the contested standards-making for land classification. In addition, this indicator proved influential in the unused land pilot project. The pilot project never collected data from satellite imagery and interviews because the project had already determined that the land in question could not be considered 'unused.'

Indicator for future land use change and avoiding fuel-food competition

According to the 'LIIB Methodology' that was the basis for the standard's development, biofuel production should take place 'in a region with an *excess potential* of unused arable [potentially cultivable] land that is equally or more suitable for the biofuel crop than the project site' (Ecofys, EPFL, WWF, 2012: 34; emphasis added). In other words, the land area targeted for biofuels must contain additional land for other uses – at least three times the area of the biofuel site, according to the methodology. Its rationale is to prevent biofuels from competing with future food production needs (Ecofys, EPFL, WWF, 2012). This indicator anticipates future land use change associated with political-economic and demographic factors, in contrast to the temporally 'static' approach of satellite imagery (Nalepa and Bauer, 2012). In addition, the requirement that the excess land be of equal or greater suitability than the biofuel site protects local food producers against being relegated to the lowest quality soils.

This indicator's emphasis on future conditions served to complicate the meaning of unused land. When the LIIB methodology was put to the test, it was this very indicator that accounted for an unexpected outcome: land initially identified as 'unused' was in fact not unused. I now turn to the pilot project where this indicator brought into relief the on-the-ground complexities of land classification.

WWF pilot project: future food demand complicates 'unused' classification

The World Wildlife Fund in Mozambique undertook a pilot project to test the viability of the unused land approach (Ecofys, EPFL, WWF, 2012). It had planned to carry out certification of a jatropha plantation situated on unused land. However, before the audit could even be conducted, WWF staff had already determined that no unused land existed in the entire region they had been investigating. Their anticipation of future food production needs was central to this determination, as the quote below illustrates. As a result, the audit for jatropha was cancelled, and thus WWF never tested the satellite imagery and interview indicators.

The project began with a two-party desk study, using 'coarse and readily available information to quickly identify the most promising areas' and then performing 'a more detailed assessment' of such areas 'to further refine the pre-selection of promising areas.' The next step involved field-based work to provide on-the-ground verification that land was unused in the region in question. WWF determined, however, that no land in the area could be classified as unused, owing to the high likelihood of the land coming under pressure for food production:

> In general, land use is scattered, and used in a low intensity. 'Unused land' doesn't seem to exist, though. Even land that is at this point in time not used, is *likely to be used in the near future*. As a result of low inputs, yields are currently declining, and with a growing population the production potential with currently applied management practices is likely to be *insufficient to provide for future needs*.
>
> (WWF International/WWF Mozambique, n.d.: 30; emphasis added)

The project in turn concluded, 'The research projects significant pressure on land in central Mozambique due to increasing demand for food' (Ecofys, EPFL, WWF, 2012: 39).

The LIIB methodology's indicator attentive to future food production helped to spotlight the messy realities on the ground. Data anticipating future land use priorities served to cast doubt upon the project's assumptions about land availability. In this context, then, the land classification scheme protected local food producers against the intrusion of large-scale biofuel cultivation. The measurements embedded in classification recognized food security as an important value in central Mozambique, thwarting the transition of ostensibly empty land to monocrop jatropha production. This finding stands out as an important contrast to critical scholars' contention that scientific studies measuring the extent of available farmland are titled in favour of industrialized export agriculture and against local practices (White *et al.*, 2012). Rather, the reverse is possible: technical measurements, in this case, *validated* the viability of local food production and *produced exclusion* of export-oriented production.

Following the cancellation of the audit for the jatropha plantation, the auditor provided comments on the challenges associated with the RSB's work on identifying unused land amenable to certification. The auditor, DNV, is a prominent certification body in the renewable energy industry. DNV's conclusion is quite revealing:

> In practice, land that can also be used for biofuel feedstock/biomass production is almost never completely unused. That does not mean that the land is not available for biofuels production … It is recommended to elaborate the methodology further to accommodate for '*underused*' land.
>
> (Ecofys, EPFL, WWF, 2012: 40; emphasis added)

Heeding this advice, the technical experts who authored the 'LIIB Methodology' – the scientific guidance behind the RSB standard – updated their report to include production on 'underused' land as a viable approach to avoid ILUC emissions. However, this category was never ultimately inserted into the RSB standard. What might explain this omission? The designation 'underused' defies easy categorization. It is not amenable to a coherent definition, but rather depends upon the context-specific political and economic dimensions of agriculture. Such variability poses a limitation for transnational standardization as a form of governance, which requires that a land category be conducive to consistent measurement procedures across contexts. Consistency is necessary to ensure that all biofuel companies are evaluated in the same manner – a criterion for the legitimacy of private standards (ISEAL, 2012; Loconto and Busch, 2010). This situation reflects Goldstein's (2012) point that the definitional challenges and murkiness associated with land classification categories undermine their institutionalization as a political project.

Revising the indicators: ignoring on-the-ground complexity

The RSB was thus faced with a conundrum: how to proceed with certifying companies interested in cultivating on 'unused land,' after the complete failure of the project that was supposed to serve as proof of concept? The roundtable responded by dispensing with the indicator that had, during the pilot study, disproven the land's 'unused' condition. By taking account of the *future* trajectory of rural development – with particular concern for the fuel vs. food conflict – the indicator had made clear that the land in question could not be considered unused. In other words, standard-setters moved to foreclose the possible emergence of data that would obfuscate the ability of agribusiness firms to simplify distant landscapes. In doing so, the RSB rejected the guidance of the experts it had sought out to contribute to the standard's development.

The indicator requiring excess available land is conspicuously absent from the RSB's standard on low ILUC biofuel production. According to a high-level staffer of the RSB Secretariat: 'The LIIB Methodology was adapted in order to ensure the practicality and the auditability. That is why there may be some differences between the LIIB [methodology] and the [RSB's] low ILUC [standard]' (personal communication, 2015). 'Practicality' and 'auditability' seem to rest on the logic that certain measurements warrant removal because they are too technically demanding for third-party auditors to verify. Yet that logic hardly holds true in this case. After all, the WWF project was *successful* in capturing this information. Most important, this data on the region's future food demand delivered a fatal blow to the project, discrediting the project's claim to have identified 'unused' land.

Seen in this context, then, the RSB's exclusion of the technical experts' guidance reflects standard-setters preoccupation with keeping the standard attractive to agribusiness companies. Given roundtables' dependence on industry's willingness to buy-in to the rules, corporations wield a tremendous amount of power in the standards-making process (Cheyns, 2011). If indicators are so rigorous that they present a high likelihood of non-compliance, then they will not be accepted by industry (Klooster, 2010). This risk of non-compliance is exactly what biofuel companies would have faced had the excess land indicator remained a requirement.

The voluntary nature of the RSB sustainability standard places the organization in the position of having to court companies to apply for certification. This imperative is amplified by the competition among standards bodies to attract companies. An RSB Secretariat staff member quoted by Ponte (2014: 267) noted that the RSB 'has been slow in selling it to companies and in preparing the tools to apply it in practice.' Another staff member claimed that the RSB planned to move away from being an 'academic standard,' referring to its idealistic nature and its struggle in finding practical applications (personal communication, 2016). A roundtable member from the biofuel producers' chamber remarked that 'it is more difficult to get RSB certification than a license to operate a nuclear plant' (Ponte, 2014: 267). Its difficulty in attracting companies for certification can be attributed to two factors. First, its inclusive structure (relative to other private standards schemes) slows down the decision-making process and leaves space for disagreement (Ponte, 2014). Second, its highly comprehensive set of social and environmental criteria makes the auditing process quite rigorous for companies.

What explains why the RSB contains these particular characteristics? The initial momentum for establishing a biofuel sustainability standard arose when activists on a wide range of social and environmental issues began raising concerns about the effects of biofuels. That is, the multitude impacts of biofuel production made such fuels a target of civil society groups focused on many different issues (i.e. food security, land rights, labour, climate change, energy, etc.). In addition, the Forest Stewardship Council served as the inspiration for having a balance between industry and civil society interests. It was the Brazilian sugarcane producers' association, UNICA that recommended an 11-chamber structure (which after one year was altered to 7) (Ponte, 2014). An RSB Steering Board member interviewed by Ponte remarked that setting the 'gold standard' for sustainability certification would exert pressure on other roundtables to be more inclusive and comprehensive.

The consternation, expressed in the quotes above, about the standard's slow uptake among industry offers insight into why the RSB tweaked its indicators to be more palatable to agribusiness companies. The technical experts' idea to account for future food production proved too threatening for companies targeting unused land in developing countries. That is, this indicator complicated – and could potentially *thwart* – their task of identifying farmland amenable to RSB certification. Had the standard required companies to undertake the more complex step of anticipating alternative land use possibilities, they would risk encountering results antagonistic to their desired conception of the land as 'unused' – and in turn fail to gain certification. Such a prospect would only exacerbate the RSB's challenges in getting companies to sign up to its standard.

Discussion: diverging meanings of sustainability

In revising its approach to land classification, the RSB profoundly altered the meaning of sustainability. Each approach had implications for the inclusion and exclusion of different social groups. The LIIB Methodology's insistence on verifying the existence of additional farmland elevates food security as a priority. It opens space for contemplating: How can this land be rendered valuable for local needs rather than 'investable' for outside companies (Goldstein and Yates, 2017; De Schutter, 2011)? This lens provides the basis for an inclusive land use planning process, exploring how political change can position smallholder farmers to utilize the land for food production. Agricultural systems that may have been made invisible through a static, reductionist vision of the landscape are thus brought into the realm of the possible. Such recognition of possible future land use patterns can function as a bulwark against externally imposed definitions of the land's condition.

The RSB's approach, on the other hand, closes down space for an inclusive land use planning process. That is, by turning a blind eye to the potential for biofuels' competition with food production, it ignores food security concerns of local populations. As a result, sustainability becomes narrowly framed as compatible with the expansion of export agriculture – with little regard for alternative land uses rooted in a more socially just, rights-based conception of sustainability.

This tension over selecting indicators illuminates how land classification is a messy political project shaped by the evolving configurations of power throughout the standard-making process. The effects of classification were highly contingent upon the knowledge forms prevailing at different junctures of the rule-making process. First, the WWF project revealed how technical experts wielded sizable power and influence. That is, the project's assessment of land use drew directly upon these experts' attention to alternative land use possibilities. Their knowledge thus shaped who would access and be excluded from the farmland in question. Yet, the RSB's omission of this data reflected a shift in who had the power to shape classification. This involved the subordination of technical experts' knowledge, in favour of an approach more conducive to agribusiness firms' ability to render legible distant land. By removing the experts' proposed measurements that contained potential for generating undesirable findings for companies, the RSB placed agribusiness firms in a position to more expediently classify land as 'unused' and receive a sustainability certification. This revision reveals quite a shift in the power dynamics of the regulatory process – from the authority of technical experts attentive to local communities' future food demands, to the industry-friendly erasure of such considerations.

Conclusion: towards an emancipatory politics of land classification?

The RSB standard-setting process raised two divergent political possibilities associated with land classification systems. In different contexts, they serve as a mechanism of *exclusion* or a vehicle for *recognition* of local livelihoods; they can either flatten or validate local difference. What relations of power contribute to such varying outcomes?

A useful lens is to consider the relationship between the actors defining the land category and those subject to regulation by the category. The RSB standard's revision revealed how the struggle to define empty land became susceptible to the uneven power relations of the decision-making process. Agribusiness companies wielded significant power because the roundtable depended upon their acceptance of the standard to ensure its relevance as a governing body. Thus when the definition of unused land posed a barrier to agribusiness accumulation, it was watered down to become easier for the companies to comply.

A key factor for explaining the social exclusion generated by land classification schemes is: what is the premise from which their creation begins? In this case, indicators were developed based on the assumption that large-scale plantation agriculture *should* take hold throughout the world, and that its sustainability requires merely spatially re-organizing its production to spare carbon-rich forests. Under this premise, there is little political space for considering whether local food production should be a higher priority than biofuel exports. As a result, the RSB's unused land definition was tailored to fit within this frame of the inevitable expansion of export-based agriculture.

Under what social relations of land classification might a more socially just outcome prevail – one that thwarts attempts by outsiders to simplify and subordinate local socio-ecological relations? Such schemes must be embedded in a political struggle committed to the values and priorities of smallholder farmers and other local producers. Contrary to the RSB case, the development of indicators must begin from the premise of protecting smallholders against dispossession.

Importantly, this bottom-up political potential is shaped by decisions over what ecological and social factors to measure on the ground. By what criteria is the land defined as unused versus under local production? These measurements must be grounded in local conceptions of land use if they are to succeed in safeguarding smallholder systems against intrusion by large-scale capitalist production.

In this regard, the WWF pilot study in Mozambique offered an example of how technical measurement proved effective in validating local food security priorities. The measurement criteria were comprehensive enough to recognize complex land use changes. Yet, the classification scheme was dis-embedded from a *political* struggle intent on contesting export-oriented agriculture. It was not part of a bottom-up articulation of sustainability valuing local control over land use decision-making. As a result, the indicators were re-worked in favour of powerful external actors.

Baird's (2014) work in Laos is insightful for showing how local actors can draw upon land classification schemes in their political mobilization against plantations. An NGO-backed coalition, the Land Issues Working Group, advocated that plantations be confined to 'degraded' land, so that local communities on highly productive soils are protected from dispossession. Interestingly, targeting 'degraded' land is the same strategy deployed by proponents of the large-scale land acquisitions *causing* dispossession (Goldstein, 2014; White *et al.*, 2012). It is striking that the same exact discourse wielded by corporations to legitimize investments is also invoked as a weapon in *opposition* to them. This paradox can be explained by the divergent political projects in which the discourse is embedded. Land concessions are justified by a development paradigm based on commercializing agriculture in areas left out of the world market; land rights advocacy groups, on the other hand, strive to bring recognition and validation to local farming practices, contesting the land's takeover by powerful interests.

Finally, developing land classification schemes in the service of an emancipatory politics warrants a greater role for political ecologists in the technical work of land categorization. Political ecologists' explicit political commitment to conferring legitimacy on local livelihood strategies can prove an asset to a socially just land classification endeavour. Their attentiveness to on-the-ground complexity ensures that technical measurements used to categorize farmland capture the land use strategies that some outsiders prefer to expediently ignore. After all, it is not the discursive construction of land categories that is the problem, but rather the unevenness in who is positioned to utilize them.

Acknowledgements

I am grateful to the many faculty and colleagues who have provided guidance on this project. This chapter was first presented as a paper at the Dimensions of Political Ecology conference at the University of Kentucky in February 2016. I thank Hilary Faxon and Ewan Robinson for inviting me to participate, and Garrett Graddy for her constructive comments as discussant. I am also grateful to Doug Constance, Maki Hatanaka, and Jason Konefal for including me in their session at the 2016 International Rural Sociological Association meeting in Toronto, and for their helpful feedback on this chapter. Josh Garoon, Daniel Kleinman, and Joe Conti at Wisconsin also offered valuable insights and mentorship.

Note

1 The RSB chambers are organized as follows. Business chambers, 1–3. Chamber 1: farmers and growers of biomass; Chamber 2: biofuel producers; Chamber 3: retailers/blenders, transportation industry, banks/investors. Civil society chambers, 4–6. Chamber 4: Rights-based NGOs and trade unions; Chamber 5: Rural development, food security, smallholder farmers, indigenous peoples, and community-based organizations; and Chamber 6: Environmental conservation, climate change, and policy organizations.

References

Bain, C. and Hatanaka, M. (2010). The practice of third-party certification: Enhancing environmental sustainability and social justice in the Global South? In Higgins V. and Larner W. (eds) *Calculating the Social*: 56–74). Palgrave Macmillan UK.

Baird, I. G. (2014). Degraded forest, degraded land and the development of industrial tree plantations in Laos. *Singapore journal of tropical geography*, 35(3): 328–344.

Borras Jr, S. and Franco, J. (2010). From threat to opportunity? Problems with the idea of a code of conduct for land-grabbing. *Yale Human Rights and Development Journal*, 13(2): 507–523.

Borras, S. M. and Franco, J. C. (2012). Global land grabbing and trajectories of agrarian change: A preliminary analysis. *Journal of Agrarian Change*, 12(1): 34–59.

Cheyns, E. (2011). Multi-stakeholder initiatives for sustainable agriculture: limits of the 'inclusiveness' paradigm. In Ponte, S., Vestergaard, J. & Gibbon, P. (eds). *Governing through standards: Origins, drivers and limits*, London: Palgrave: 318–354.

Cheyns, E. and Riisgaard, L. (2014). Introduction to the symposium. *Agriculture and Human Values*, 31(3): 409–423.

De Schutter, O. (2011). How not to think of land-grabbing: three critiques of large-scale investments in farmland. *The Journal of Peasant Studies*, 38(2): 249–279.

Djama, M., Fouilleux, E. and Vagneron, I. (2011). Standard-setting, certifying and benchmarking. In Ponte, S., Vestergaard, J. and Gibbon, P. (eds) *Governing through standards: Origins, drivers and limits*, London: Palgrave: 1840209. Ecosys, EPFL and WWF International. (2012). Low Indirect Impact Biomass (LIIB) Methodology Version 0. www.ecofys.com/files/files/12-09-03-liib-methodology-version-0-july-2012.pdf. (Accessed 18 April 2018). Fortin, E. and Richardson, B. (2013). Certification schemes and the governance of land: enforcing standards or enabling scrutiny? *Globalizations*, 10(1): 141–159.

Geisler, C. (2012). New Terra Nullius Narratives and the Gentrification of Africa's 'Empty Lands,' *Journal of World-Systems Research*, 18(1): 15–29.

Goldstein, J. E. (2014). The afterlives of degraded tropical forests: New value for conservation and development. *Environment and Society*: 124–140.

Goldstein, J. E. and Yates, J. S. (2017). Introduction: Rendering land investable. *Geoforum*, 82: 209–211.

ISEAL Alliance (2012). *Assuring Compliance with Social and Environment Standards*. London. www.isealalliance.org/sites/default/files/resource/2018-02/ISEAL_Assurance_Code_Version_2.0.pdf. (Accessed 18 April 2018).

Klooster, D. (2010). Standardizing sustainable development? The Forest Stewardship Council's plantation policy review process as neoliberal environmental governance. *Geoforum*, 41(1): 117–129.

Li, T. M. (2014). What is land? Assembling a resource for global investment. *Transactions of the Institute of British Geographers*, 39(4): 589–602.

Loconto, A. and Busch, L. (2010). Standards, techno-economic networks, and playing fields: Performing the global market economy. *Review of International Political Economy*, 17(3): 507–536.

Merry, S. E. (2011). Measuring the world: Indicators, human rights, and global governance. *Current Anthropology*, 52(S3): S83–S95.

Nalepa, R. A. and Bauer, D. M. (2012). Marginal lands: the role of remote sensing in constructing landscapes for agrofuel development. *The Journal of Peasant Studies*, 39(2): 403–444.

Ponte, S. (2014). 'Roundtabling' sustainability: lessons from the biofuel industry. *Geoforum*, 54: 261–271.

Ponte, S. and Daugbjerg, C. (2015). Biofuel sustainability and the formation of transnational hybrid governance. *Environmental Politics*, 24(1): 96–114.

RSB (2015). *Low ILUC Risk Biomass Criteria and Compliance Indicators* Geneva Switzerland. https://rsb.org/wp-content/uploads/2017/02/RSB-STD-04-001-ver0.3RSBLowiLUCCriteriaIndicators.pdf. (Accessed 18 April 2018).

RSB Secretariat (2015). Personal communication.

RSB Secretariat (2016). Personal communication.

Sapp, M. (2013). RSB agrees to 'extra credit' for low ILUC risk biofuels. *Biofuels Digest*. www.biofuelsdigest.com/bdigest/2013/03/25/rsb-agrees-to-extra-credit-for-low-iluc-risk-biofuels/. (Accessed 20 April 2018).

Scott, J. C. (1998). *Seeing like a state: How certain schemes to improve the human condition have failed*. Yale University Press.

Searchinger, T., Heimlich, R., Houghton, R. A., Dong, F., Elobeid, A., Fabiosa, J., Tokgoz, S., Hayes, D. and Yu, T. H. (2008). Use of US croplands for biofuels increases greenhouse gases through emissions from land-use change. *Science*, 319(5867): 1238–1240.

Selfa, T., Bain, C. and Moreno, R. (2014). Depoliticizing land and water "grabs" in Colombia: the limits of Bonsucro certification for enhancing sustainable biofuel practices. *Agriculture and Human Values*, 31(3): 455–468.

Silva-Castañeda, L. (2012). A forest of evidence: third-party certification and multiple forms of proof – a case study of oil palm plantations in Indonesia. *Agriculture and Human Values*, 29(3): 361–370.

Timmermans, S. and Epstein, S. (2010). A world of standards but not a standard world: toward a sociology of standards and standardization. *Annual review of Sociology*, 36: 69–89.

White, B., Borras Jr, S. M., Hall, R., Scoones, I. and Wolford, W. (2012). The new enclosures: critical perspectives on corporate land deals. *The Journal of Peasant Studies*, 39(3–4): 619–647.

WWF International/WWF Mozambique (n.d.). *Identifying Responsible Cultivation Areas in Mozambique* www.hcvnetwork.org/resources/folder.2006-09-29.6584228415/identifying-responsible-cultivation-areas-in-mozambique. (Accessed 18 April 2018).

15 Justifying the standardization of sustainability impact[1]

Allison Marie Loconto

Introduction

Since 2015, a new global agenda for the meta-governance of sustainable development has come into force with an astounding list of 17 goals, 169 targets and 232 individual indicators.[2] There is an interesting thing that happens with Goal 12; in some documents (and on the icon) the goal is referred to as 'Responsible Consumption and Production,' while the goal itself is to 'ensure sustainable consumption and production patterns' (UN, 2015: 22). This blurring of responsibility with sustainability becomes clearer when we examine two of SDG12's targets: 'By 2030, achieve the sustainable management and efficient use of natural resources' (12.2) and 'Encourage companies, especially large and transnational companies, to adopt sustainable practices and to integrate sustainability information into their reporting cycle (12.6)' (UN, 2015: 22). This is a clear statement made by State actors that corporations do indeed have a responsibility to deliver sustainability impact.

Within the business ethics literature, taking responsibility often centres on ensuring that actors along the value chain are held responsible for good economic, environmental and social performance (Utting, 2002, Utting, 2007, Dolan and Rajak, 2011, Temri *et al.*, 2015). The most common means to ensure, or govern, this is through the use of voluntary social and environmental (also referred to as sustainability) standards (Cashore *et al.*, 2004, Vogel, 2010). These standards use science-based methods (Bain *et al.*, 2010, Ransom *et al.*, 2017) to determine compliance with specific sets of values developed by a range of international organizations, which make claims about the ability of these standards to influence the sustainability of practices (Reinecke, 2010). Indeed, the indicators used to measure the two abovementioned SDG targets are the publishing of sustainability/corporate social responsibility (CSR) reports and material footprints, which are often fundamental components of sustainability standards. We might characterize this as the closing of a circle which began in the private and civic sector, whereby now all societal actors are beginning to use codes of conduct, voluntary guidelines, and multi-actor dialogues to *responsibilize* actors both individually and collectively (e.g. Visser, 2010, Dorbeck-Jung and Shelley-Egan, 2013, Lindner *et al.*, 2016).

With the rising importance of sustainability for business activity and delivering on global societal goals, there has been an explosion of standards available to be complied with, for example, the Ecolabel Index counts 465.[3] Due to the increased multiplication of these standards, competing schemes cause 'an unnecessary duplication of efforts, undermine the stringency of standards programs, and may exacerbate third party concerns regarding the credibility and legitimacy of private standards setting' (Derkx, 2011: 1). Both actors and scholars have pointed to concerns about the legitimacy of specific, individual standards related to who created them (Constance and Bonanno, 2000, Bartley, 2003, O'Rourke, 2005) and how multiple stakeholders are included (Cheyns, 2011, Fransen and Kolk, 2007). The existence of conflicts of interest in certification practices (Lytton, 2014) is critiqued, as is the way standards are used by mainstream multi-national companies (Arnold, 2010) or how diverse types of knowledge are used at different times during the standardization process to justify a specific definition of sustainability (Auld and Bull, 2003, Loconto and Hatanaka, 2017). Questioning of the fundamental legitimacy of individual standards as a way to achieve the 'right impacts' has been questioned because of the use of different logics in how they develop their systems (Auld *et al.*, 2015, Renard and Loconto, 2013) and the ability of actors to hold standard-setters and companies to account for their claims (Marx, 2013). Attacks on the legitimacy of these standards are simultaneously attacks on the visions of sustainability and responsibility that they are able to deliver, which is why concrete efforts have been taken to reduce these contestations.

In response to the competitive nature of multiple sustainability standards and the threats to their legitimacy, an international, member-based organization (the ISEAL Alliance[4]) emerged at the turn of this century to create meta-standards, or meta-regulations, that govern the 'credibility' of voluntary standards, so to allow their creators and users to make claims of social and environmental responsibility. Bernstein (2011) argues that the meta-governance initiative set up by ISEAL meets the input legitimacy requirements – such as stakeholder access, transparency, deliberation and North-South equity – that are most often under threat from individual standards. What remains unclear is whether those mechanisms that have been created at a meta-governance level actually do ensure the credibility of this system of governance (i.e. the complex system of 'objective' oversight) and are enough to establish this as a legitimate way to demonstrate users' responsibility for delivering on sustainability goals (Fransen, 2015). The ability of meta-standards to resolve these contests about sustainability and responsibility is the focus of this chapter, as I explore the structure and function of a meta-standard that aims to reduce conflict about how to measure the impact of sustainability standards.

Case method

Voluntary sustainability standards have been referred to as: 'one of the most innovative and startling institutional designs of the past 50 years' (Cashore *et al*., 2004). Therefore, it is fitting that a social inquiry considers how these standards take form and enable claims of responsibility. This chapter does this through a case study (Yin, 1984) of the ISEAL Alliance. The ISEAL Alliance is a non-profit membership organization that was officially created in 2002. Its four goals are to: 'improve the impacts of standards, define credibility for sustainability standards, increase the uptake of credible sustainability standards, and improve the effectiveness of standards' (ISEAL, 2012a; 1). The core normative work of ISEAL is what they call 'defining credibility.' This currently consists of three outputs: (1) credibility principles where public consultation is used to define what a credible standard is, (2) codes of good practice, which are the meta-standards that lay out requirements for members related to setting-standards, evaluating impacts and assuring compliance (the Standards Code, the Impacts Code and the Assurance Code), and (3) making accurate claims, which is a public consultation on the creation of a taxonomy of claims about sustainability (ISEAL, 2013c).

Loconto and Fouilleux (2014) argue that the ISEAL Alliance is an institutional entrepreneur that is actively working to institutionalize standards as the main governance instruments of an organizational field of sustainability. The focus on credibility is a technique used by ISEAL to imbue their standards with input legitimacy. Since 2014, however, there has been a shift in rhetoric towards output legitimacy and questions of responsibility – particularly related to the expectations that are being set for the ISEAL members. An ISEAL manager explains the changing responsibilities as follows:

> As Credibility Director, I continued to focus on developing a shared understanding of how standards and certification can be more effective, with responsibilities for our Codes of Good Practice and Credibility Principles, and supporting members and emerging initiatives directly to develop and improve their programmes. Now, as Director, Innovations, I am leading our work to ensure members are well-positioned to drive transformation to full sustainability in the sectors where they work.[5]

Within this argument, the Impacts Code is considered to be a means to contribute to ISEAL's credibility as a regulator by codifying tools that will provide output legitimacy. This means that legitimacy is judged based on the effectiveness of how an objective is achieved (Scharpf, 1999). As we are reminded,

> an actor grows with the number of relationships he or she can put, as we say, in black boxes. A black box contains that which no longer needs to be reconsidered, those things whose contents have become a matter of indifference.
>
> (Callon and Latour, 1981: 285)

The Impacts Code is one of three meta-standards that the ISEAL Alliance created and its purpose is to encourage its members to monitor and evaluate the impact that their standard is having on sustainability (cf. Loconto and

Fouilleux, 2014). In this chapter, I argue that as a meta-standard that black-boxes specific metrics, the ISEAL Impacts Code is a 'good' form for governing innovation, but the justifications of the 'goodness' of the form require more care and attention as the structures and functions of meta-standards are still highly contested. I arrive at this conclusion by exploring: *what measures have been developed to 'prove' the 'responsibility' of stakeholders in 'impacting' sustainable practices?*

The empirical data used in this chapter draws from document analysis of the ISEAL Impacts Code, internal communications, ISEAL community blogs, and background papers. These were collected during participant observation as an ISEAL subscriber between 2010 and 2013. Observations consisted of attendance at ISEAL events dedicated to the Impacts Code, participation in online training and webinars (4) and participating in the Assurance Code steering and technical committees (average of four 2–3 day meetings per year). I also conducted interviews with ISEAL staff, members and stakeholders who participated in developing the Impacts Code (four interviews), which were recorded through field notes. ISEAL events are run using the *Chatham House Rule*,[6] therefore, verbatim transcripts are not available from the observations or the informal interviews carried out during these events. An alternative approach to direct quotation from participant observation was used in this chapter, which was to rely upon quotations found in written public documents that were in line with the quotations recorded in field notes.

The next section presents my analytical framework which focuses on how standards are forms for establishing value and how these forms then enable the evaluation of the values at stake. I then analyse the ISEAL Alliance's Impacts Code using this analytical framework in order to expose the values and justifications that are legitimately at stake. I conclude with reflections on how the investment in these forms sheds light on the contested nature of proving the responsibility of actors to have an impact on sustainability.

Forms, standards and their evaluation

How does one prove responsibility, or any other value, that is not clearly visible to the observer but relies instead upon interpretation? This is a question that goes back to Plato's theory of forms, and perhaps beyond.[7] In the *Republic*, Plato argued that what we describe with our language is our perceptions, not of reality, but of the forms that translate reality (cf. Callon and Latour, 1981). The form, in this sense, is an idea of reality that our perceptions and observations must be measured against in order to be able to describe what we see (Geach, 1956). The *Third Man Argument* explains these forms, in that there must be a third person, or form, against which the original form and its participants (people and objects) must be measured (Cohen, 1971, Vlastos, 1954). In this sense, forms are multi-level paradigms, where the three principles of 'one over many,' 'self-predication' and 'non-identity' are all present (Sellars, 1955, Vlastos, 1954). What this means is that a form must be the sum of its parts, explain both its parts and the sum of its parts, and be an idea that is different but can be compared to the sum of its parts. Philosophers have equated this notion of the form to a standard (Geach, 1956; Sellars, 1955), in order to more easily explain how forms can be used for judgement (both technical and moral) and how actors can gain a reflective understanding of reality (Busch, 2012).

Yet standards, unlike Plato's forms, do not remain fixed, but can change over time (Dolev, 2007, Wittgenstein, 1968). Wittgenstein (1968) famously claimed 'there is one thing of which one can say neither that it is one metre long, nor that it is not one metre long, and that is the standard metre in Paris' (1968: 50). This is because of the problem of materiality; material standards are affected by environmental elements that erode them over time and thus there is a need for a third standard to compare it to. ISEAL's standards are less material than they are organizational and cognitive. Sustainability standards are essentially management standards (Mutersbaugh, 2005). They describe the processes, policies and procedures that must be in place for a firm to be able to make a claim of being in compliance with the standard. However, these standards are also changing because of the increased participation of users and other stakeholders in their formulation (Renard and Loconto, 2013; Cheyns, 2011; Ponte and Cheyns, 2013).

As a result, meta-standards, or standards for standards, must be sought in order to re-establish the relation to the original idea that the standard is meant to represent (Dendler, 2013). In this sense, they represent the regression of Plato's forms, as they are *ratcheted-up* through the creation of new hierarchical organizational and institutional structures (Bartley, 2011, Reinecke *et al.*, 2012). These institutions provide a means to continue using the standards, or forms, through time and space beyond those interactions that created them (Callon and Latour, 1981; Loconto and Busch, 2010). This is also how standards can be used to govern the judgements of persons, processes and things (Busch and Bingen, 2006). Thus, standardization, or rather the construction of the form, is a way to create equivalencies and to establish power. In this way, standards are the joining of conformity and information (Thévenot, 2009). Yet at the same time the black-boxing nature of standardization means that many of the debates and concerns have been locked away and concealed from public view.

In order to re-open the black-box to understand the relationship of the form to the values it translates, we must pose more general questions about *what makes a 'good' form and what is the 'goodness' of the form's purpose?* An answer to these two interrelated questions requires an analysis of (1) the 'provability' of the form and (2) the justifications used to explain why it is good. I draw upon the French school of pragmatic sociology to do this (cf. Glerup and Horst, 2014). Thévenot (1984) argues that standard forms permit the 'establishment of equivalents – i.e. the stable and economic articulation of these forms' (1984: 8). To achieve this requires a costly investment 'to establish a *stable* relation *with a certain lifespan*' and an area of *validity* (Thévenot, 1984). A form's '"provability" can then be evaluated based on [its] *material equipment*' (Thévenot, 2009). Lampland and Star (2009) argue that in the case of standards, the content of the form may vary from instance to instance, but the shape of the form becomes the primary human-capital investment. These aspects stabilize certain properties and enable specific claims to be made as to the 'goodness' of the form's purpose. In sum, 'a "good form" is one which fulfils the requirement of reproducing, with the minimum of human intervention, a state of relations between standard objects' (Thévenot, 1984: 10).

To understand the 'goodness' of the form, I rely upon Boltanski and Thévenot's (2006 [1991]) plural 'orders of worth' to analyse how the standard is justified. This is not considered an individual or strategic process of framing an issue for social mobilization (e.g. Benford and Snow, 2000). Rather, it is a way for a collective to justify fairness and procedural requirements for the public space (Thévenot, 2007).[8] Each legitimate polity, to which a justification is made, emphasizes different values as being more salient for the good of society and requires the presentation of different types of evidence. In this sense, it provides the justification for how objectives can be achieved (i.e. output legitimacy). First, the order of *inspiration* values those who break the bonds with the earthly world to find inspiration. Second, the *domestic* polity finds value in tradition and kinship ties. The third order of *fame* draws upon Hobbe's notion of honour and values public opinion. Fourth, the *civic* order of worth is found in collective solidarity as exemplified in Rousseau's social contract. Competition is the main value in the *market* order of worth, while according to the *industrial* polity the value of efficiency takes precedence and is achieved through a functional system of rules that are determined and implemented through scientific methods and technological objects. It is important to understand that these are plural orders, meaning that they are co-existing in society and at times overlapping. There may be a need to appeal to more than one form of 'the good' to justify actions in different spaces and time. The analysis of the goodness of the form will provide a means to understand how voluntary standards contribute to the responsible governance of innovation.

Investing in the form: the Impacts Code

ISEAL was originally formed to increase communication between competing sustainability standards, but today its work is more entrepreneurial in its efforts to expand the markets of its members as well as increase the credibility legitimacy of standards as a tool for achieving sustainability, which is left purposely undefined at a global level (Loconto and Fouilleux, 2014). This latter work requires greater coordination between members and more solid definitions of what their standards can do (cf. Fransen, 2015). For example, ISEAL argues that the nature of 'credence' attributes (Caswell, 1998) that characterize the claims made by sustainability standards organizations makes it difficult for buyers and consumers alike to compare standards and to understand what impact they are having on sustainable production (ISEAL, 2011b). This is exacerbated by weak or non-existent monitoring and evaluation systems for ISEAL members.[9] ISEAL itself claimed that:

> Recent experiences show that we are beginning to move beyond anecdotal evidence by applying methodologies and tools that provide a more accurate picture of social and environmental change. However, these initiatives are nascent and there is a sense of urgency in reaching a shared understanding of how to set up and implement methodologies for measuring impacts. This shared understanding can inform existing processes for measuring impacts, as well as influence the development of new and emerging standards initiatives.
>
> (ISEAL, 2010b: 1)

The urgency in the above quote was the increasing demand from consumers, governments, companies and other stakeholders to prove that voluntary standards are having a positive impact – and thus justify their use by these stakeholders. This provided significant motivation for ISEAL to invest in a form that could establish equivalents between standards systems.

The *stable relation* of this investment is found in the Code of Good Practice for Assessing the Impacts of Standards Systems (Impacts Code) that was the result of a 22-month technical consultation process with two 60 day periods of public consultation (2008–2010). The code development was governed by a steering committee, an issues committee and a methodologies committee and consisted of 29 participants. Forty-one per cent of the participants came from ISEAL member organizations, the remainder (and majority) came from NGOs, research, industry (including a certification body) and one public donor. Seventeen staff members from ISEAL members were included in the confidential committee communications and 13 in-person or web meetings were held during the period (ISEAL, 2010b).

With this code, ISEAL pushes members 'to define and document the changes they intend to bring about and collect the data that will allow them and others to assess their contributions toward that change' (ISEAL, 2011a: 2). This quote illustrates that ISEAL is using the code to establish output legitimacy for those who comply with it. The Impacts Code sets out a structure for members to measure change and to understand how this change contributes to long-term impacts. It includes requirements and guidance on which issues to measure and how best to measure them. This standard is highly procedural, focusing on methodology rather than on content. This is noted in the standard, and in lieu of extensive prescriptions, ISEAL suggested principles that should guide impact assessment: appropriate scope, practical focus, quality, openness (transparency), effective communication, broad participation, learning and improving, and institutional capacity.

In this way, this meta-standard establishes a requirement for all credible standards systems to measure and demonstrate their contributions to social and environmental impacts using consistent methodologies. Practically, only members of the ISEAL Alliance are required to comply with the Impacts Code and thus, according to ISEAL and its members, only they can be considered 'credible.' This sets the *spatial validity* of the form to ISEAL members in good standing and creates a boundary around who and what can participate in the form. The initial *temporal validity* of this standard was two years with a review conducted in 2012. Subsequently, the standard must be reviewed and revised based on member and other stakeholder input every four years (ISEAL, 2010a). These reflexive loops are meant to encourage learning processes and accommodate changes that the participants to the form undergo; they do not mean that the temporal validity of the form will expire every four years. Rather the temporal validity is fixed against the spatial validity of the form in that as long as a standards development organization is a member of ISEAL, they must comply with the code. This first short temporal validity of the standard relates to the difficulties that ISEAL members had in implementing the Impacts Code.[10] Current members were given three years to comply with the standard, but it became clear that setting up a monitoring and evaluation (M&E) system, as was required by the standard, was more difficult than originally envisioned (ISEAL, 2012b). As a result, ISEAL invested more time, resources and effort into making the material equipment of the standard more solid. Through these efforts we can see some of the contestations that are black-boxed into the standard.

The solidity of the material equipment

The *solidity* of the material equipment of the form refers to the extent to which the form is objectified (Thévenot, 1984). What this means is that to stabilize a form, there must be investment in anonymous tools (e.g. technological, legal, scientific, management) that can be applied objectively across the space and time of the form. This means that these implements derive from and contribute to fixing and diffusing the form. I argue that this is done in the Impacts Code through its choice of using the 'theory of change' metric, which is a tool of management science (Zand and Sorensen, 1975). What this metric does is effectively fix the generic, objective form, which allows for specificity at the level of each member. In this way it enables the Impacts Code to serve as a unifying form for the circulation of particulars (O'Connell, 1993).

ISEAL determined that much of the difficulty that members faced in proving their impacts in the past was due to ambiguity concerning what they were measuring. They found that members often conflated specific standards interventions, or even compliance to individual indicators, to an aggregate overall impact of the organization's work. Therefore, ISEAL established the following two specific research questions as the means to gauge impacts: (1) To what extent does compliance with the standards lead to the intended impacts? (2) How effective are the strategies employed by the standards system at increasing the use of standards-compliant practices? (ISEAL, 2010a) In effect, this approach allowed ISEAL to define impact by separating it into the impact OF what and the impact ON what (Rusillo, 2009). This differentiation was further clarified by introducing a 'Theory of Change' approach to impact assessment:

> This involves developing an assessment plan for how an organization intends to create change in the world, and then determining how to assess whether that change had occurred. Standard setters need to articulate their goals and their strategies to achieve them. This includes the implementation of standards as well as other activities such as influencing policy and capacity building. This process helps them to determine what indicators are important to monitor in order to assess whether the anticipated change has occurred
>
> (ISEAL, 2009: 1)

The above theory-guided programme evaluation framework was first introduced by Weiss (1972, 1995) and defines a theory of change as a theory of how and why an initiative works. It explains both the interim steps that lead to the long-term goal and the connections between activities and outcomes at each step. It introduces a logical model for evaluation. This theory has been developed over the past 40 years and has become popular in the evaluation of development projects and community programmes. The adoption of the theory of change by ISEAL effectively forces its member to implement a 'results-based framework' of management, not just for the impacts their standards are having, but for themselves as organizations.

This requires ISEAL members to clearly define goals (impacts) and objectives (outcomes), as well as the road map to get there. This model also requires standard setters to develop a monitoring and evaluation programme, which will ensure that there are procedures for collecting data on the indicators to be assessed. The notion of feedback loops is very important to the model that ISEAL has developed, as it speaks to the stakeholder participation that is required of ISEAL members. The data collected through the monitoring and evaluation exercise will then be evaluated and turned into knowledge that can be used to improve the impact of the standards systems and as contribution to impact evaluation studies (ISEAL, 2009).

Promoters of the 'theory of change' argue that it is an 'approach to evaluation; it is not an evaluation method that stands on its own' (Connell and Kubisch, 1998). This means that for ISEAL members to measure their impact, they need to investment in additional implements. Whether or not investment at this level might be a collective project is contested and the classic social science dilemma between the relative worth of qualitative and quantitative data is at stake. For example, Raggo (2009) cites the ISEAL background report that 'while the positive impacts are apparent and supported by qualitative data gathered by those who have worked in different fields of activity, it is challenging to translate these impacts into cross-sector, comparable, and quantitative data' (Raggo, 2009: 143).

This dilemma was presumably resolved by the introduction of the 'theory of change' into the code and is explained in the public summary by the recommendation that:

> Standard systems should shift the focus from proving impact as an end in itself to using the evaluation process for learning about changes instigated

> and influenced. Impact assessment should be considered within a monitoring and evaluation system as one level of evaluation, with monitoring and evaluation taking place at multiple levels (outputs/outcomes/impacts). The focus of impact assessment then shifts from proving impacts to becoming part of a learning and improving process.
>
> (Rusillo, 2009: 3)

However, discussions at the 2011 Annual meeting during the Public Day session entitled 'Measuring the Impacts of Certification: Towards a Focus for Future Research' and the members only session entitled 'Towards Compliance with the ISEAL Impacts Code: Member Case Studies' demonstrated that this issue was not yet resolved (ISEAL, 2011d). For example, during the Public Day session the 'state of the art' in impact research was presented and there was discussion around the lack of impact studies that used 'credible counterfactuals' (cf. Blackman and Rivera, 2010, International Trade Centre (ITC), 2011). The presentation focusing on the future of impact research presented the 'gold standard' of impact assessment – Repeatable, Random, Controlled Experiments. This proposal raised a number of methodological, feasibility and ethical questions from the audience. For example, the comparison was made to the pharmaceutical industry's clinical drug trials and the speaker proposed that the ISEAL community should be doing similar trials, including paying participants, in order to determine the causal impacts of their standards.

Many participants found this proposal to be counterproductive to the debate. Indeed, during the members only session the same people were again involved in discussing the experiences of three ISEAL members who had begun to implement the Impacts Code. During this session, it was evident that members had not adopted the experimental design for impact research and are instead engaging a range of different methods, both quantitative and qualitative, to determine the impact of their standards. What was new was the type of equipment that was being brought into the impacts frameworks. Some focused on the use of geographic positioning systems (GPS), others on lifecycle assessment, others on qualitative participatory audits, and still others on simple risk-analysis tools (ISEAL, 2011c). At the 2013 annual conference, even more scientific and management tools were proposed, as was a renewed call for 'quasi-experimental designs' and 'common data exchange formats and frameworks' (Kennedy, 2013: 8). It was clear that 'proving' impact was still important for ISEAL members, yet the qualitative/quantitative debate persisted. Nonetheless, ISEAL members were encouraged to collaborate even further: 'Each standard system has a different theory of how change occurs and employs different strategies to promote sustainable livelihoods and reduce poverty, and yet there are many common elements in the approaches and goals across systems' (ISEAL, 2012c: 1). This commonality, while promoted through the form of the Impacts Code, is maintained through collective projects and services offered by ISEAL (e.g. webinars and online community blogs). Through these collaborative efforts, we can see how the Impacts Code as a 'good form' is being solidified and how the 'goodness' of the form is being justified.

The 'goodness' of the form, or what is responsible about meta-standardizing sustainability impact?

The French school of pragmatic sociology has been used in the past to understand how the concepts of 'fair price' (Reinecke, 2010), 'evidence in third-party certification' (Silva-Castañeda, 2012) and 'inclusiveness' (Cheyns, 2011) are constructed and justified. The contribution of this chapter is to explore how this 'good' form reproduces a notion of 'responsibility' for the owners and users of voluntary standards.

Responsibility can be understood both in terms of taking responsibility and being held accountable for actions. It is about incorporating democratic principles of governance and focusing on the 'right impacts' (Owen *et al.*, 2012), thus bridging the observed divide between input and output legitimacy (Fuchs *et al.*, 2011). For the international agri-business community, the 'right impacts' are increasingly tied into CSR programmes and are linked to proving sustainability[11] in production (Loconto, 2016; Mejlgaard and Griessler, 2016). Responsibility can also be thought of as responsiveness, which refers to collective responsibility (how individuals become responsive to each other) and ethical choices in practice (Gorgoni, 2009; Walhout and Kuhlmann, 2013). The collective projects of ISEAL, including communications strategies and training, show how the Impacts Code is giving form to multiple justifications of responsibility.

First, the justification of responsibility is linked to the responsibility of ISEAL members to prove to their stakeholders that they are indeed having an impact on sustainability. As ISEAL reminds its members, 'the culture of impact and demonstrating results is here to stay' (ISEAL, 2013g: 1). ISEAL members can thus improve their responsiveness by setting up a monitoring and evaluation system (M&E). The reasons why an M&E system is needed are explained to members in an online training session as follows:

> External: stakeholders and users want to see evidence of results. Is the standard we support making a difference? Is it worth our time and worth the extra cost?
>
> Internal: clarity about goals and expected results improves program and standard design measuring and monitoring results drives and enables learning and improvement.
>
> (ISEAL, 2013e: 1)

These justifications illustrate that ISEAL places an importance on demonstration and learning as a way to act responsibly (Komives, 2013b). This is facilitated collectively through collaboration via an M&E 'community of practice' (ISEAL, 2013h), where one of the collaborative efforts has been on developing a 'set of shared, or common, indicators' (ISEAL Alliance, 2013: 1). Quoting David Suzuki from his United Nations address at the Convention for Biological Diversity in 2004, one of ISEAL's members explained that 'when there is no interconnectedness there is no sense of shared responsibility' (Kennedy, 2013: 11). This quote was used to justify the need to 'start simple, focus on measures that we all need (Shared Values = Shared Responsibility)' (Kennedy, 2013: 12). These shared values are found in traditional geographic, demographic and agronomic metrics, whose forms are relatively well stabilized. Nonetheless, the creation of a collective and the pre-competitive forum within which ISEAL members are trying to work offer insights into how this form is allowing the creation of shared responsibility and ISEAL also hopes that this will fuel innovation in evaluation techniques (ISEAL, 2013f).

Second, the Impacts Code helps ISEAL members to focus their end goal on broader social and public goods, rather than on a strict internal strategy of corporate responsibility. This is a way to *responsibilize* ISEAL members, as illustrated by how ISEAL defines what its members' standards should be:

> Given that international standards should be used as references for the development of national and regional standards, it is important that they are placed in the public domain and are available without cost. National and regional standards that focus on social and environmental issues should also be placed in the public domain as they are in the *public interest and public good.*
>
> (ISEAL, 2010c: 12)

Through training webinars and the project supported by the Ford Foundation, ISEAL and its agriculture and forestry members are widening the focus of impact beyond the idea of firms being responsible for their operations (which is the focus of most ISEAL member standards) towards the responsibility of ISEAL members to help firms and farmers to contribute to human well-being:

> The impact goal: The end goal in the conceptual framework is significant and sustained improvement in *human well-being at the household level.* Human well-being is multi-dimensional, and includes economic, environmental, human, social, and political dimensions. Higher incomes are part of the picture, but equally important are control over resources, quality of life, and empowerment.
>
> (ISEAL, 2012c: 1)

> The success of a standard is not measured solely by increases in certified area or the uptake of better practices. What really determines the impact of a standard is the tangible difference it makes to social and environmental sustainability.
>
> (ISEAL, 2013d: 1)

In the two quotes above, ISEAL argues that the impacts of its members should be evaluated, despite their individual theories of change, within the light of broader public goods. These are clearly appeals to a civic notion of worth, whereby the collective good is the desired outcome of a *responsibilization* process.

However, this sense of responsibility is further justified by those ISEAL members who make commitments to evaluation and investment in the form, which reflect an appeal to the industrial order of worth:

> The true test of a standard is whether it actually improves sustainability. Credible standards have systems to collect information such as whether the standard is leading to the desired impacts or whether it is addressing important environmental and social challenges. Credible standards also contribute to deeper evaluations to know if the standard itself is responsible for changes observed. It takes time and commitment to do this accurately, but credible standard-setters make that commitment and by doing so, are able to prove their impacts while also improving their standard.
>
> (ISEAL, 2013d: 1)

Third, how the Impacts Code is justified in public communication positions the actors towards discourse about values that are considered to be fundamental to responsible innovation. As ISEAL explains to its members, the ability of outsiders to believe their impacts data relies upon transparency:

> Be transparent about results: post and share studies and monitoring reports. Report good and bad results. Be accurate. Discuss learning and how you will respond.
>
> (ISEAL, 2013e: 1)

> Honesty and transparency are key here in communications and marketing terms. We need to be open, not just about what we're getting right, but about the challenges and complexities we face.
>
> (ISEAL, 2013a: 1)

> And while it is important to show positive impacts, we also have to be open to recognize weaknesses and address these in a pro-active manner, both technically, but also through transparent and honest communication.
>
> (ISEAL, 2013b: 1)

Appealing to honesty and transparency, particularly in public communication, reveals appeals to goodness found in the fame order of worth. They are also the foundational ethics of what scholars are calling responsible innovation (Lindner *et al.*, 2016).

However, despite strong rhetoric within ISEAL, they have not achieved their ideal of honest, transparent and pre-competitive collaboration on impacts tools. Why? Because of the fact that it is not simple to 'align without full transparency' and ISEAL members remain 'afraid of being compared to others' (Komives, 2013a: 1). Thus, despite emerging claims to civic and public orders of worth, the industrial order of worth remains an important justification of the form, particularly when it interacts with the market ideal of competition and investment (cf. Ponte, 2009). Described in this way, it is clear that despite the existence of stable, standardized forms, the justifications of their value for proving responsibility remain unstable and still highly contested. If sustainability standards are to be considered as tools for proving responsibility for sustainability impact, then there must be more work done to understand the plurality of justifications used to define actions as responsible.

Conclusions

In this chapter, I argue that the ISEAL Alliance's meta-standard, the Impacts Code, is a 'good' form for establishing sustainability standards as governance tools in responsible innovation. However, the 'goodness' of this form in terms of the value of 'responsibility' is not clear regarding the definition of who becomes responsible for what. By drawing upon the French school of pragmatic sociology my analysis shows that the process of standardization attempts to bound responsibility for impact according to civic, fame, industrial and market notions of worth i.e. those responsible standards are those that can prove a positive impact on society. The standardization of impact therefore introduces new justifications into the realm of 'provable' forms and shows that analyses of sustainability standards based solely upon a market logic where market competition will 'prove' which standard is best are limited in their explanations of responsibility for sustainability impact. Likewise, those that focus solely on civic notions of goodness will also fall short.

The 'good' forms that are promoted by the Impacts Code are tools of industrial worth as they are standardized metrics that provide scientific and managerial evidence that can prove impact, contribute to ISEAL's project of defining credibility and, as a result, begin to provide ISEAL members with the foundations of output legitimacy. However, the industrial forms of proof clash with some of the claims of the 'goodness' of these standards, as further exploration of ISEAL members' 'theories of change' have illustrated (Komives, 2013a). As a result, ISEAL, its members and international donors simultaneously made additional investments in collaboration and communication that are part of efforts to 'compromise' for the common good (Boltanski and Thévenot, 2006 (1991)). This analysis reveals significant tensions between the 'good' form and the 'goodness' of the form, which highlights the plurality and co-existence of conflicting justifications for what is sustainable and who is responsible.

The effects of this investment in form may be longstanding and have influence on both the practices of sustainability standards and how knowledge about their impacts is produced. As ISEAL members are increasingly beginning to collect impacts data about their systems, they are simultaneously constructing databases that will format their data according to each member's 'theory of change.' However, if the collective project works, isomorphism will take hold and we will have increasingly standardized indicators that are being collected across all of the systems. For ISEAL, their members and consumers, this is great news. It means that it would be easier to compare across standards systems and the transparency that is promised should prove the responsibility and accountability of ISEAL members.

On the other hand, it also means that independent knowledge about the impacts of these systems will be harder to find. As the ISEAL M&E manager notes: 'I frequently remind researchers that they should not be surprised if they receive well-founded constructive criticism on their research designs and reports from the standard systems they study' (ISEAL, 2013h: 1). But what happens if researchers find different data and impacts from that which is held in the increasingly large databases of the standards systems? Which results will be considered more legitimate? Alternatively, what if researchers don't find different data and impacts; does this then prove the responsibility of standards development organizations or the sustainability of their practices? Perhaps it only proves that the investment has been successful in standardizing the form, but not the substance. In either case, it illustrates that there are many aspects of meta-governance, including its forms, which we must continue to explore.

Notes

1 Funding: This work was supported by the Res-AGorA project (Responsible Research and Innovation in a Distributed Anticipatory Governance Frame. A Constructive Socio-normative Approach) under the European Union's Seventh Framework Programme for research and technological development and demonstration [321427]; and the French Institute for Research Innovation and Society (IFRIS).
2 https://sustainabledevelopment.un.org/.
3 www.ecolabelindex.com/. (Accessed 15 September 2017).
4 ISEAL originally stood for International Social and Environmental Accreditation and Labelling Alliance. Its current members include: the 4C Association, Accreditation Services International (ASI), the Alliance for Water Stewardship, Aquaculture Stewardship Council (ASC), Bonsucro, Equitable Origin, Fairtrade International (FLO), the Forest Stewardship Council (FSC), the Global Sustainable Tourism Council (GSTC), Golf Environment Organization (GEO), GoodWeave, International Organic Accreditation Services (IOAS), Linking Environment and Farming (LEAF), the Marine Stewardship Council (MSC), the Responsible Jewellery Council (RJC), the Roundtable on Sustainable Biomaterials (RSB), the Roundtable on Sustainable Palm Oil (RSPO), Social Accountability Accreditation Services (SAAS), the Rainforest Alliance/Sustainable Agriculture Network (SAN), the Union for Ethical BioTrade (UEBT), UTZ Certified.
5 www.isealalliance.org/online-community/people/patrick-mallet. (Accessed 5 April 2016).
6 'When a meeting, or part thereof, is held under the Chatham House Rule, participants are free to use the information received, but neither the identity nor the affiliation of the speaker(s), nor that of any other participant, may be revealed.' www.chathamhouse.org/about-us/chathamhouserule. (Accessed 10 March 2013).
7 I want to thank Lawrence Busch for bringing up this point. See: Busch, L. (2017). Standards and Their Problems: From Technical Specifications to World-Making. In: Miele, M., Higgins, V., Bjørkhaug, H. and Truninger, M. (eds) *Transforming the Rural*. Bingley, UK: Emerald Publishing Limited.
8 Thévenot describes this as the *justifiable action engagement regime*, which is oriented towards a public and is characterized by generality and legitimacy. This is one of three regimes of engagement found in society, the other two are the regime of familiar engagement and the regime of engagement in a plan.

9 Interview with ISEAL staff member March 2011.
10 These were discussed during the impacts sessions at the ISEAL Annual Meetings in 2011 and 2012.
11 Typically envisioned along the three pillars of economic, environmental and social sustainability. See: Brundtland G. H. C. (1987). Our common future. *World Commission on Environment and Development*. Oxford, England: Oxford University Press.

References

Arnold, C. (2010). *Fairtrade vs. Rainforest Alliance. Sara Lee vs. Organics. It's one big green fight* [Online]. Brand Republic. Available: http://community.brandrepublic.com/blogs/arnold_on_ethical_marketing/archive/2010/03/01/fairtrade-vs-rainforest-alliance-sara-lee-vs-organics-it-s-one-big-green-fight.aspx. (Accessed 7 November 2010).

Auld, G. and Bull, G. Q. (2003). The institutional design of forest certification standards initiatives and its influence on the role of science: the case of forest genetic resources. *Journal of Environmental Management*, 69: 47–62.

Auld, G., Renckens, S. and Cashore, B. (2015). Transnational private governance between the logics of empowerment and control. *Regulation & Governance*, 9: 108–124.

Bain, C., Ransom, E. and Worosz, M. (2010). Constructing Credibility: Using Technoscience to Legitimate Strategies in Agrifood Governance. *Journal of Rural Social Science*, 25: 160–192.

Bartley, T. (2003). Certifying Forests and Factories: States, Social Movements, and the Rise of Private Regulation in the Apparel and Forest Product Fields. *Politics and Society*, 31: 433–464.

Bartley, T. (2011). Certification as a mode of social regulation. In: Levi-Faur, D. (ed.) *Handbook on the Politics of Regulation*. Cheltenham, UK and Northampton, MA: Edward Elgar.

Benford, R. D. and Snow, D. A. (2000). Framing Processes and Social Movements: An Overview and Assessment. *Annual Review of Sociology*, 26: 611–639.

Bernstein, S. (2011). Legitimacy in intergovernmental and non-state global governance. *Review of International Political Economy*, 18: 17–51.

Blackman, A. and Rivera, J. (2010). *The evidence base for environmental and socioeconomic impacts of 'sustainable' certification*. Washington, DC: Resources for the future.

Boltanski, L. and Thévenot, L. (2006) [1991]. *On Justification: Economies of Worth*, Princeton, NJ: Princeton University Press.

Brundtland G. H. C. (1987). Our common future. *World Commission on Environment and Development*. Oxford, England: Oxford University Press.

Busch, L. (2012). Standards and Their Problems: From Technical Specifications to World-Making. *XIII World Congress of Rural Sociology 2012 – Working Group 4: Standards and Innovation Transitions: Practising and Knowing Sustainable Rural Futures*. Lisbon, Portugal.

Busch, L. (2017). Standards and Their Problems: From Technical Specifications to World-Making. In: Miele, M., Higgins, V., Bjørkhaug, H. and Truninger, M. (eds) *Transforming the Rural*. Bingley, UK: Emerald Publishing Limited.

Busch, L. and Bingen, J. (2006). Introduction: A New World of Standards. In: Bingen, J. and Busch, L. (eds) *Agricultural Standards: The Shape of the Global Food and Fiber System*. Dordrecht, NL: Springer.

Callon, M. and Latour, B. (1981). Unscrewing the big Leviathan: how actors macro-structure reality and how sociologists help them to do so. In: Knorr-Cetina, K. and

Cicourel, A. V. (eds) *Advances in social theory and methodology: Toward an integration of micro and macro-sociologies*. Boston: Routledge and Kegan Paul.

Cashore, B. W., Auld, G. and Newsom, D. (2004). *Governing through markets: Forest certification and the emergence of non-state authority*. New Haven: Yale University Press.

Caswell, J. A. (1998). How labelling of safety and process attributes affects markets for food. *Agricultural and Resource Economics Review*, October: 151–158.

Cheyns, E. (2011). Multi-stakeholder Initiatives for Sustainable Agriculture: The Limits of the 'Inclusiveness' Paradigm. In: Ponte, S., Gibbon, P. and Vestergaard, J. (eds) *Governing Through Standards*. Houndmills, Basingstoke, Hampshire; New York: Palgrave Macmillan.

Cohen, S. M. (1971). The Logic of the Third Man. *The Philosophical Review*, 80: 448–475.

Connell, J. P. and Kubisch, A. C. (1998). *Applying a Theory of Change Approach to the Evaluation of Comprehensive Community Initiatives: Progress, Prospects, and Problems*. Washington, DC: The Aspen Institute.

Constance, D. H. and Bonanno, A. (2000). Regulating the global fisheries: The World Wildlife Fund, Unilever, and the Marine Stewardship Council. *Agriculture and Human Values*, 17: 125–139.

Dendler, L. (2013). Sustainability Meta Labelling: an effective measure to facilitate more sustainable consumption and production? *Journal of Cleaner Production*, 63: 74–83.

Derkx, B. (2011). Metagovernance in the Realm of Private Sustainability Standards Setting. MSc MSc Thesis, University of Utrecht.

Dolan, C. and Rajak, D. (2011). Introduction: Ethnographies of corporate ethicizing. *Focaal*, 2011: 3–8.

Dolev, Y. (2007). Mission Impossible and Wittgenstein's Standard Metre. *Philosophical Investigations*, 30: 127–137.

Dorbeck-Jung, B. and Shelley-Egan, C. (2013). Meta-Regulation and Nanotechnologies: The Challenge of Responsibilisation Within the European Commission's Code of Conduct for Responsible Nanosciences and Nanotechnologies Research. *NanoEthics*, 7: 55–68.

Fransen, L. (2015). The politics of meta-governance in transnational private sustainability governance. *Policy Sciences*, 48(3): 298–317.

Fransen, L. W. and Kolk, A. (2007). Global rule-setting for business: A critical analysis of multi-stakeholder standards. *Organization*, 14: 667–684.

Fuchs, D., Kalfagianni, A. and Havinga, T. (2011). Actors in private food governance: the legitimacy of retail standards and multistakeholder initiatives with civil society participation. *Agriculture and Human Values*, 28: 353–367.

Geach, P. T. (1956). The Third Man Again. *The Philosophical Review*, 65: 72–82.

Glerup, C. and Horst, M. (2014). Mapping 'social responsibility' in science. *Journal of Responsible Innovation*, 1: 31–50.

Gorgoni, G. (2009). La responsabilità come progetto. Primi elementi per un'analisi dell'idea giuridica di responsabilità prospettica. *Diritto e Società*, 2: 243–292.

International Trade Centre (ITC) (2011). The impacts of private sustainability standards on producers in developing countries. *Literature review on the impacts of private sustainability standards; Part II*. Geneva: International Trade Centre (ITC).

ISEAL (2009). *The ISEAL Impacts Code: Implications for Standards Systems*. London: ISEAL Alliance.

ISEAL (2010a). *Assessing the Impacts of Social and Environmental Standards Systems v1.0 ISEAL Code of Good Practice*. London: ISEAL Alliance.

ISEAL (2010b). *Final Report of the Impacts Code Project*. London: ISEAL Alliance.

ISEAL (2010c). *Setting Social and Environmental Standards v5.0: ISEAL Code of Good Practice*. London: ISEAL Alliance.

ISEAL (2011a). *An Introduction to ISEAL: Making Good Standards Better*. London: ISEAL Alliance.

ISEAL (2011b). *The ISEAL 100: A Survey of Thought Leader Views on Sustainability Standards*. London: ISEAL Alliance.

ISEAL (2011c). ISEAL Member-Only Conference Days 2011: Resources. *ISEAL community only blog*. ISEAL Alliance.

ISEAL (2011d). Measuring the Impacts of Certification. *Online community blog*. ISEAL Alliance.

ISEAL (2012a). *About Us* [Online]. Available: www.isealalliance.org/content/about-us. (Accessed 18 February 2012).

ISEAL (2012b). *Assessing compliance with the ISEAL Impacts Code: For all organisations that become ISEAL Associate Members after December 31, 2010*. Version 3.1 – April 2012. London: ISEAL Alliance.

ISEAL (2012c). *How Standards Contribute to Human Well-Being: A Conceptual Framework*. London: ISEAL Alliance.

ISEAL (2013a). Community Day of ISEAL conference to showcase innovation and collaboration. *ISEAL community only blogs*. ISEAL Alliance.

ISEAL (2013b). Conference Sneak Peek 3 Questions With WWF's Nina Haase On Communicating Impacts. *ISEAL community only blogs*. ISEAL Alliance.

ISEAL (2013c). *Defining credibility* [Online]. Available: www.isealalliance.org/our-work/defining-credibility. (Accessed 11 August 2013).

ISEAL (2013d). Standards Show The Way: The Demonstrating Impacts Video. *Online Community Blogs*. ISEAL Alliance.

ISEAL (2013e). Standards Systems Essentials: Good Practice in Monitoring and Evaluation. *ISEAL community only blogs*. ISEAL Alliance.

ISEAL (2013f). Talking Impacts Collaboration with the Ford Foundation. *ISEAL community only blogs*. ISEAL Alliance.

ISEAL (2013g). Things to consider when communicating about impacts. *ISEAL community only blogs*. ISEAL Alliance.

ISEAL (2013h). What has changed in two years of working on impacts? *ISEAL community only blog*. ISEAL Alliance.

ISEAL Alliance (2013). *ISEAL & Impacts: Working Together to Demonstrate & Improve Poverty Impacts*. London: ISEAL Alliance.

Kennedy, E. (2013). Digging for data … together: Rainforest Alliance sustainable agriculture network experiences and aspirations. *ISEAL Annual Conference 2013: Impacts and Innovation*. London.

Komives, K. (2013a). Digging for data … together: Introduction. *ISEAL Annual Conference 2013: Impacts and Innovation*. London.

Komives, K. (2013b). Fairtrade's monitoring report tracks impact for over one million farmers and workers. *Online Community News*. ISEAL Alliance.

Lampland, M. and Star, S. L. (2009). *Standards and their stories: how quantifying, classifying, and formalizing practices shape everyday life*, Ithaca: Cornell University Press.

Lindner, R., Kuhlmann, S., Randles, S., Bedsted, B., Gorgoni, G., Griessler, E., Loconto, A. and Mejlgaard, N. (eds) (2016). *Navigating Towards Shared Responsibility in Research and Innovation. Approach, Process and Results of the Res-AGorA Project* Karlsruhe, DE: Fraunhofer Institute for Systems and Innovation Research (ISI).

Loconto, A. (2016). Case Study 4: Critical Organisations – Multi-national corporations. In: Lindner, R., Kuhlmann, S., Randles, S., Bedsted, B., Gorgoni, G., Griessler, E., Loconto, A. and Mejlgaard, N. (eds) *Navigating Towards Shared Responsibility in Research and Innovation. Approach, Process and Results of the Res-AGorA Project*. Karlsruhe, DE: Fraunhofer Institute for Systems and Innovation Research (ISI).

Loconto, A. and Busch, L. (2010). Standards, techno-economic networks, and playing fields: Performing the global market economy. *Review of International Political Economy*, 17: 507–536.

Loconto, A. and Fouilleux, E. (2014). Politics of Private Regulation: ISEAL and the shaping of transnational sustainability governance. *Regulation & Governance*, 8: 166–185.

Loconto, A. and Hatanaka, M. (2017). Participatory Guarantee Systems: Alternative Ways of Defining, Measuring, and Assessing 'Sustainability.' *Sociologia Ruralis*, 00: 1–21.

Lytton, T. D. (2014). Competitive Third-Party Regulation: How Private Certification Can Overcome Constraints That Frustrate Government Regulation. *Theoretical Inquiries in Law*, 15: 539.

Marx, A. (2013). Varieties of legitimacy: a configurational institutional design analysis of eco-labels. *Innovation: The European Journal of Social Science Research*, 26: 268–287.

Mejlgaard, N. and Griessler, E. (2016). Monitoring RRI in Europe: approach and key observations. In: Lindner, R., Kuhlmann, S., Randles, S., Bedsted, B., Gorgoni, G., Griessler, E., Loconto, A. and Mejlgaard, N. (eds) *Navigating Towards Shared Responsibility in Research and Innovation. Approach, Process and Results of the Res-AGorA Project*. Karlsruhe, DE: Fraunhofer Institute for Systems and Innovation Research (ISI).

Mutersbaugh, T. (2005). Just-in-space: Certified rural products, labor of quality, and regulatory spaces. *Journal of Rural Studies*, 21: 389–402.

O'Connell, J. (1993). Metrology: The Creation of Universality by the Circulation of Particulars. *Social Studies of Science*, 23: 129–173.

O'Rourke, D. (2005). Market Movements: Nongovernmental Organization Strategies to Influence Global Production and Consumption. *Journal of Industrial Ecology*, 9: 1–14.

Owen, R., Macnaghten, P. and Stilgoe, J. (2012). Responsible research and innovation: From science in society to science for society, with society. *Science and Public Policy*, 39: 751–760.

Ponte, S. (2009). Governing through Quality: Conventions and Supply Relations in the Value Chain for South African Wine. *Sociologia Ruralis*, 49: 236–257.

Ponte, S. and Cheyns, E. (2013). Voluntary standards, expert knowledge and the governance of sustainability networks. *Global Networks*, 13: 459–477.

Raggo, P. (2009). The ISEAL Alliance: Setting standards for social and environmental change. In: Leipziger, D. (ed.) *SA8000 The First Decade: Implementation, Influence and Impact*. Sheffield, UK: Greenleaf Publishing Limited.

Ransom, E., Hatanaka, M., Konefal, J. and Loconto, A. (2017). Science and Standards. In: Tyfield, D., Lave, R., Randalls, S. and Thorpe, C. (eds) *The Routledge Handbook of the Political Economy of Science*. New York: Routledge.

Reinecke, J. (2010). Beyond a subjective theory of value and towards a 'fair price': an organizational perspective on Fairtrade minimum price setting. *Organization*, 17: 563–581.

Reinecke, J., Manning, S. and Von Hagen, O. (2012). The Emergence of a Standards Market: Multiplicity of Sustainability Standards in the Global Coffee Industry. *Organization Studies*, 33: 791–814.

Renard, M.-C. and Loconto, A. (2013). Competing Logics in the Further Standardization of Fair Trade: ISEAL and the Símbolo de Pequeños Productores. *International Journal of Sociology of Agriculture and Food*, 20: 51–68.

Rusillo, A. (2009). *State of the Art in Measuring the Impacts of Social and Environmental Standard Systems R081 – Public Summary – January 2009 An overview of the two ISEAL reports: 'Key Impact Issues: WHAT is Being Measured' and 'Evaluation Approaches: HOW It Is Being Measured.'* London: ISEAL Alliance.

Scharpf, F. W. (1999). *Governing in Europe: Effective and Democratic?* Oxford: Oxford University Press.

Sellars, W. (1955). Vlastos and 'The Third Man.' *The Philosophical Review*, 64: 405–437.

Silva-Castañeda, L. (2012). A forest of evidence: third-party certification and multiple forms of proof—a case study of oil palm plantations in Indonesia. *Agriculture and Human Values*, 29: 361–370.

Temri, L., Giordano, G. and Kessari, M.-E. (2015). Innovation et responsabilité sociale des entreprises (RSE) dans les entreprises agroalimentaires du Languedoc-Roussillon: le rôle de la performance économique. *Innovations*, 46: 115–139.

Thévenot, L. (1984). Rules and implements: investment in forms. *Social Science Information*, 23: 1–45.

Thévenot, L. (2007). The Plurality of Cognitive Formats and Engagements. *European Journal of Social Theory*, 10: 409–423.

Thévenot, L. (2009). Governing life by standards: A view from engagements. *Social Studies of Sciences*, 39: 793–813.

UN (2015). Transforming our world: the 2030 Agenda for Sustainable Development. Resolution adopted by the General Assembly on 25 September 2015. In: Assembly, U. N. G. (ed.) *A/RES/70/1* New York: United Nations.

Utting, P. (2002). Regulating Business via Multistakeholder Initiatives: A Preliminary Assessment. In: NGLS (ed.) *Voluntary approaches to corporate responsibility: readings and a resource guide*. Geneva, Switzerland: UN Non-Governmental Liaison Service.

Utting, P. (2007). CSR and equality. *Third World Quarterly*, 28: 697–712.

Visser, W. (2010). CSR 2.0: From the Age of Greed to the Age of Responsibility. In: Sun, W., Stewart, J. and Pollard, D. (eds) *Reframing Corporate Social Responsibility: Lessons from the Global Financial Crisis*. Bingley, UK: Emerald.

Vlastos, G. (1954). The Third Man Argument in the Parmenides. *The Philosophical Review*, 63: 319–349.

Vogel, D. (2010). The Private Regulation of Global Corporate Conduct. *Business & Society*, 49: 68–87.

Walhout, I. I. A. M. and Kuhlmann, P. D. S. (2013). In search of a governance framework for responsible research and innovation. *2013 IEEE International Technology Management Conference & 19th ICE Conference, 24–26 June 2013, the Hague*. The Hague.

Weiss, C. H. (1972). *Evaluation research: methods for assessing program effectiveness*, Englewood Cliffs, N.J.: Prentice-Hall.

Weiss, C. H. (1995). Nothing as Practical as Good Theory: Exploring Theory-based Evaluation for Comprehensive Community Initiatives for Children and Families. In: Connell, J., Kubisch, A. C., Schorr, L. B. and Weiss, C. H. (eds) *New Approaches to Evaluating Community Initiatives: Concepts, Methods, and Contexts*. Washington, DC: Aspen Institute.

Wittgenstein, L. (1968). *Philosophical investigations*, Oxford: Basil Blackwell.

Yin, R. K. (1984). *Case study research: Design and methods*, Newbury Park, CA: Sage.

Zand, D. E. and Sorensen, R. E. (1975). Theory of Change and the Effective Use of Management Science. *Administrative Science Quarterly*, 20: 532–545.

Part V

Conclusion

16 Fault lines in agricultural sustainability

Contestation, cooptation, reform, and transformation

Jason Konefal

Sustainability transitions are in vogue today. It is widely acknowledged that 'long term, multi-dimensional and fundamental transformation processes' that shift socio-technical systems, such as agriculture, towards 'more sustainable modes of production and consumption' are both necessary and possible (Markard, Raven and Truffer, 2012: 956). Over the last two decades there has been a surge in activity focused on accelerating the sustainability of agriculture. This has included research and policy reports, governmental actions, and private initiatives. Consequently, today there are a multitude of actors and initiatives that claim to be committed to and working towards sustainability. However, as the chapters in this volume indicate, not all these actors and initiatives are working towards the same vision of sustainable agriculture. Thus, while agriculture may be in the midst of a sustainability transition, there is substantial disagreement regarding the path(s) of this transition.

This volume echoes Hinrichs' (2014: 151) reminder that we cannot take 'sustainability as an unproblematic and uncontested goal.' Rather, researchers need to interrogate the meaning(s) of agricultural sustainability itself. Different ways of conceptualizing sustainability privilege some values, interests, and actors, and marginalize others (Gaspartos, 2010; Thompson, 2010). In other words, much is at stake in the codification of sustainability. As Gertler, Jaffe, and Beckie (this volume) observe, how sustainability is conceptualized affects 'regional ecologies, scientific practices, development trajectories, markets, and the moral high ground in terms food security and sustainability.'

This book has sought to elucidate current agricultural sustainability discourses. Specifically, it identifies and analyses a host of sustainability discourses, including sustainable intensification (SI), agroecology, food security, and food sovereignty, among others. As these multiple discourses demonstrate, the idea of sustainable agriculture remains a messy and contested idea. Building on the analyses presented in this volume, the first part of this concluding chapter identifies the key tensions between the myriad of sustainability discourses. Specifically, seven key divergences are identified: technology, social innovation, experts, democratization, markets, nature, and the extent of change.

The second part of this chapter examines the effects of governance and institutionalization on both sustainability discourses and their enactment. It is through governance processes that sustainability is transformed from a general idea to more tangible forms, such as reports, policies, and standards. While the transitions literature acknowledges that governance is an important component of sustainability transitions, the effects of governance processes on understandings of sustainability are under-theorized. Building on analyses presented in this volume, governance processes are found to have a moderating effect on sustainability transitions.

Tensions in sustainability discourses

Surveying the sustainability discourses presented throughout this volume, seven key divergences are identified in how they frame agricultural sustainability transitions: technology, social innovation, experts, democratization, markets, nature, and the extent of change. Each of these divergences is briefly discussed below.

Technology

What role should technology play in agricultural sustainability transitions? While all stakeholders acknowledge that technology needs to be a component of sustainability transitions in agriculture, there is disagreement as to what kinds of technologies and how central a role technology should play. On the one hand, you have discourses, most notably SI and livestock intensification, in which technology is advanced as the primary mechanism for accelerating sustainability. From this perspective, capital-intensive technologies that intensify production and increase eco-efficiencies are central to agricultural sustainability transitions. On the other hand, there are discourses that tend to view technologies more critically. For example, agroecologists see many agricultural technological innovations as key drivers of current unsustainabilities in agriculture. Hence, while not anti-technological, agroecologists tend to promote low-input and low-cost technologies.

Social innovation

What is the role of social innovation in agricultural sustainability transitions? Innovation is widely acknowledged to be part of sustainability transitions. However, different sustainability discourses emphasize different kinds of innovations. Namely, whereas some approaches, such as SI, substantially emphasize technological innovations, other approaches place more emphasis on innovation in social relations. In alignment with the idea of social innovation (Moulaert, MacCallum, and Hillier, 2013), sustainability discourses, such as agroecology and food sovereignty, view social, cultural, political, and economic changes as necessary for sustainability transitions.

Experts

What is the role of experts in sustainability transitions? Agricultural sustainability discourses vary in how they conceptualize the role of experts. On the one hand, discourses that emphasize technological innovations tend to privilege scientists. For example, SI largely frames sustainability transitions in agriculture as science-based. In alignment with the classic adoption and diffusion approach (Rogers, 2003), scientists' development of new technologies that are then diffused to farmers is viewed as a key part of transitioning agriculture. Thus, in SI discourses, scientists are positioned as core actors in sustainability transitions. On the other hand, other discourses, such as agroecology, are critical of privileging scientists and expert knowledge. Prominent agroecologists Altieri and Toledo (2011: 588) note that innovations have to be 'based on techniques that are not delivered top-down but developed on the basis of farmers' knowledge and experimentation.' To accomplish this, agroecologists contend that 'farmer-to-farmer research and grassroots extension approaches' need to be vital components of sustainability transitions (Altieri and Toledo, 2011: 588).

Democratization

To what extent should sustainability transitions be democratic? All sustainability discourses stress the need for democratic governance in sustainability transitions. However, there is considerable variation in terms of the extent to which the governance of transitions should be democratic. Discourses, such as food sovereignty and agroecology, advocate for strong forms of participatory governance that include diverse actors and epistemologies (e.g. local and experiential knowledge). In contrast, those discourses that favour experts and technological innovations tend to be less inclusive and democratic in how they frame governance in sustainability transitions. This includes both the range of stakeholders that participate, and the issues open for discussion and debate. For example, through their principles and objectives, multi-stakeholder initiatives often eliminate certain topics from consideration, such as debate over the sustainability of plantation-style agriculture or genetically-modified organisms (see Chapters 12 and 14).

Markets

What is the role of markets in sustainability transitions? There is considerable disagreement as to the role that markets can, and/or should, play in sustainability transitions. On the one hand, some sustainability discourses frame market processes as key drivers of sustainability transitions. Here, congruent with neoliberal discourse, market pressures are advanced as having the potential to drive increased (eco)efficiencies. For example, proponents of SI view market incentives as key drivers of technological innovations and farmer adoption of such innovations. On the other hand, other sustainability discourses see market processes as drivers of unsustainabilities at worst, or as limited in their capacity to incentivize sustainability transitions at best. For example, in alternative agri-food, agroecological, and food sovereignty discourses, changes in markets, such as stronger anti-trust legislation, redistributive policies (e.g. land reform and wages), and more governance of trade, are viewed as a necessary part of sustainability transitions (Holt Gimenez and Shattuck, 2011). Additionally, these discourses also advance alternative economic arrangements, such as localism, cooperatives, and worker-owned businesses, which are more inclusive of non-economic values (e.g. ecology and community), as central components of sustainability transitions.

Nature

What is the relationship between sustainability and nature in sustainability transitions? While the chapters in this book demonstrate that a key component of agricultural sustainability transitions is environmental conservation, there is disagreement as to how this is best achieved. Agricultural sustainability discourses can be divided into two positions. The first are those discourses that seek to advance environmental conservation through further separating agriculture from the environment. For example, in both SI and livestock intensification discourses, technological innovations are framed as a way to further reduce agriculture's dependence on natural resources. In other words, increased environmental sustainability is to be accomplished through further disembedding agriculture from nature. This will, in turn, allow for more natural resources and land to be conserved (i.e. land sparing). A second set of discourses, most clearly articulated by agroecology, advocates for (re)embedding agriculture in nature. In this approach, environmental conservation is advanced through the use of low-impact practices that enable the co-existence of agriculture and ecologies (i.e. land sharing).

Extent of change

To what extent do sustainability transitions entail transformative changes? Sustainability discourses diverge with respect to whether they advocate incremental or transformative change. Discourses such as SI, food security, and livestock intensification focus on stabilizing current agricultural systems by ameliorating challenges that threaten the food and agriculture sectors, such as resource scarcities, population growth, and nutrition transitions. Hence, these are reformist discourses that envision sustainability transitions in agriculture as possible through incremental changes to current agricultural systems. In contrast to such reformist discourses, sustainability discourses such as civil commons, agrocecology and food sovereignty frame sustainability as entailing transformative changes. For example, in this volume, Sumner proposes linking sustainability to the idea of the 'civil commons' – 'a co-operative, not competitive ... mode of engagement.' This is an oppositional discourse that frames sustainability transitions as entailing the wholesale transformation of agricultural systems. Similarly, agrocecology and food sovereignty discourses frame sustainability transitions as requiring systemic transformation of current industrialized and capitalistic agricultural systems.

Discussion

The above analysis highlights key divergences across agricultural sustainability discourses. These divergences illustrate that the meaning and processes of sustainability transitions in agriculture remain very much contested. This includes in terms of who should participate in sustainability transitions and in what capacity, how they should be governed, and the character and extent of changes.

In much of current thinking about agricultural sustainability transitions, different discourses are juxtaposed against each other, such as agroceology versus SI and food security versus food sovereignty. The incommensurabilities and tensions between these competing discourses are often highlighted. In teasing out specific points of contention between conflicting discourses, the analysis presented here seeks to reframe thinking about sustainability discourses beyond dualistic juxtapositions. Given the diversity of agricultural systems, and the political economic realities of food and agriculture, sustainability transitions will be multifaceted and diverse. This means that experts, technology, participatory governance, social innovation, incremental reforms and transformative change all have a role to play. Hence, research is needed on the potential and limitations of each of these in sustainability transitions, as well as the ways they privilege and marginalize different stakeholders and interests.

Governance and institutionalization

A key focus of many of the chapters in this volume is the effect that governance and institutionalization are having on sustainability transitions. In transitions theory, a key part of accelerating sustainability is the institutionalization of innovations at the regime level. However, how this processes works in practice, and how it affects sustainability transitions have been under-theorized in the transitions literature. Markard, Raven and Truffer (2012: 962), in their review article, note that transitions theory has largely neglected 'issues of power and politics.' Marsden (2013: 132) also argues that 'how power, both economic and political, is differentially allocated and mediated' is under-theorized. This volume begins to address this research gap with a focus on how governance and institutionalization processes affect sustainability discourses and their enactment.

A key finding of this volume is that sustainability discourses are being reformulated in governance processes towards more moderate discourses. We see this with agroecology, where a variety of actors, including governments and international bodies, are now advancing agroecology as an important component of accelerating the sustainability of agriculture. However, as several chapters in this volume indicate (see Chapters 2, 3, 4, and 9), the ways that agroecology is conceptualized has changed with its institutionalization. For example, while, in France, agroecology has been officially incorporated into agricultural policy, it is 'a technical vision of agroecology' that largely excludes the 'social dimensions of agroecology' (Thomas, this volume). This application of agroecology is very different from how agroecology has long-been envisioned. Thus, what has occurred in France is the selective appropriation of some aspects of agroecology as part of a sustainability transition, namely those that fit best with existing practices.

We also see similar moderation of sustainability discourses as they are scaled upwards from the national to global level. In her analysis of discourses of hunger and sustainability, Smith (this volume) demonstrates how the Zero Hunger programme has shifted from a fairly progressive initiative to one that is more reformist with its transition from a Brazilian to a global development initiative. As a national programme in Brazil, it sought 'high involvement of civic actors, redistribution via smallholder farmer cooperatives, social protection programmes and an emphasis on agroecology and justice' (Smith, this volume). However, with its inclusion in the United Nations Sustainable Development Goals, Smith (this volume) observes that Zero Hunger has been reformulated in ways that favour market-based measures, established institutional actors, and 'narratives of modernization.' Thus, as Zero Hunger became part of 'high level diplomatic language and negotiation processes' it was reformulated to better align with dominant development discourses and the interests of powerful stakeholders.

In addition to (re)formulating sustainability discourses, governance processes can also affect the legitimacy of sustainability discourses. We see this in terms of SI, which has gone from a nearly non-existent idea at the start of the millennium to, perhaps, the dominant sustainable agriculture discourse today. This has happened through the adoption of SI by governmental and international bodies, as well as the emergence of multi-stakeholder initiatives that advance SI. Analyses of SI in this volume find that lead actors in the current agrifood regime are a key force behind this legitimation (see Chapters 2, 3, 4, and 13). Given that SI is more in alignment with the current agrifood regime than other agricultural sustainability discourses, lead actors have used their influence and resources to legitimize SI as a viable path for agricultural sustainability transitions.

In brief, the cases presented in this volume largely suggest that current forms of governance have a moderating affect on sustainability transitions. Lead actors in agrifood regimes are able to exert influence in governance processes to moderate sustainability discourse in ways that 'fit-and-conform to the incumbent regime' (Levidow, this volume). In other words, current governance processes appear to result in a channeling of sustainability transitions towards incremental reforms that are accommodative of current political, economic, and socio-cultural relations. These findings raise questions as to current governance arrangements – both public and private – to facilitate transformative sustainability transitions in agrifood systems.

Conclusion

This volume contributes to research on agricultural sustainability transitions on two fronts. First, it provides an overview of agricultural sustainability discourses.

In particular, the chapters in the volume call attention to the contested meaning of sustainable agriculture. They demonstrate that how sustainability is understood matters significantly. Depending on the vision(s) of sustainability put into practice, the future of agriculture, as well as rural communities and ecologies, will look very different.

While competing visions of sustainability are often juxtaposed today – most notably agroecology and SI – the reality of agricultural sustainability transitions is likely to be more messy and complex (Lamine, 2015). Given the diversity of agricultural systems, and the political economic realities of food and agriculture, there is not a one-size fits all sustainability transition for agriculture (Spaargaren, Loeber, and Oosterveer, 2012). Put differently, there are likely to be multiple sustainability transitions in agriculture, which, in all likelihood, will combine elements from different sustainability discourses. Recognizing this diversity, this chapter has sought to shift analyses of sustainability from a focus on competing discourses to the specific tensions among stakeholders.

The second contribution of this volume is with respect to the governance of sustainability transitions. Current governance processes are shown to have a moderating effect on sustainability transitions. In other words, current governance processes are channeling sustainability transitions in the direction of incremental changes in agricultural sustainability. This path is in the interests of many stakeholders, including some farmers, agribusiness corporations, and consumers. However, questions remain regarding whether incremental changes are sufficient to overcome the myriad of challenges facing food and agriculture. For actors who see more transformative changes as necessary, sustainability transitions will require substantial innovation in governance.

References

Altieri, M. A. and Toledo, V. M. (2011). 'The agroecological transition in Latin America: Rescuing nature, ensuring food sovereignty and empowering peasants,' *The Journal of Peasant Studies*, vol. 38: 587–612.

Gaspartos, A. (2010). 'Embedded value systems in sustainability assessment tools and their implications,' *Journal of Environmental Management*, vol. 91, no. 8: 1613–1622.

Hinrichs, C. C. (2014). 'Transitions to sustainability: A change in thinking about food systems change?' *Agriculture and Human Values*, vol. 31, no. 1: 143–155.

Holt Giménez, E. and Shattuck, A. (2011). 'Food crises, food regimes and food movements: rumblings of reform or tides of transformation?' *The Journal of Peasant Studies*, vol. 38, no. 1: 109–144.

Lamine, C. (2015). 'Sustainability and Resilience in Agrifood Systems: Reconnecting Agriculture, Food, and the Environment,' *Sociological Ruralis*, vol. 55, no. 1: 41–61.

Markard, J., Raven, R. and Truffer, B. (2012). 'Sustainability transitions: An emerging field of research and its prospects,' *Resource Policy*, vol. 41, no. 6: 955–967.

Marsden, T. (2013). 'From post-productionism to reflexive governance: Contested transitions in securing more sustainable food futures,' *Journal of Rural Studies*, vol. 29, no. 1: 123–134.

Moulaert, F., MacCallum, D. and Hillier, J. (2013). 'Chapter 1: Social innovation: intuition, precept, concept, theory, and practice.' In: F. Moulaert, D. MacCallum, A. Mehmood and A. Hamdouch (eds), *The International Handbook on Social Innovation: Collective Action, Social Learning, and Transdisciplinary Research*, Edward Elgar Publishing: Cheltenham, UK: 13–24.

Rogers, E. M. (2003). *Diffusion of Innovation*. Free Press: New York.

Spaargaren, G., Loeber, A. and Oosterveer, P. (2012). 'Food Futures in the Making.' In G. Spaargaren, A. Loeber and P. Oosterveer (eds), *Food Practices in Transition: Changing Food Consumption, Retail and Production in the Age of Reflexive Modernity*, Routledge: New York: 312–338.

Thompson, P. B. (2010). *The Agrarian Vision: Sustainability and Environmental Ethics*, University of Kentucky Press: Lexington, KY.

Index

Page numbers in **bold** denote tables, those in *italics* denote figures.

For Product Safety Concerns and Information please contact our EU representative GPSR@taylorandfrancis.com
Taylor & Francis Verlag GmbH, Kaufingerstraße 24, 80331 München, Germany

www.ingramcontent.com/pod-product-compliance
Ingram Content Group UK Ltd.
Pitfield, Milton Keynes, MK11 3LW, UK
UKHW022012180926
475701UK00003B/323
9780367589363